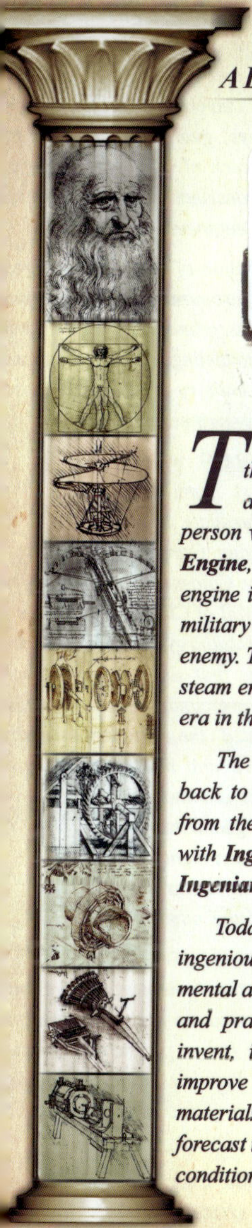

A Brief History of "Engineering"

The Archimedes Screw

*The term **Engineering** is derived from the word **Engineer** which dates back to around 1300 when an **Engine(er)** was a person who literally operated / constructed an **Engine**, which in those days meant a military engine i.e. a mechanical device which gave the military some mechanical advantage over its enemy. The word **Engine** was later taken over by steam engines and the like during the industrial era in the late sixteenth century.*

*The word **Engineer** itself can be traced back to the eleventh century and was derived from the Latin word **Ingeniator**, meaning one with **Ingenium**, the ingenious (clever) one and **Ingeniare** which means 'to devise / invent'*

*Today, we still look upon an **Engineer** as an ingenious person who uses his cleverness and mental ability of mathematics, science, economic and practical knowledge in order to devise, invent, innovate, design, build, maintain and improve machines, tools, systems, components, materials, structures, processes, solutions and forecast their behaviour under specific operating conditions and intended functions.*

Engineering *has existed since ancient times, when practical artisans and craftsmen developed their skills mainly by trial and error and over time devised fundamental inventions such as the wheel, pulleys, gears, wedges, levers, levels, etc. Each of these inventions is essentially consistent with the modern definition of engineering.*

Lever

Later on, as the design and construction of civilian structures such as aqueducts, bridges, canals, harbours, lighthouses and buildings advanced as a technical discipline, the term ***Civil Engineering*** *entered the scene as a way to distinguish between those specializing in the construction of non-military projects and those involved in the older discipline of military engineering.*

<u>The Ancient Era (3500BC - 250AD)</u> *which includes - the Pottery wheel, the first Dam, the Archimedes Screw, the Roman Aqueducts, the Pyramid of Djoser in Egypt, the Hanging Gardens of Babylon, the Parthenon in Greece, the Colosseum, the Inca and Aztec Empires, the Great Wall of China, etc. The birth of metallurgy, and the resultan construction of the first siege weapons suc as ladders for overcoming enemy walls clearly demonstrated the skill and ingenuit of the ancient civil and military engineers. It was during this era that Pythagoras developed the famous geometric formula* $c^2 = a^2 + b^2$ *which states that in a right-angle triangle, the square of the hypotenuse is equal to the sum of the squares of the other two sides.*

Pottery Wheel

Pyramid of Djoser

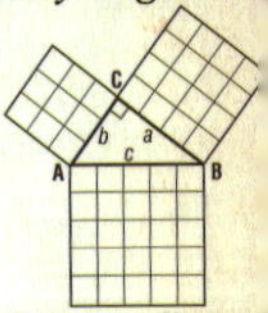

Pythagoras Theoren

<u>The Renaissance Era</u> *- following the Ancient era, there wa a period of relative quiet where there were no major innovation in engineering until 1421 when the first Italian patent was grante by the Republic of Florence to Filippo Brunelleschi for the desig of a special shipping barge with hoisting gear that carried marb along the Arno River in Italy.*

During this same period around 1500, a series of sketches an manuscripts were produced by various artists / engineers, the mo famous being Leonardo da Vinci, who was given the official title of ***"Ingegnere Generale"****. He is today still considered one of the most prolific inventors in history and dreamed up inventions and innovations across a variety of fields. Whether designing weapons of war, flying machines, water systems or work tools,*

Leonardo Da Vinci

Leonardo da Vinci the inventor, much like Leonardo da Vinci the artist, was never afraid to look beyond traditional thinking. His sketchbooks provide ample evidence that he had envisioned many ideas long before the technology to build them actually existed.

Flying Machine

Around the same time, other inventions such as the first spring-activated pocket watch by Peter Henlein, the pendulum clock by Christiaan Huygens, the microscope by Hans and Zacharias Janssen, the telescope by Hans Lippershey and the first mechanical calculator called Pascaline by Blaise Pascal, who has a popular computer programming language named in his honour today called 'Pascal'.

Blaise Pascal

The first steam engine was built in 1698 by a mechanical engineer called Thomas Savery and this gave rise to the Industrial Revolution in the coming decades.

The Industrial Revolution *started around the 17th century and included the invention of the first Pressure Pot by Denis Papin. Other pioneers of the steam engine and subsequent steam locomotive were Thomas Savery, Guillaume Amontons, Thomas Newcomen, John Calley and further improved by James Watt in 1776. During this period many other inventions came into being which made it possible to advance the first cotton textile factories in England in 1771and the US in 1789.*

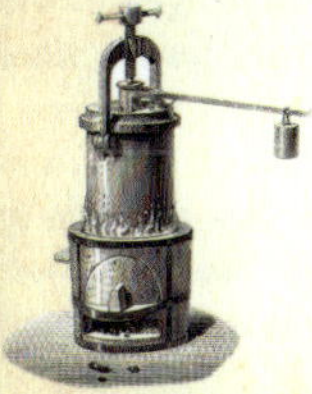

Pressure Pot

The Second Industrial Revolution *started in the 19th century when greater availability of raw materials and increase of steel production was brought about by the invention of the Bessemer Converter by Henry Bessemer in 1855. Also, during this period Reinforced Concrete was developed by Joseph Monier in 1849, who designed the first concrete bridge in 1857.*

Henry Bessemer

As a result of the development of steel and reinforced concrete, buildings began to grow in height and so came the first elevator designed by engineer Elisha Otis in 1857. The continuous rapid development of steel resulted in the construction of the Eiffel Tower, built by Gustave Eiffel in Paris in 1889.

Eiffel Tower

Alexander Graham Bell is credited with the invention of the telephone in 1876.

Michael Faraday, George Ohm and Nikola Tesla were involved in the development of the Electric Motor between 1872 and 1882. Thomas Edison invented the electric light bulb in 1879. All of the above and more gave rise to modern Mechanical and Electrical Engineering.

Following World War II, the automation of machine tools took off at a rapid pace, first through the use of cams, servos, etc, leading up to the development of the first Numerical Control (NC) machine credited to John Parsons around 1942. The first NC machines in production used Punch Cards as a means of programming. As computing and networking evolved, so did Direct Numerical Control (DNC) and this was quickly followed by the development of the Computer Numerical Control (CNC) machines which used a mini computer using G-Code as the most common method of programming.

CNC Machine

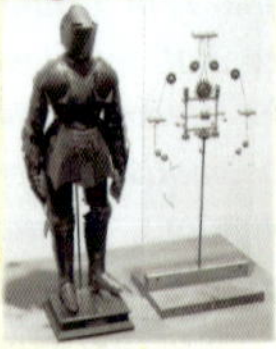
Model of Leonardo Da Vinci's Robot with inner working

Digitally Controlled Industrial Robots and Robots making use of artificial intelligence have been built since the 1960s and are common place today in most of the large industrial factories. However, looking into the future, we human beings still need to remember that the craft of a Skilled Machinist and their Machine Tools will always be responsible for most of these new inventions to have occurred.

<u>The Information/Automation/Digital/Media Age</u> *is the* ***New Age*** *we live in and is characterized by the shift from traditional industry that the Industrial Revolution brought about through industrialization, to an economy based on information and automation. With the development of the smart phone, it seems that nothing is beyond the reach of the human being with many un-imaginable developments to come in the future.*

Smart Phone

About the Book

Welcome to the ***Engineers Black Book (EBB) - 3rd Edition (INCH)***. It should be noted that the previous edition to this was in fact called the EBB - USA Edition or popularly referred to as the INCH Edition, first released in 2005.

In keeping with the soon to be released ***EBB - 3rd Edition (METRIC)***, the publisher decided that in the interest of uniformity, it was necessary for the **INCH** Edition to skip a generation (2nd Edition) and call this release, the ***EBB - 3rd Edition (INCH)*** and therefore bypass the ***EBB - 2nd Edition (INCH)*** which in fact was never released.

As a means of differentiating between Inch and Metric information, the author has maintained the **INCH** and **METRIC** layout.

The EBB - 3rd Edition (INCH) **contains hard to come by information on:**

- *Drill Size & Decimal Equivalents*
- *Counterboring/Countersinking*
- *Common Thread Forms*
- *Tapping Drill Formulae*
- *Engineering Plastics*
- *Conversion Factors*
- *Torque Conversion*
- *Tapers and Angles*
- *Spanner Selection*
- *Tungsten Carbide*
- *Grinding Wheels*
- *Drill Sharpening*
- *Speeds & Feeds*
- *Bolts and Nuts*
- *Flange Details*

and other technical information too numerous to list.

The motivation that led to the compilation of this book was the need as the author saw it, to amalgamate the plethora of information on standard technical information that is used on a day to day basis by most engineering workshops and present it in a very basic, convenient and user-friendly pocket-book format.

A great deal of thought has gone into the development of this ***EBB - 3rd Edition (INCH)***, to ensure that it provides a better understanding by way of pictorially depicting standard engineering information alongside relevant dimensional details where possible.

It should be noted that this book is not intended to be a technical or specification manual, but a quick reference (for identification purposes) to dimensions and other relevant information on the more commonly used workshop tools and equipment. Technical data and specifications were researched and compiled from numerous industry sources for the sole purpose of providing basic information.

It is not the intention of the publisher through the medium of this book to provide solutions to problems that might arise in manufacturing processes as this is the realm of qualified engineers.

The layout of information in this book follows the style of the popular Engineers Black Book - USA Edition first released in 2005 by the same author. As was the case in the earlier compilation, this ***Engineers Black Book - 3rd Edition (INCH)*** too has been designed to provide a wealth of information intended to assist engineers, apprentices, trainees, technicians, etc., all of whom would be using engineering information on a routine basis.

CONTENTS

CONTENTS

CONTENTS

CONTENTS

COMMON-SENSE SAFETY AT WORK

Most tool-rooms and machine shops are equipped with a variety of machines like lathes, milling machines, grinders, drill presses, cut-off saws, welding machines, etc. These machines are either manually operated or are fully automatic. No matter how a machine operates, it very likely creates safety hazards during its operation that could cause serious injury or even death to operators or other employees in the immediate area.

The most important reason for accidents at work is apathy. The benefits of safe practice at work are not usually appreciated until after an accident occurs and then it is too late. The objective of these two pages is to provide some common-sense safety do's and dont's when dealing with machinery as well as act as a reminder never to adopt the ***"I've done it before"*** approach as regards to safety in the work place.

The following suggestions are offered as guidelines for basic operator safety in the work place:

Proper Clothing: The rotating spindle of a machine tool can snag loose fitting clothing, jewelry or long hair. Never wear jewelry, long sleeves, neckties, gloves or anything else that could become caught when operating a machine tool. Long hair must be restrained or netted to prevent it from becoming entangled in a rotating spindle.

Safety Shoes: Steel toed boots should always be worn in any machine environment.

Proper Eye Protection: Always wear good quality safety glasses with side shields to protect your eyes from flying particles such as flying chips and sparks.

Proper Work Piece Fixturing: Never hold the work piece or the vise by hand. The work piece(or vise) must always be clamped firmly to the table of the machine so that it cannot move, rotate or lift.

Be Alert: Always be aware of the potential hazards of a machining operation. Sometimes working with your machine can seem routine. This is what is meant by apathy. A feeling of false security can develop which may lead to serious injury. Again, this is the most important reason for accidents at work - the second being machine guarding.

Always:

- Keep hands, body parts, clothing, jewelry and hair out of the areas of operation, when the machine spindle is rotating. Wait for the machine spindle to completely stop. Never work in wet areas around machines or slippery floors.
- Operate the machine only when the guards surrounding the power transmission such as belts, pulleys, motor shafts, leadscrews and chains, lathe chucks, grinding wheels, etc. are present.

COMMON-SENSE SAFETY AT WORK

- Be aware of exposed, damaged or frayed electrical insulation on electrical parts especially when wet conditions exist or coolants are used. Burning odor may indicate overheated insulation. This particularly refers to portable electrical tools such as angle grinders, drilling machines, etc.
- Ensure proper lighting conditions exist especially where there are exposed rotating machinery parts such as lathe chucks, milling machine spindles, grinding wheels, etc.
- Make sure that operator controls such as emergency-stop devices, disconnect switches are functional.
- Never operate dirty or greasy equipment, because potential hazards may be covered up from sight - have the machine cleaned before use.
- Use proper lifting techniques (keep back straight, bend from knees, grasp object firmly)
- Ensure that work-rests on bench grinders are maintained at the correct maximum opening of 1/8" or 3mm to prevent the workpiece from being jammed between the wheel and the work-rest, which may cause wheel breakage.
- Never work under stress related to urgency - this is when most operator errors occur.
- Be aware of safety signs around the machine and all around the work place - they are there for a reason.
- Familiarize yourself with electricity shut-offs (kill switches), first-aid supplies and contact details of emergency medical help - hopefully, you will never need it.

In Case of Electrical Shock

- If the victim usually gets stuck to the source of the electricity, it is important that you first separate him from the electrical source. If safely possible, shut off the electrical current. Unplug the cord, remove the fuse from the fuse box, or turn off the circuit breakers if possible.
- If the current cannot be turned off, use a non-conducting object, such as a broom, chair, rug, or rubber doormat to push the victim away from the source of the current. Do not use a wet or metal object.
- Once the victim is free from the source of electricity, check the victim's airway, breathing and pulse. If either has stopped or seems dangerously slow or shallow, initiate first aid (CPR - Cardio Pulmonary Resuscitation).
- Call for medical help.

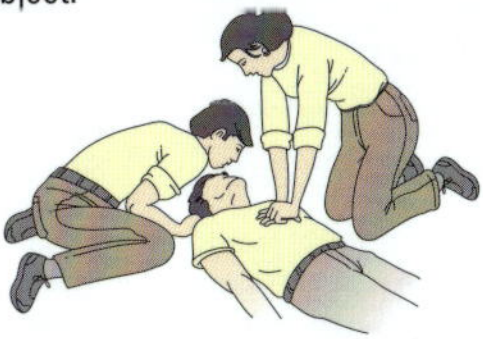

Cardio Pulmonary Resuscitation (CPR)

COMMON CONVERSION FACTORS
IMPERIAL TO METRIC

TO CONVERT	INTO	MULTIPLY BY
Acres	Hectares	0.4047
Atmospheres	Kilopascals	101.3250
British Thermal Units	Kilojoules	1.0551
Bushels	Cubic Meters	0.0364
Cubic Feet per Minute	Liters per Second	0.4719
Cubic Feet	Cubic Meters	0.0283
Cubic Inches	Cubic Centimeters	16.3871
Cubic Inches	Fluid Ounces	0.5767
Cubic Inches	Liters	0.0164
Cubic Yards	Cubic Meters	0.7646
Cycles per Second	Hertz	1.0000
Degrees Fahrenheit	Degrees Celsius	- 32 x 5 ÷ 9
Feet per Second	Miles per Hour	0.6819
Feet	Meters	0.3048
Fluid Ounces	Cubic Inches	1.7339
Fluid Ounces	Milliliters	28.4130
Fluid Ounces	Liters	0.0284
Foot Pound	Newton-meter	1.3558
Furlongs	Meters	201.0000
Furlongs	Kilometers	0.2010
Gallons per Minute	Liters per Second	0.0758
Gallons per Hour	Liters per Second	0.0013
Gallons (UK)	Liters	4.5461
Gallons (US)	Liters	3.7854
Grains	Grams	0.0648
Horse Power (Electric)	Kilowatts	0.7460
Hundredweights	Kilograms	50.8032
Inches of Mercury	Millibars	33.7410
Inches	Millimeters	25.4000
Inches	Centimeters	2.5400
Kilowatt Hours	Megajoules	3.6000
Miles per Hour	Feet per Second	1.4667
Miles per Hour	Kilometers per Hour	1.6093
Miles	Kilometers	1.6093
Miles	Nautical Miles	0.8690
Ounces, Avoirdupois	Grams	28.3495
Pints	Liters	0.5683
Pounds-Force	Newtons	4.4482
Pounds per Square Inch	Bar	0.0689
Pounds per Square Inch	Kilopascals	6.8948
Pounds per Square Inch	Megapascals	0.0069
Pounds	Grams	453.5920
Pounds	Kilograms	0.4536
Square Inches	Square Millimeters	645.1612
Square Inches	Square Centimeters	6.4516
Square Feet	Square Centimeters	929.0310
Square Feet	Square Meters	0.0929
Square Yards	Square Meters	0.8361
Square Miles	Square Kilometers	2.5899
Ton-Force	Kilonewtons	9.9640
Tons-Long (2,240 pounds)	Tonnes	1.0161
Tons-Short (2,000 pounds)	Tonnes	0.9072
Tons per Square Inch	Megapascals	15.4000
Yards	Meters	0.9144
Yards	Kilometers	0.0009

COMMON CONVERSION FACTORS

METRIC TO IMPERIAL

TO CONVERT	INTO	MULTIPLY BY
Bar	Pounds per Square Inch	14.5038
Centimeters	Inches	0.3937
Cubic Centimeters	Cubic Inches	0.0610
Cubic Meters	Bushels	27.4962
Cubic Meters	Cubic Feet	35.3147
Cubic Meters	Cubic Yards	1.3079
Degrees Celsius	Degrees Fahrenheit	x 9 ÷ 5 + 32
Grams	Ounces	0.0353
Grams	Grains	15.4324
Grams	Ounces, Avoirdupois	0.0353
Grams	Pounds	0.0022
Hectares	Acres	2.4710
Hertz	Cycles per Second	1.0000
Kilograms	Hundredweights	0.0197
Kilograms	Pounds	2.2046
Kilogram Force	Newton	9.8066
Kilometers per Hour	Miles per Hour	0.6214
Kilometers	Yards	1093.6132
Kilometers	Furlongs	4.9750
Kilometers	Miles	0.6214
Kilopascals	Pounds per Square Inch	0.1450
Kilopascals	Atmospheres	0.0099
Kilojoules	British Thermal Units	0.9478
Kilonewtons	Ton-Force	0.1004
Kilowatts	Horse Power (Electric)	1.3410
Liters per Second	Cubic Feet per Minute	2.1190
Liters per Second	Gallons per Minute	13.1980
Liters per Second	Gallons per Hour	791.8900
Liters	Cubic Inches	60.9362
Liters	Fluid Ounces	35.2113
Liters	Pints	1.7598
Liters	Gallons (UK)	0.2200
Liters	Gallons (US)	0.2642
Megajoules	Kilowatt Hours	0.2778
Megapascals	Pounds per Square Inch	145.0389
Megapascals	Tons per Square Inch	0.0649
Meters	Feet	3.2808
Meters	Furlong	0.0050
Meters	Yards	1.0936
Millibars	Inches of Mercury	0.0296
Milliliters	Fluid Ounces	0.0352
Millimeters	Inches	0.0394
Nautical Miles	Miles	1.1508
Newtons	Pounds-Force	0.2248
Newton	Kilogram-Force	0.1020
Newton-meter	Foot Pound	0.7376
Square Centimeters	Square Inches	0.1550
Square Centimeters	Square Feet	0.0011
Square Meters	Square Feet	10.7639
Square Meters	Square Yards	1.1959
Square Millimeters	Square Inches	0.0015
Square Kilometers	Square Miles	0.3861
Tonnes	Long Tons (2,240 pounds)	0.9842
Tonnes	Short Tons (2,000 pounds	1.1020

14 DRILL SIZE AND DECIMAL EQUIVALENT CHART

(*INCLUDES TAPPING DRILL SIZES FOR THREAD CUTTING TAPS)

Fract.	inch	mm	Gauge	Thread Size
	0.00394	**0.100**		
	0.00591	0.150	97	
	0.00630	0.160	96	
	0.00669	0.170	95	
	0.00709	0.180	94	
	0.00748	0.190	93	
1/128	0.00781	0.198		
	0.00787	**0.200**	92	
	0.00827	0.210	91	
	0.00866	0.220	90	
	0.00906	0.230	89	
	0.00945	0.240	88	
	0.00984	0.250		
	0.01000	0.254	87	
	0.01024	0.260		
	0.01063	0.270	86	
	0.01102	0.280	85	
	0.01142	0.290	84	
	0.01181	**0.300**		
	0.01201	0.305	83	
	0.01220	0.310		
	0.01248	0.317	82	
	0.01260	0.320		
	0.01299	0.330	81	
	0.01339	0.340		
	0.01350	0.343	80	
	0.01378	0.350		
	0.01417	0.360		
	0.01449	0.368	79	
	0.01457	0.370		
	0.01496	0.380		
	0.01535	0.390		
1/64	0.01563	0.397		
	0.01575	**0.400**		
	0.01598	0.406	78	
	0.01614	0.410		
	0.01654	0.420		
	0.01693	0.430		
	0.01732	0.440		
	0.01772	0.450		
	0.01799	0.457	77	
	0.01811	0.460		
	0.01850	0.470		
	0.01890	0.480		
	0.01929	0.490		
	0.01969	**0.500**		
	0.02000	0.508	76	
	0.02008	0.510		
	0.02047	0.520		
	0.02087	0.530		
	0.02098	0.533	75	
	0.02126	0.540		
	0.02165	0.550		
	0.02205	0.560		
	0.02244	0.570		
	0.02252	0.572	74	
	0.02283	0.580		
	0.02323	0.590		
	0.02362	**0.600**		
	0.02402	0.610	73	
	0.02441	0.620		
	0.02480	0.630		
	0.02500	0.635	72	
	0.02520	0.640		
	0.02559	0.650		
	0.02598	0.660	71	
	0.02638	0.670		
	0.02677	0.680		
	0.02717	0.690		
	0.02756	**0.700**		
	0.02795	0.710		
	0.02799	0.711	70	
	0.02835	0.720		
	0.02874	0.730		
	0.02913	0.740		
	0.02921	0.742	69	
	0.02953	0.750		**M1.0 x 0.25 (C)**
	0.02992	0.760		
	0.03031	0.770		
	0.03071	0.780		
	0.03098	0.787	68	
	0.03110	0.790		
1/32	0.03125	0.794		
	0.03150	**0.800**		
	0.03189	0.810		
	0.03201	0.813	67	
	0.03228	0.820		
	0.03268	0.830		
	0.03299	0.838	66	
	0.03307	0.840		
	0.03346	0.850		**M1.1 x 0.25 (C)**
	0.03386	0.860		
	0.03425	0.870		
	0.03465	0.880		
	0.03500	0.889	65	
	0.03504	0.890		
	0.03543	**0.900**		
	0.03583	0.910		

(C)=Coarse Thread

*Also refer pages 36-47 for Tapping Drill Sizes

DRILL SIZE AND DECIMAL EQUIVALENT CHART

(*INCLUDES TAPPING DRILL SIZES FOR THREAD CUTTING TAPS)

Fract.	inch	mm	Gauge	Thread Size
	0.03598	0.914	64	
	0.03622	0.920		
	0.03661	0.930		
	0.03701	0.940	63	
	0.03740	0.950		M1.2 x 0.25 (C)
	0.03780	0.960		
	0.03799	0.965	62	
	0.03819	0.970		
	0.03858	0.980		
	0.03898	0.990		
	0.03902	0.991	61	
	0.03937	**1.000**		
	0.04000	1.016	60	
	0.04098	1.041	59	
	0.04134	1.050		
	0.04201	1.067	58	
	0.04299	1.092	57	
	0.04331	1.100		M1.4 x 0.30 (C)
	0.04528	1.150		
	0.04650	1.181	56	#0 - 80 UNF
3/64	0.04688	1.191		
	0.04724	1.200		
	0.04921	1.250		M1.6 x 0.35 (C)
	0.05118	1.300		
	0.05201	1.321	55	
	0.05315	1.350		M1.7 x 0.35 (SP)
	0.05500	1.397		
	0.05512	1.400		
	0.05709	1.450		M1.8 x 0.35 (C)
	0.05906	1.500		
	0.05949	1.511	53	#1 - 64 UNC #1 - 72 UNF
	0.06102	1.550		M2.0 x 0.45 (SP)
1/16	0.06250	1.588		
	0.06299	1.600		M2.0 x 0.40 (C)
	0.06350	1.613	52	
	0.06496	1.650		
	0.06693	1.700		
	0.06701	1.702	51	
	0.06890	1.750		M2.2 x 0.45 (C)
	0.07000	1.778	50	#2 - 56 UNC #2 - 64 UNF
	0.07087	1.800		
	0.07283	1.850		
	0.07299	1.854	49	
	0.07480	1.900		M2.3 x 0.40 (SP)
	0.07598	1.930	48	
	0.07677	1.950		
5/64	0.07813	1.984		
	0.07850	1.994	47	#3 - 48 UNC
	0.07874	**2.000**		
	0.08071	2.050		M2.5 x 0.45 (C)
	0.08098	2.057	46	
	0.08201	2.083	45	#3 - 56 UNF
	0.08268	2.100		
	0.08465	2.150		M2.6 x 0.45 (SP)
	0.08598	2.184	44	
	0.08661	2.200		
	0.08858	2.250		
	0.08902	2.261	43	#4 - 40 UNC
	0.09055	2.300		
	0.09252	2.350		
	0.09350	2.375	42	#4 - 48 UNF
3/32	0.09375	2.381		
	0.09449	2.400		M3.0 x 0.60 (SP)
	0.09598	2.438	41	
	0.09646	2.450		
	0.09799	2.489	40	
	0.09843	2.500		M3.0 x 0.50 (C)
	0.09949	2.527	39	
	0.10039	2.550		
	0.10150	2.578	38	#5 - 40 UNC
	0.10236	2.600		
	0.10402	2.642	37	#5 - 44 UNF
	0.10433	2.650		M3.0 x 0.35 (F)
	0.10630	2.700		
	0.10650	2.705	36	#6 - 32 UNC
	0.10827	2.750		
7/64	0.10938	2.778		
	0.11000	2.794	35	
	0.11024	2.800		
	0.11098	2.819	34	
	0.11220	2.850		
	0.11299	2.870	33	#6 - 40 UNF
	0.11417	2.900		M3.5 x 0.60 (C)
	0.11598	2.946	32	
	0.11614	2.950		
	0.11811	**3.000**		
	0.12000	3.048	31	
	0.12205	3.100		
1/8	0.12500	3.175		
	0.12598	3.200		M4.0 x 0.75 (SP)
	0.12850	3.264	30	
	0.12992	3.300		M4.0 x 0.70 (C)
	0.13386	3.400		
	0.13598	3.454	29	#8 - 32 UNC #8 - 36 UNF
	0.13780	3.500		M4.0 x 0.50 (F)

(C)=Coarse Thread (F)=Fine Thread (SP)=Special Pitch Thread

*Also refer pages 36-47 for Tapping Drill Sizes

DRILL SIZE AND DECIMAL EQUIVALENT CHART

(*INCLUDES TAPPING DRILL SIZES FOR THREAD CUTTING TAPS)

Fract.	inch	mm	Gauge	Thread Size
	0.14051	3.569	28	
9/64	0.14063	3.572		
	0.14173	3.600		
	0.14402	3.658	27	
	0.14567	3.700		M4.5 x 0.75 (C)
	0.14701	3.734	26	
	0.14949	3.797	25	#10 - 24 UNC
	0.14961	3.800		
	0.15201	3.861	24	
	0.15354	3.900		
	0.15402	3.912	23	
5/32	0.15625	3.969		
	0.15701	3.988	22	
	0.15748	4.000		
	0.15902	4.039	21	#10 - 32 UNF
	0.16098	4.089	20	
	0.16142	4.100		
	0.16535	4.200		M5.0 x 0.80 (C)
	0.16598	4.216	19	
	0.16929	4.300		
	0.16949	4.305	18	
11/64	0.17188	4.366		
	0.17299	4.394	17	
	0.17323	4.400		
	0.17701	4.496	16	#12 - 24 UNC
	0.17717	4.500		M5.0 x 0.50 (F)
	0.18000	4.572	15	
	0.18110	4.600		
	0.18201	4.623	14	#12 - 28 UNF
	0.18504	4.700	13	
3/16	0.18750	4.763		
	0.18898	4.800	12	
	0.19098	4.851	11	
	0.19291	4.900		
	0.19350	4.915	10	
	0.19598	4.978	9	
	0.19685	5.000		M6.0 x 1.00 (C)
	0.19902	5.055	8	
	0.20079	5.100		
	0.20098	5.105	7	1/4 - 20 UNC
13/64	0.20313	5.159		
	0.20402	5.182	6	
	0.20472	5.200		M6.0 x 0.75 (F)
	0.20551	5.220	5	
	0.20866	5.300		
	0.20902	5.309	4	
	0.21260	5.400		
	0.21299	5.410	3	1/4 - 28 UNF
	0.21654	5.500		
7/32	0.21875	5.556		

Fract.	inch	mm	Gg. Let	Thread Size
	0.22047	5.600		
	0.22098	5.613	2	
	0.22441	5.700		
	0.22799	5.791	1	
	0.22835	5.800		
	0.23228	5.900		
	0.23402	5.944	A	
15/64	0.23438	5.953		
	0.23622	6.000		M7.0 x 1.00 (C)
	0.23799	6.045	B	
	0.24016	6.100		
	0.24201	6.147	C	
	0.24409	6.200		M7.0 x 0.75 (F)
	0.24598	6.248	D	
	0.24803	6.300		
1/4	0.25000	6.350	E	
	0.25197	6.400		
	0.25591	6.500		
	0.25701	6.528	F	5/16 - 18 UNC
	0.25984	6.600		
	0.26098	6.629	G	
	0.26378	6.700		
17/64	0.26563	6.747		
	0.26598	6.756	H	
	0.26772	6.800		M8.0 x 1.25 (C)
	0.27165	6.900		
	0.27201	6.909	I	5/16 - 24 UNF
	0.27559	7.000		M8.0 x 1.00 (F)
	0.27701	7.036	J	
	0.27953	7.100		
	0.28098	7.137	K	
9/32	0.28125	7.144		
	0.28346	7.200		M8.0 x 0.75 (SP)
	0.28740	7.300		
	0.29000	7.366	L	
	0.29134	7.400		
	0.29500	7.493	M	
	0.29528	7.500		
19/64	0.29688	7.541		
	0.29921	7.600		
	0.30201	7.671	N	
	0.30315	7.700		
	0.30709	7.800		M9.0 x 1.25 (C)
	0.31102	7.900		
5/16	0.31250	7.938		3/8 - 16 UNC
	0.31496	8.000		M9.0 x 1.00 (SP)
	0.31598	8.026	O	
	0.31890	8.100		
	0.32283	8.200		
	0.32299	8.204	P	

(C)=Coarse Thread (F)=Fine Thread (SP)=Special Pitch Thread

*Also refer pages 36-47 for Tapping Drill Sizes

DRILL SIZE AND DECIMAL EQUIVALENT CHART

(*INCLUDES TAPPING DRILL SIZES FOR THREAD CUTTING TAPS)

Fract.	inch	mm	Letter	Thread Size
	0.32677	8.300		
21/64	0.32813	8.334		
	0.33071	8.400		
	0.33201	8.433	Q	3/8 - 24 UNF
	0.33465	8.500		M10.0 x 1.50 (C)
	0.33858	8.600		
	0.33902	8.611	R	
	0.34252	8.700		
11/32	0.34375	8.731		
	0.34646	8.800		M10.0 x 1.25 (F)
	0.34799	8.839	S	
	0.35039	8.900		
	0.35433	9.000		M10.0 x 1.00 (SP)
	0.35799	9.093	T	
	0.35827	9.100		
23/64	0.35938	9.128		
	0.36220	9.200		M10.0 x 0.75 (SP)
	0.36614	9.300		
	0.36799	9.347	U	7/16 - 14 UNC
	0.37008	9.400		
	0.37402	9.500		M11.0 x 1.50 (C)
3/8	0.37500	9.525		
	0.37701	9.576	V	
	0.37795	9.600		
	0.38189	9.700		
	0.38583	9.800		
	0.38598	9.804	W	
	0.38976	9.900		
25/64	0.39063	9.922		7/16 - 20 UNF
	0.39370	10.000		
	0.39701	10.084	X	
	0.39764	10.100		
	0.40157	10.200		M12.0 x 1.75 (C)
	0.40402	10.262	Y	
	0.40551	10.300		
13/32	0.40625	10.319		
	0.40945	10.400		
	0.41299	10.490	Z	
	0.41339	10.500		M12.0 x 1.50 (SP)
	0.41732	10.600		
	0.42126	10.700		
27/64	0.42188	10.716		1/2 - 13 UNC
	0.42520	10.800		M12.0 x 1.25 (F)
	0.42913	10.900		
	0.43307	11.000		M12.0 x 1.00 (SP)
	0.43701	11.100		
7/16	0.43750	11.113		
	0.44094	11.200		
	0.44488	11.300		
	0.44882	11.400		

Fraction	inch	mm	Thread Size
	0.45276	11.500	
29/64	0.45313	11.509	1/2 - 20 UNF
	0.45669	11.600	
	0.46063	11.700	
	0.46457	11.800	
	0.46850	11.900	
15/32	0.46875	11.906	
	0.47244	12.000	M14.0 x 2.00 (C)
	0.47638	12.100	
	0.48031	12.200	
	0.48425	12.300	
31/64	0.48438	12.303	9/16 - 12 UNC
	0.48819	12.400	
	0.49213	12.500	M14.0 x 1.50 (F)
	0.49606	12.600	
1/2	0.50000	12.700	
	0.50394	12.800	M14.0 x 1.25 (SP)
	0.50787	12.900	
	0.51181	13.000	
33/64	0.51563	13.097	9/16 - 18 UNF
	0.51575	13.100	
	0.51969	13.200	
	0.52362	13.300	
	0.52756	13.400	
17/32	0.53125	13.494	5/8 - 11 UNC
	0.53150	13.500	
	0.53543	13.600	
	0.53937	13.700	
	0.54331	13.800	
35/64	0.54688	13.891	
	0.54724	13.900	
	0.55118	14.000	M16.0 x 2.00 (C)
	0.56102	14.250	
9/16	0.56250	14.288	
	0.57087	14.500	M16.0 x 1.50 (F)
37/64	0.57813	14.684	5/8 - 18 UNF
	0.58071	14.750	
	0.59055	15.000	M16.0 x 1.00 (SP)
19/32	0.59375	15.081	
	0.60039	15.250	
39/64	0.60938	15.478	
	0.61024	15.500	M18.0 x 2.50 (C)
	0.62008	15.750	
5/8	0.62500	15.875	
	0.62992	16.000	
	0.63976	16.250	
41/64	0.64063	16.272	
	0.64961	16.500	M18.0 x 1.50 (F)
21/32	0.65625	16.669	3/4 - 10 UNC
	0.65945	16.750	

(C) = Coarse Thread (F) = Fine Thread (SP) = Special Pitch Thread

*Also refer pages 36-47 for Tapping Drill Sizes

DRILL SIZE AND DECIMAL EQUIVALENT CHART
(*INCLUDES TAPPING DRILL SIZES FOR THREAD CUTTING TAPS)

Fraction	inch	mm	Thread Size
	0.66929	17.000	
43/64	0.67188	17.066	
	0.67913	17.250	
11/16	0.68750	17.463	3/4 - 16 UNF
	0.68898	17.500	M20.0 x 2.50 (C)
	0.69882	17.750	
45/64	0.70313	17.859	
	0.70866	18.000	M20.0 x 2.00 (SP)
	0.71850	18.250	
23/32	0.71875	18.256	
	0.72835	18.500	M20.0 x 1.50 (F)
47/64	0.73438	18.653	
	0.73819	18.750	
	0.74803	19.000	M20.0 x 1.00 (SP)
3/4	0.75000	19.050	
	0.75787	19.250	
49/64	0.76563	19.447	7/8 - 9 UNC
	0.76772	19.500	M22.0 x 2.50 (C)
	0.77756	19.750	
25/32	0.78125	19.844	
	0.78740	20.000	
51/64	0.79688	20.241	
	0.79724	20.250	
	0.80709	20.500	M22.0 x 1.50 (F)
13/16	0.81250	20.638	7/8 - 14 UNF
	0.81693	20.750	
	0.82677	21.000	M24.0 x 3.00 (C)
53/64	0.82813	21.034	
	0.83333	21.250	
27/32	0.84375	21.431	
	0.84646	21.500	
	0.85630	21.750	
55/64	0.85938	21.828	
	0.86614	22.000	M24.0 x 2.00 (F)
7/8	0.87500	22.225	1 - 8 UNC
	0.87598	22.250	
	0.88583	22.500	M24.0 x 1.50 (SP)
57/64	0.89063	22.622	
	0.89567	22.750	
	0.90551	23.000	
29/32	0.90625	23.019	
	0.91535	23.250	
59/64	0.92188	23.416	1 - 12 UNF
	0.92520	23.500	M25.0 x 1.50 (SP)
	0.93504	23.750	
15/16	0.93750	23.813	1 - 14 UNS
	0.94488	24.000	M27.0 x 3.00 (C)
61/64	0.95313	24.209	
	0.95472	24.250	
	0.96457	24.500	
31/32	0.96875	24.606	

Fraction	inch	mm	Thread Size
	0.97441	24.750	
	0.98425	25.000	M27.0 x 2.00 (F)
63/64	0.98438	25.003	1 1/8 - 7 UNC
	0.99409	25.250	
1	1.00000	25.400	
	1.00394	25.500	
	1.01378	25.750	
1 1/64	1.01563	25.797	
	1.02362	26.000	
1 1/32	1.03125	26.194	
	1.03346	26.250	
	1.04331	26.500	M30.0 x 3.50 (C)
1 3/64	1.04688	26.591	1 1/8 - 12 UNF
	1.05315	26.750	
1 1/16	1.06250	26.988	
	1.06299	27.000	
	1.07283	27.250	
1 5/64	1.07813	27.384	
	1.08268	27.500	
	1.09252	27.750	
1 3/32	1.09375	27.781	
	1.10236	28.000	M30.0 x 2.00 (F)
1 7/64	1.10938	28.178	1 1/4 - 7 UNC
	1.11220	28.250	
	1.12205	28.500	
1 1/8	1.12500	28.575	
	1.13189	28.750	
1 9/64	1.14063	28.972	
	1.14173	29.000	
	1.15157	29.250	
1 5/32	1.15625	29.369	
	1.16142	29.500	M33.0 x 3.50 (C)
	1.17126	29.750	
1 11/64	1.17188	29.766	1 1/4 - 12 UNF
	1.18110	30.000	
1 3/16	1.18750	30.163	
	1.19094	30.250	
	1.20079	30.500	M32.0 x 1.50 (SP)
1 13/64	1.20313	30.559	
	1.21063	30.750	
1 7/32	1.21875	30.956	1 3/8 - 6 UNC
	1.22047	31.000	
	1.23031	31.250	
1 15/64	1.23438	31.353	
	1.24010	31.500	
1 1/4	1.25000	31.750	
	1.25984	32.000	M36.0 x 4.00 (C)
1 17/64	1.26563	32.147	
	1.27953	32.500	
1 9/32	1.28125	32.544	
1 19/64	1.29688	32.941	1 3/8 - 12 UNF

(C)=Coarse Thread (F)=Fine Thread (SP)=Special Pitch Thread

*Also refer pages 36-47 for Tapping Drill Sizes

DRILL SIZE AND DECIMAL EQUIVALENT CHART
(*INCLUDES TAPPING DRILL SIZES FOR THREAD CUTTING TAPS)

Fraction	inch	mm	Thread Size
	1.29921	**33.000**	
1 5/16	1.31250	33.338	
	1.31890	33.500	
1 21/64	1.32813	33.734	
	1.33858	**34.000**	
1 11/32	1.34375	34.131	1 1/2 - 6 UNC
	1.35827	34.500	
1 23/64	1.35938	34.528	
1 3/8	1.37500	34.925	
	1.37795	**35.000**	M39.0 x 4.00 (C)
1 25/64	1.39063	35.322	
	1.39764	35.500	
1 13/32	1.40625	35.719	
	1.41732	**36.000**	
1 27/64	1.42188	36.116	1 1/2 - 12 UNF
	1.43701	36.500	
1 7/16	1.43750	36.513	
1 29/64	1.45313	36.909	
	1.45669	**37.000**	
1 15/32	1.46875	37.306	
	1.47638	37.500	M42.0 x 4.50 (C)
1 31/64	1.48438	37.703	
	1.49606	**38.000**	
1 1/2	1.50000	38.100	
1 33/64	1.51563	38.497	
	1.51575	38.500	
1 17/32	1.53125	38.894	
	1.53543	**39.000**	
1 35/64	1.54688	39.291	
	1.55512	39.500	
1 9/16	1.56250	39.688	1 3/4 - 5 UNC
	1.57480	**40.000**	
1 37/64	1.57813	40.084	
1 19/32	1.59375	40.481	
	1.59449	40.500	M45.0 x 4.50 (C)
1 39/64	1.60938	40.878	
	1.61417	**41.000**	
1 5/8	1.62500	41.275	
	1.63386	41.500	
1 41/64	1.64063	41.672	
	1.65354	**42.000**	
1 21/32	1.65625	42.069	
1 43/64	1.67188	42.466	
	1.67323	42.500	
1 11/16	1.68750	42.863	
	1.69291	**43.000**	M48.0 x 5.00 (C)
1 45/64	1.70313	43.259	
	1.71260	43.500	
1 23/32	1.71875	43.656	
	1.73228	**44.000**	
1 47/64	1.73438	44.053	

Fraction	inch	mm	Thread Size
1 3/4	1.75000	44.450	
	1.75197	44.500	
1 49/64	1.76563	44.847	
	1.77165	**45.000**	
1 25/32	1.78125	45.244	2 - 4 1/2 UNC
	1.79134	45.500	
1 51/64	1.79688	45.641	
	1.81102	**46.000**	
1 13/16	1.81250	46.038	
1 53/64	1.82813	46.434	
	1.83071	46.500	
1 27/32	1.84375	46.831	
	1.85039	**47.000**	M52.0 x 5.00 (C)
1 55/64	1.85938	47.228	
	1.87008	47.500	
1 7/8	1.87500	47.625	
	1.88976	**48.000**	
1 57/64	1.89063	48.022	
1 29/32	1.90625	48.419	
	1.90945	48.500	
1 59/64	1.92188	48.816	
	1.92913	**49.000**	
1 15/16	1.93750	49.213	
	1.94882	49.500	
1 61/64	1.95313	49.609	
	1.96850	**50.000**	
1 31/32	1.96875	50.006	
1 63/64	1.98438	50.403	
	1.98819	50.500	M56.0 x 5.50 (C)
2	2.00000	50.800	
	2.00787	**51.000**	
2 1/32	2.03125	51.594	2 1/4 - 4 1/2 UNC
	2.04724	**52.000**	
2 1/16	2.06250	52.388	
	2.07283	52.650	
	2.08661	**53.000**	
2 3/32	2.09375	53.181	
2 1/8	2.12500	53.975	
	2.12598	**54.000**	
2 5/32	2.15625	54.769	
	2.16535	**55.000**	
2 3/16	2.18750	55.563	
	2.20472	**56.000**	
2 7/32	2.21875	56.356	
	2.24409	**57.000**	
2 1/4	2.25000	57.150	2 1/2 - 4 UNC
2 9/32	2.28125	57.944	
	2.28346	**58.000**	M64.0 x 6 (C)
	2.30314	58.500	
2 5/16	2.31250	58.738	
	2.32283	**59.000**	

(C)=Coarse Thread

*Also refer pages 36-47 for Tapping Drill Sizes

DRILL SIZE AND DECIMAL EQUIVALENT CHART
(*INCLUDES TAPPING DRILL SIZES FOR THREAD CUTTING TAPS)

Fraction	inch	mm	Thread Size
2 11/32	2.34375	59.531	
	2.36220	**60.000**	
2 3/8	2.37500	60.325	
	2.40157	**61.000**	
2 13/32	2.40625	61.119	
2 9/16	2.43750	61.913	
	2.44094	**62.000**	**M72.0 x 6 (C)**
2 15/32	2.46875	62.706	
	2.48031	**63.000**	
2 1/2	2.50000	63.500	**2 3/4 - 4 UNC**
	2.51969	**64.000**	
2 17/32	2.53125	64.294	
	2.54921	64.750	
	2.55906	**65.000**	
2 9/16	2.56250	65.088	
2 19/32	2.59375	65.881	
	2.59843	**66.000**	
2 5/8	2.62500	66.675	
	2.63780	**67.000**	
2 21/32	2.65625	67.469	
	2.67717	**68.000**	
2 11/16	2.68750	68.263	
	2.71654	**69.000**	
2 23/32	2.71875	69.056	
2 3/4	2.75000	69.850	**3 - 4 UNC**
	2.75591	**70.000**	
2 25/32	2.78125	70.644	
	2.79528	**71.000**	
	2.79921	71.100	
2 13/16	2.81250	71.438	
	2.83465	**72.000**	
2 27/32	2.84375	72.231	
	2.87402	**73.000**	
2 7/8	2.87500	73.025	
2 29/32	2.90625	73.819	
	2.91339	**74.000**	**M80.0 x 6 (C)**
2 15/16	2.93750	74.613	
	2.95276	**75.000**	
2 31/32	2.96875	75.406	
	2.99213	**76.000**	
3	3.00000	76.200	**3 1/4 - 4 UNC**
3 1/32	3.03125	76.994	
	3.03150	**77.000**	
	3.04921	77.450	
3 1/16	3.06250	77.788	
	3.07087	**78.000**	
3 3/32	3.09375	78.581	
	3.11024	**79.000**	
3 1/8	3.12500	79.375	
	3.14961	**80.000**	

Fraction	inch	mm	Thread Size
3 5/32	3.15625	80.169	
3 3/16	3.18750	80.963	
	3.18898	**81.000**	
3 7/32	3.21875	81.756	
	3.22835	**82.000**	
3 1/4	3.25000	82.550	**3 1/2 - 4 UNC**
	3.26772	**83.000**	
3 9/32	3.28125	83.344	
	3.29921	83.800	
	3.30709	**84.000**	**M90.0 x 6 (C)**
3 5/16	3.31250	84.138	
3 11/32	3.34375	84.931	
	3.34646	**85.000**	
3 3/8	3.37500	85.725	
	3.38583	**86.000**	
3 13/32	3.40625	86.519	
	3.42520	**87.000**	
3 7/16	3.43750	87.313	
	3.46457	**88.000**	
3 15/32	3.46875	88.106	
3 1/2	3.50000	88.900	**3 3/4 - 4 UNC**
	3.50394	**89.000**	
3 17/32	3.53125	89.694	
	3.54331	**90.000**	
	3.54921	90.150	
3 9/16	3.56250	90.488	
	3.58268	**91.000**	
3 19/32	3.59375	91.281	
	3.62205	**92.000**	
3 5/8	3.62500	92.075	
3 21/32	3.65625	92.869	
	3.66142	**93.000**	
3 11/16	3.68750	93.663	
	3.70079	**94.000**	**M100.0 x 6 (C)**
3 23/32	3.71875	94.456	
	3.74016	**95.000**	
3 3/4	3.75000	95.250	**4 - 4 UNC**
	3.77953	**96.000**	
3 25/32	3.78125	96.044	
	3.79921	96.500	
3 13/16	3.81250	96.838	
	3.81890	**97.000**	
3 27/32	3.84375	97.631	
	3.85827	**98.000**	
3 7/8	3.87500	98.425	
	3.89764	**99.000**	
3 29/32	3.90625	99.219	
	3.93701	**100.000**	
3 15/16	3.93750	100.013	
4	4.00000	101.600	

(C)=Coarse Thread

*Also refer pages 36-47 for Tapping Drill Sizes

SPANNER SELECTION CHART

UNF / UNC

Thread Size (inch)	Spanner Size A/F (inch)	Spanner Opening min (inch)	Spanner Opening max (inch)
1/4"	7/16	0.440	0.446
5/16"	1/2	0.504	0.510
3/8"	9/16	0.566	0.573
7/16"	5/8	0.629	0.636
1/2"	3/4	0.755	0.763
9/16"	13/16	0.818	0.826
5/8"	15/16	0.944	0.953
3/4"	1 1/8	1.132	1.142
7/8"	1 5/16	1.320	1.331
1"	1 1/2	1.508	1.520
1 1/8"	1 11/16	1.696	1.708
1 1/4"	1 7/8	1.885	1.898
1 3/8"	2 1/16	2.074	2.088
1 1/2"	2 1/4	2.262	2.277
1 3/4"	2 5/8	2.639	2.656
2"	3	3.016	3.035
2 1/4"	3 3/8	3.393	3.414
2 1/2"	3 3/4	3.770	3.793
2 3/4"	4 1/8	4.147	4.172
3"	4 1/2	4.524	4.550

BS (Whitworth)

Thread Size BS	Spanner Size A/F Whit	Spanner Size A/F (inch)	Spanner Size A/F (mm)	Nearest Fitting Spanner Size A/F (inch)	Nearest Fitting Spanner Size A/F (mm)
1/4"	3/16 W	0.445	11.30	7/16	12 mm
5/16"	1/4 W	0.525	13.33	17/32	14 mm
3/8"	5/16 W	0.600	15.24	19/32	16 mm
7/16"	3/8 W	0.710	18.03	20/32	18 mm
1/2"	7/16 W	0.820	20.83	13/16	21 mm
9/16"	1/2 W	0.920	23.37	15/16	24 mm
5/8"	9/16 W	1.010	25.65	1	26 mm
3/4"	11/16 W	1.200	30.48	1 1/4	31 mm
7/8"	3/4 W	1.300	33.02	1 5/16	33 mm
1"	7/8 W	1.480	37.59	1 1/2	38 mm
1 1/8"	1 W	1.670	42.42	1 11/16	43 mm
1 1/4"	1 1/8 W	1.860	47.24	1 7/8	48 mm
1 3/8"	1 1/4 W	2.050	52.07	2 1/16	52 mm
1 1/2"	1 3/8 W	2.220	56.39	2 1/4	57 mm
1 3/4"	1 5/8 W	2.580	65.53	2 9/16	66 mm
2"	1 3/4 W	2.760	70.10	2 3/4	70 mm

NOTE: Whitworth (W) = Old Standard
BS = New Standard

METRIC

Thread Size (mm)	Spanner Size A/F (mm)	Spanner Opening min (mm)	Spanner Opening max (mm)
M 1.6	3.2	3.22	3.28
M 2.0	4.0	4.02	4.12
M 2.5	5.0	5.02	5.12
M 3.0	5.5	5.52	5.62
M 4.0	7.0	7.03	7.15
M 5.0	8.0	8.03	8.15
M 6.0	10.0	10.04	10.19
M 7.0	11.0	11.04	11.19
M 8.0	13.0	13.04	13.24
M 10.0	17.0	17.05	17.30
M 12.0	19.0	19.06	19.36
M 14.0	22.0	22.06	22.36
M 16.0	24.0	24.06	24.36
M 18.0	27.0	27.08	27.48
M 20.0	30.0	30.08	30.48
M 22.0	32.0	32.08	32.48
M 24.0	36.0	36.10	36.60
M 27.0	41.0	41.10	41.60
M 30.0	46.0	46.10	46.60
M 33.0	50.0	50.10	50.60
M 36.0	55.0	55.12	55.72
M 39.0	60.0	60.12	60.72
M 42.0	65.0	65.12	65.72
M 45.0	70.0	70.12	70.72
M 48.0	75.0	75.15	75.85
M 52.0	80.0	80.15	80.85
M 56.0	85.0	85.15	85.85
M 60.0	90.0	90.15	90.85
M 64.0	95.0	95.15	95.85
M 68.0	100.0	100.15	100.85
M 72.0	105.0	105.20	106.00
M 76.0	110.0	110.20	111.00
M 80.0	115.0	115.20	116.00
M 85.0	120.0	120.20	121.00
M 90.0	130.0	130.20	131.00
M 95.0	135.0	135.20	136.00
M 100.0	145.0	145.20	146.00
M 105.0	150.0	150.25	151.25
M 110.0	155.0	155.25	156.25

SOCKET SIZING CROSS REFERENCE CHART
INCH TO METRIC

Socket Sizes A/F (inch)	Metric Equivalent (mm)	Nearest Alternative Socket Size A/F (mm)	Diff. in (mm)
3/16"	4.762	5mm	+0.238
7/32"	5.556	5.5mm	−0.056
1/4"	6.350		
9/32"	7.144		
5/16"	7.937	8mm	+0.063
11/32"	8.731	9mm	+0.269
3/8"	9.525	10mm	+0.475
7/16"	11.112	11mm	−0.112
15/32"	11.906	12mm	+0.094
1/2"	12.700	13mm	+0.300
17/32"	13.494	14mm	+0.506
9/16"	14.287		
19/32"	15.081	15mm	−0.081
5/8"	15.875	16mm	+0.125
11/16"	17.462		
3/4"	19.050	19mm	−0.050
25/32"	19.844	20mm	+0.156
13/16"	20.637	21mm	+0.363
7/8"	22.225	22mm	−0.225
15/16"	23.812	24mm	+0.188
1"	25.400	26mm	+0.600
1 1/16"	26.987	27mm	+0.013
1 1/8"	28.575	29mm	+0.425
1 3/16"	30.162	30mm	−0.162
1 1/4"	31.750	32mm	+0.250
1 5/16"	33.337	33mm	−0.337
1 3/8"	34.925	35mm	+0.075
1 7/16"	36.512	37mm	+0.488
1 1/2"	38.100	38mm	−0.100
1 9/16"	39.687	40mm	+0.313
1 5/8"	41.275	41mm	−0.275
1 11/16"	42.862	43mm	+0.138
1 3/4"	44.450	45mm	+0.550
1 13/16"	46.037	46mm	−0.037
1 7/8"	47.625	48mm	+0.375
1 15/16"	49.212	49mm	−0.212
2"	50.800	50mm	+0.200
2 1/16"	52.387	52mm	−0.387
2 1/8"	53.975	54mm	+0.025
2 3/16"	55.562	56mm	+0.438
2 1/4"	57.150	57mm	−0.150
2 5/16"	58.737	59mm	+0.263
2 3/8"	60.325	60mm	−0.325
2 7/16"	61.912	62mm	+0.088
2 1/2"	63.500	63mm	−0.500
2 9/16"	65.087	65mm	−0.087
2 5/8"	66.675	67mm	+0.325
2 11/16"	68.262		
2 3/4"	69.850	70mm	+0.150

Socket Sizes A/F (inch)	Metric Equivalent (mm)	Nearest Alternative Socket Size A/F (mm)	Diff. in (mm)
2 13/16"	71.437		
2 7/8"	73.025		
2 15/16"	74.612	75mm	+0.388
3"	76.200		
3 1/16"	77.787		
3 1/8"	79.375	80mm	+0.625
3 3/16"	80.962		
3 1/4"	82.550		
3 5/16"	84.137		
3 3/8"	85.725		
3 7/16"	87.312		
3 1/2"	88.900		
3 9/16"	90.487	90mm	−0.487
3 5/8"	92.075		
3 3/4"	95.250	95mm	−0.250
3 13/16"	96.837		
3 7/8"	98.425		
3 15/16"	100.012	100mm	−0.012
4"	101.600		
4 1/16"	103.187		
4 1/8"	104.775	105mm	+0.225
4 3/16"	106.362		
4 1/4"	107.950		
4 5/16"	109.537	110mm	+0.463
4 3/8"	111.125		
4 7/16"	112.712		
4 1/2"	114.300	115mm	+0.700
4 5/8"	117.475		
4 3/4"	120.650	120mm	−0.650
4 7/8"	123.825		
5"	127.000		
5 1/8"	130.175	130mm	−0.175
5 1/4"	133.350		
5 3/8"	136.525		
5 7/16"	138.112		
5 1/2"	139.700	140mm	+0.300
5 5/8"	142.875		
5 3/4"	146.050		
5 7/8"	149.225		
6"	152.400	155mm	+2.600
6 1/8"	155.575	155mm	−0.575
6 1/4"	158.750	160mm	+1.250
6 3/8"	161.925		
6 1/2"	165.100	165mm	−0.100
6 5/8"	168.275		
6 3/4"	171.450		
6 7/8"	174.625		
7"	177.800		

SOCKET SIZING CROSS REFERENCE CHART
METRIC TO INCH

Socket Hex Sizes A/F (mm)	Decimal Equivalent (inch)	Nearest Alternative Socket Size A/F (inch)	Diff. in (mm)
3.5mm	0.138"		
4mm	0.157"		
4.5mm	0.177"		
5mm	0.197"		
5.5mm	0.216"	7/32"	+0.056
6mm	0.236"		
7mm	0.275"		
8mm	0.315"	5/16"	−0.063
9mm	0.354"		
10mm	0.394"		
11mm	0.433"	7/16"	+0.112
12mm	0.472"	15/32"	−0.094
13mm	0.512"		
14mm	0.551"	9/16"	+0.287
15mm	0.590"	19/32"	+0.081
16mm	0.630"	5/8"	−0.127
17mm	0.669"		
18mm	0.709"		
19mm	0.748"	3/4"	+0.051
20mm	0.787"	25/32"	−0.157
21mm	0.827"		
22mm	0.866"	7/8"	+0.226
23mm	0.905"		
24mm	0.945"	15/16"	−0.188
25mm	0.984"	1"	+0.400
26mm	1.024"		
27mm	1.063"	1 1/16"	−0.012
28mm	1.102"		
29mm	1.142"		
30mm	1.181"	1 3/16"	+0.162
31mm	1.220"		
32mm	1.260"	1 1/4"	−0.250
33mm	1.299"	1 5/16"	+0.337
34mm	1.338"		
35mm	1.378"	1 3/8"	−0.075
36mm	1.417"	1 7/16"	+0.512
37mm	1.457"		
38mm	1.496"	1 1/2"	+0.100
39mm	1.535"		
40mm	1.575"	1 9/16"	−0.312
41mm	1.614"	1 5/8"	+0.275
42mm	1.653"		
43mm	1.693"	1 11/16"	−0.137
44mm	1.732"		
45mm	1.772"		
46mm	1.811"	1 13/16"	+0.037
47mm	1.850"		
48mm	1.890"		
49mm	1.929"		
50mm	1.968"		
52mm	2.047"	2 1/16"	+0.387
54mm	2.126"	2 1/8"	−0.025
55mm	2.165"	2 3/16"	+0.562
56mm	2.205"	2 3/16"	−0.437
57mm	2.244"	2 1/4"	+0.150
58mm	2.283"		
60mm	2.362"	2 3/8"	+0.325
62mm	2.441"	2 7/16"	−0.087
63mm	2.480"	2 1/2"	+0.500
65mm	2.559"	2 9/16"	+0.087
67mm	2.638"	2 5/8"	−0.325
70mm	2.756"	2 3/4"	−0.150
75mm	2.953"	2 15/16"	−0.387
80mm	3.150"		
85mm	3.346"	3 3/8"	+0.725
90mm	3.543"	3 9/16"	+0.487
95mm	3.740"	3 3/4"	+0.250
100mm	3.937"	3 15/16"	+0.012
105mm	4.134"	4 1/8"	−0.225
110mm	4.331"		
115mm	4.527"	4 1/2"	−0.700
120mm	4.724"	4 3/4"	+0.650
125mm	4.921"		
130mm	5.118"	5 1/8"	+0.175
135mm	5.315"		
140mm	5.512"	5 1/2"	−0.300
145mm	5.709"		
150mm	5.905"		
155mm	6.102"	6 1/8"	+0.575
160mm	6.299"		
165mm	6.496"	6 1/2"	+0.100
170mm	6.693"		

TORQUE CONVERSION CHART

1 lb ft = 1.355817948 Nm   1 Nm = 0.737562149 lb ft

From pound-foot to Newton-meter				From Newton-meter to pound-foot			
lb ft ⇨	Nm	lb ft ⇨	Nm	Nm ⇨	lb ft	Nm ⇨	lb ft
1	1.36	51	69.14	1	0.74	51	37.62
2	2.71	52	70.50	2	1.48	52	38.35
3	4.07	53	71.86	3	2.21	53	39.09
4	5.42	54	73.21	4	2.95	54	39.83
5	6.78	55	74.57	5	3.69	55	40.57
6	8.13	56	75.93	6	4.43	56	41.30
7	9.49	57	77.28	7	5.16	57	42.04
8	10.85	58	78.64	8	5.90	58	42.78
9	12.20	59	79.99	9	6.64	59	43.52
10	13.56	60	81.35	10	7.38	60	44.25
11	14.91	61	82.70	11	8.11	61	44.99
12	16.27	62	84.06	12	8.85	62	45.73
13	17.63	63	85.42	13	9.59	63	46.47
14	18.98	64	86.77	14	10.33	64	47.20
15	20.34	65	88.13	15	11.06	65	47.94
16	21.69	66	89.48	16	11.80	66	48.68
17	23.05	67	90.84	17	12.54	67	49.42
18	24.40	68	92.20	18	13.28	68	50.15
19	25.76	69	93.55	19	14.01	69	50.89
20	27.12	70	94.91	20	14.75	70	51.63
21	28.47	71	96.26	21	15.49	71	52.37
22	29.83	72	97.62	22	16.23	72	53.10
23	31.18	73	98.97	23	16.96	73	53.84
24	32.54	74	100.33	24	17.70	74	54.58
25	33.90	75	101.69	25	18.44	75	55.32
26	35.25	76	103.04	26	19.18	76	56.05
27	36.61	77	104.40	27	19.91	77	56.79
28	37.96	78	105.75	28	20.65	78	57.53
29	39.32	79	107.11	29	21.39	79	58.27
30	40.67	80	108.47	30	22.13	80	59.00
31	42.03	81	109.82	31	22.86	81	59.74
32	43.39	82	111.18	32	23.60	82	60.48
33	44.74	83	112.53	33	24.34	83	61.22
34	46.10	84	113.89	34	25.08	84	61.96
35	47.45	85	115.24	35	25.81	85	62.69
36	48.81	86	116.60	36	26.55	86	63.43
37	50.17	87	117.96	37	27.29	87	64.17
38	51.52	88	119.31	38	28.03	88	64.91
39	52.88	89	120.67	39	28.76	89	65.64
40	54.23	90	122.02	40	29.50	90	66.38
41	55.59	91	123.38	41	30.24	91	67.12
42	56.94	92	124.74	42	30.98	92	67.86
43	58.30	93	126.10	43	31.72	93	68.59
44	59.66	94	127.45	44	32.45	94	69.33
45	61.01	95	128.80	45	33.19	95	70.07
46	62.37	96	130.16	46	33.93	96	70.81
47	63.72	97	131.51	47	34.67	97	71.54
48	65.08	98	132.87	48	35.40	98	72.28
49	66.44	99	134.23	49	36.14	99	73.02
50	67.79	100	135.58	50	36.88	100	73.76

ABBREVIATIONS (SYMBOLS) USED FOR STANDARD THREADS

Symbol	Reference
ACME-G	Acme Thread – General Purpose
AMO	American Standard Microscope Objective Thread
ANPT	Aeronautical National Pipe Taper Thread
BA	British Association Standard Thread
BSB	British Standard Brass (26 tpi) Thread
BSC	British Standard Cycle Thread
BS CON	British Standard Conduit Thread
BSF	British Standard Fine Thread
BS GAS	British Standard Gas Thread
BSPP	British Standard Pipe Parallel Thread
BSPT	British Standard Pipe Taper Thread
BSW	British Standard Whitworth (Coarse) Thread
M	ISO Metric (Coarse Pitch) Thread
ME	Model Engineers Thread
MF	ISO Metric (Fine Pitch) Thread
N	National 8, 12 and 16 Thread Series (8N, 12N, 16N)
N BUTT	National Buttress Thread
NC	National Coarse Thread Series
NEF	National Extra-Fine Thread Series
NF	National Fine Thread Series
NGO	National Gas Outlet Thread (specify right or left hand)
NGS	National Gas Straight Thread
NGT	National Gas Taper Thread (See also "SGT")
NH	National Hose (Garden & Fire Hose Coupling) Thread
NPT	National Pipe Taper Thread
NPTF	National Pipe Taper Fuel (Fuel & Oil) Thread
NPTR	National Pipe Taper Railing (Railing Joints) Thread
NPS	National Pipe Straight Thread
NPSC	National Pipe Straight Coupling (Pipe Couplings) Thread
NPSF	National Pipe Straight Fuel (Fuel & Oil) Thread
NPSH	National Pipe Straight Hose (Hose Couplings) Thread
NPSI	National Pipe Straight Intermediate Thread
NPSL	National Pipe Straight Lock (Lock Nut) Thread
NPSM	National Pipe Straight Mechanical Thread
NS	National Special Thread
PG	Panzerrohr-Gewinde (Steel Conduit Thread)
SB	Stove Bolt Standard Thread
SGT	Special Gas Taper Thread
STI	Special Thread Insert (for Helical Coil Wire Screw Thread Inserts)
STUB ACME	Stub Acme Thread
TR	Trapezoidal Thread
UN	Unified National (Constant-Pitch) Thread Series
UNC	Unified National Coarse Thread
UNEF	Unified National Extra-Fine Thread Series
UNF	Unified National Fine Thread
UNJ	Unified Nat. Thread Series with a 0.1501p to 0.1804p Controlled Root Radius (external threads)
UNM	Unified National Miniature Thread Series
UNR	Unified National Round (round root) Thread with a 0.1082p to 0.1443p Controlled Root Radius
UNS	Unified National Special Thread
V	A 60° "V" Thread with Truncated Crest and Root.
WHIT	Whitworth Form (Special) Thread

COMMON THREAD FORMS USED IN THE ENGINEERING INDUSTRY

GENERAL SYMBOLS

Where:

	P = Pitch	H = Height of Fundamental Triangle
D = Major Dia., Internal Thread	D_1 = Minor Dia., Internal Thread	D_2 = Pitch Dia., Internal Thread
d = Major Dia., External Thread	d_1 = Minor Dia., External Thread	d_2 = Pitch Dia., External Thread

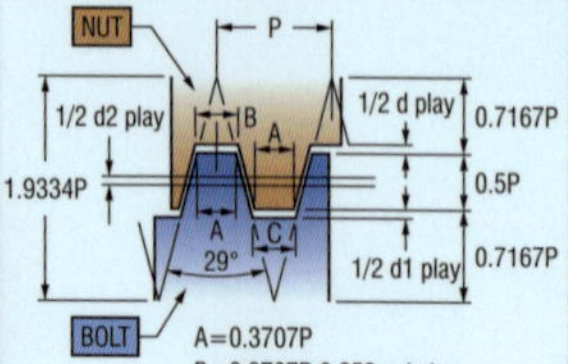

A=0.3707P
B=0.3707P-0.259 x d play
C=0.3707P-0.259 (d1 play-d2 play)

ACME

Use : Acme-General is used in the mechanical industry for motion transmission screws.

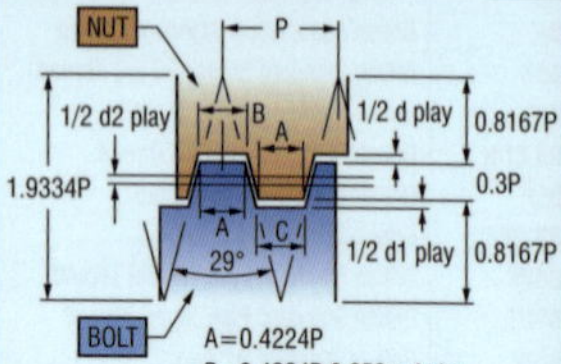

A=0.4224P
B=0.4224P-0.259 x d play
C=0.4224P-0.259 (d1 play-d2 play)

STUB ACME

Use : Stub Acme is used for motion transmission screws where normal Acme is too deep.

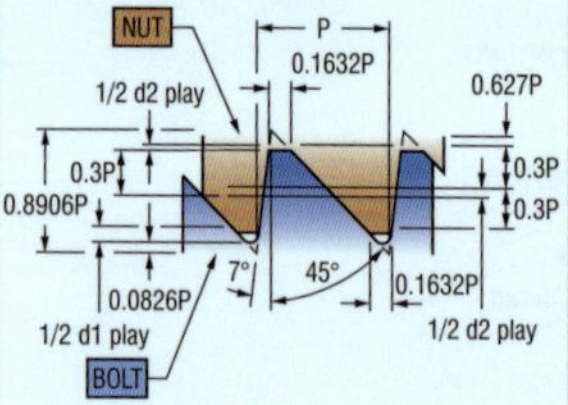

AMERICAN BUTTRESS

Use : Mechanical industry for motion transmission screws when the load is acting from one direction, e.g., in presses, rolling mills, etc.

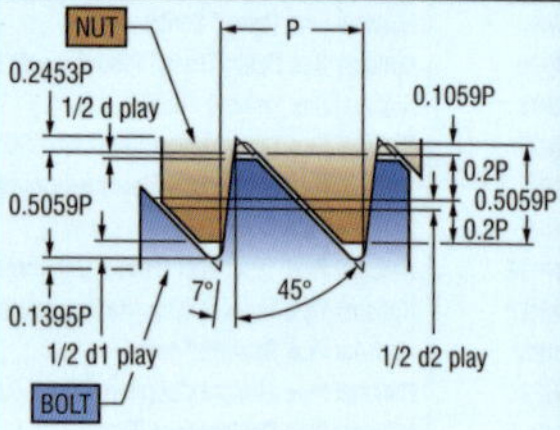

ENGLISH BUTTRESS

Use : Mechanical industry for motion transmission screws when the load is acting from one direction, e.g., in presses, rolling mills, etc.

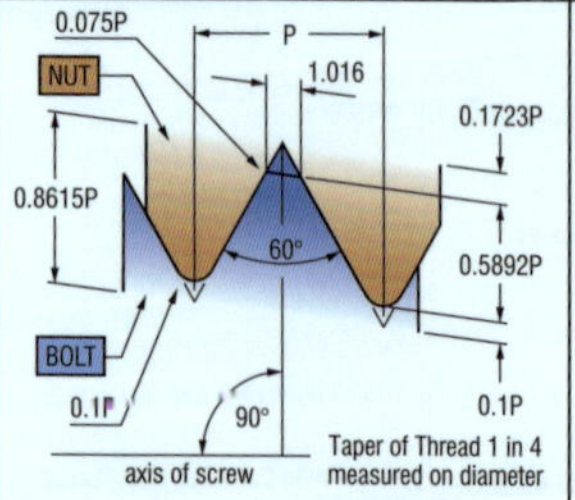

API : TAPER 1:4 (V-0.040" FLAT)

Use : Oil industry.

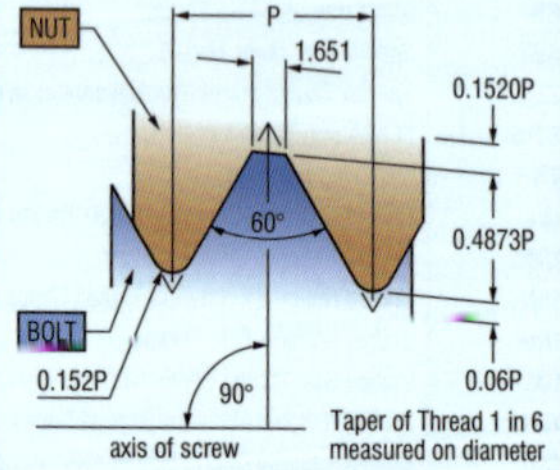

API : TAPER 1:6 (V-0.038"R)

Use : Oil industry.

COMMON THREAD FORMS USED IN THE ENGINEERING INDUSTRY

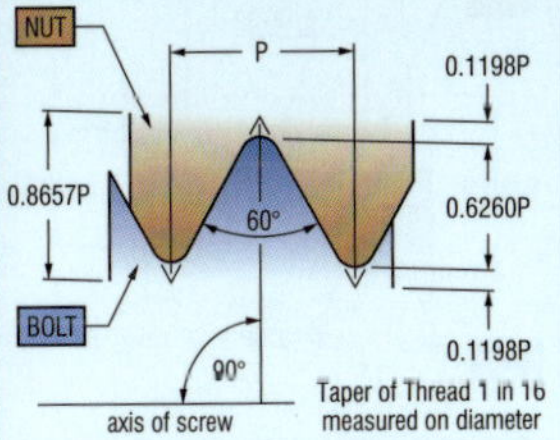

API : TAPER 1:16 (ROUNDED)

Use : Oil industry.

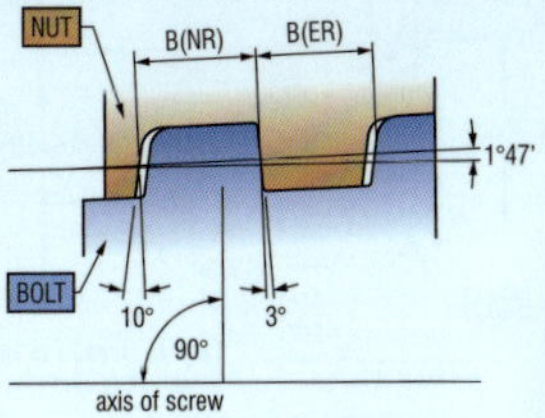

API BUTTRESS

Use : Oil industry for casing.

Pipe size : 4 1/2" to 13 3/8"

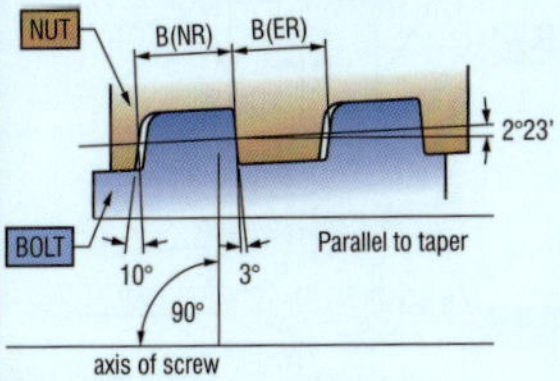

API BUTTRESS

Use : Oil industry for casing.

Pipe size : 16" and larger

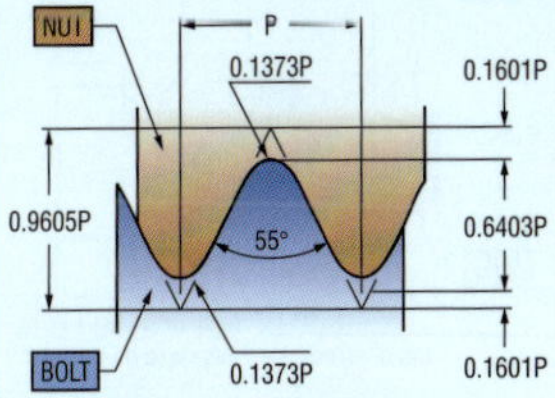

BSB (BRITISH STANDARD BRASS - 26 TPI)

Use : Brass fittings and pipe couplings.

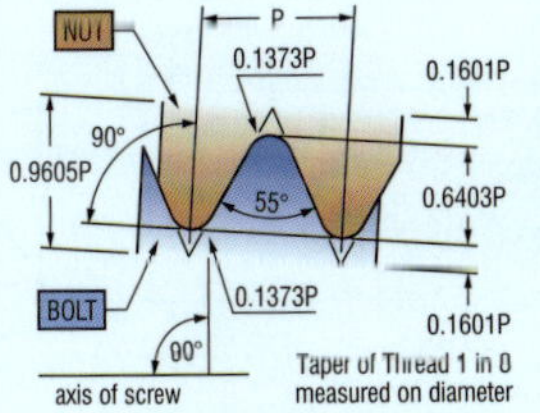

BS GAS (BRITISH STANDARD GAS)

Use : Valves for gas cylinders and similar pressure vessels.

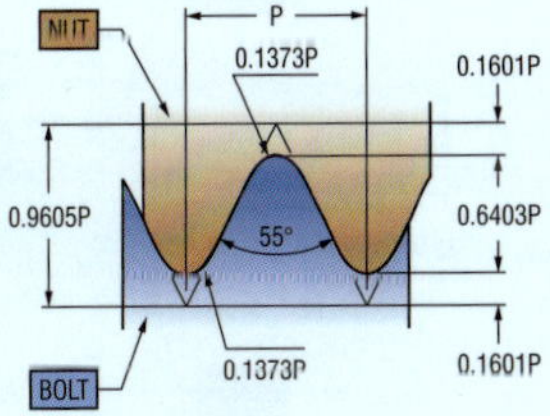

BSP (G SERIES) (BRITISH STD. PIPE PARALLEL)

Use : Fittings and pipe couplings for gas, water and sewer lines.

COMMON THREAD FORMS USED IN THE ENGINEERING INDUSTRY

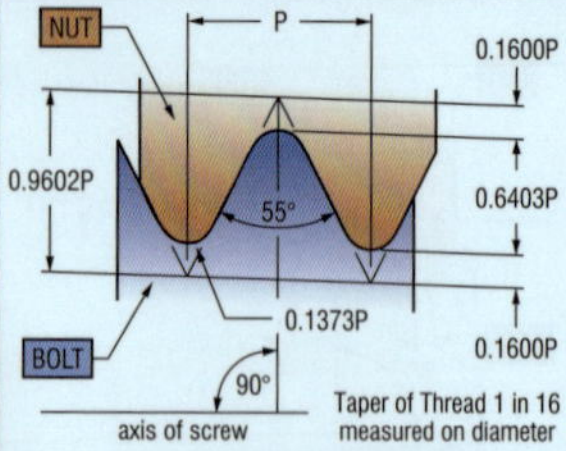

BSPT (RC SERIES) (BRITISH STD. PIPE TAPER)

Use : Pipe thread for steam, gas and water lines.

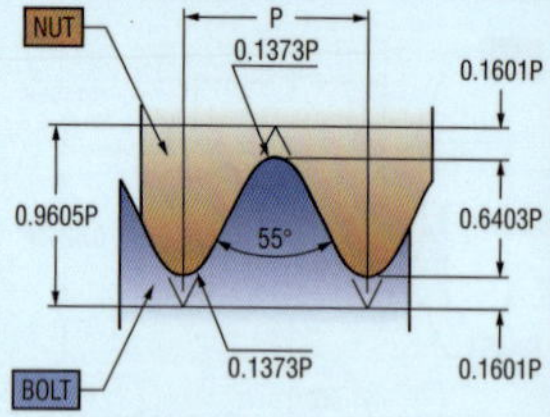

BSW (BRITISH STD. WHITWORTH) BSF (BRITISH STD. FINE)

Use : As a general machine thread.

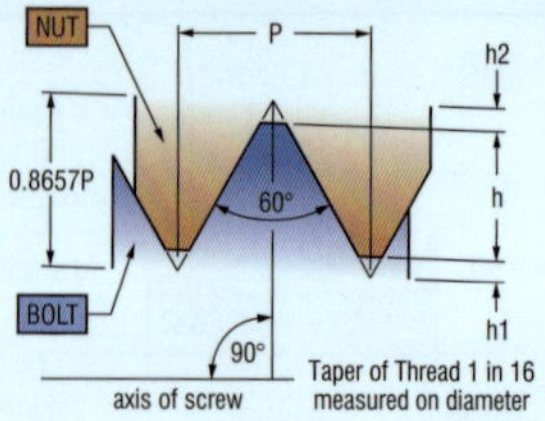

DRYSEAL

Use : Widely used for fluid-tight pipe couplings in the motor industry.

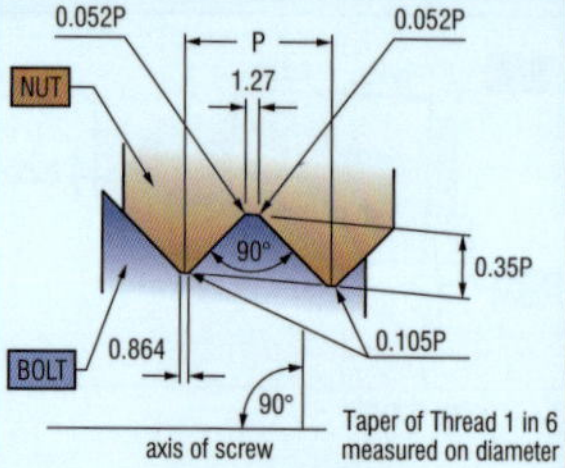

HUGHES H-90

Use : Oil industry.

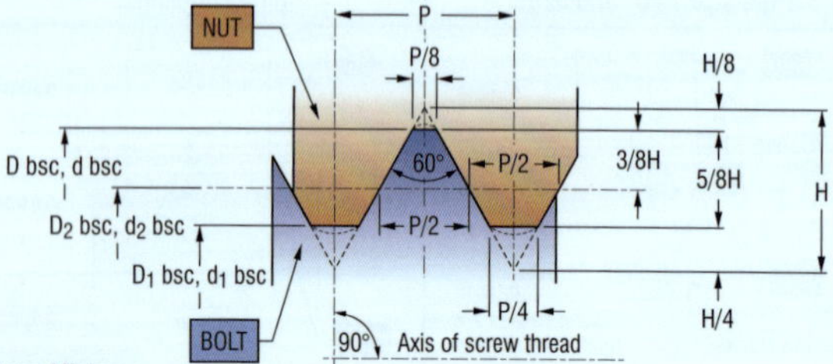

General Notes:

$H = \sqrt{3} / 2 \times P = 0.866025\ P$

$H/8 = 0.108253\ P$

$H/4 = 0.216506\ P$

$3/8H = 0.324760\ P$

$5/8H = 0.541266\ P$

M (ISO METRIC)

according to ISO 68-1 Basic Profile

Use : All branches of mechanical industry.

COMMON THREAD FORMS USED IN THE ENGINEERING INDUSTRY

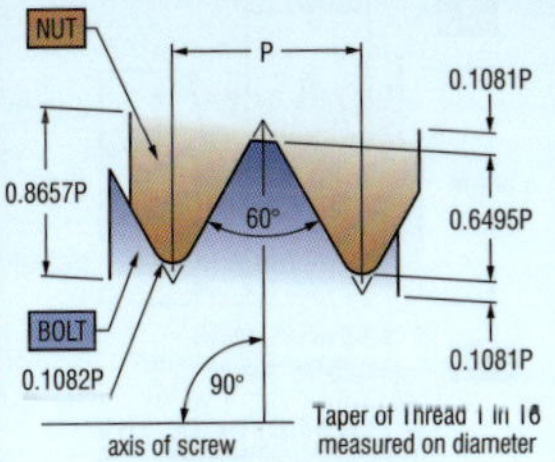

METRIC TAPER

Use : Grease fittings and pipe couplings.

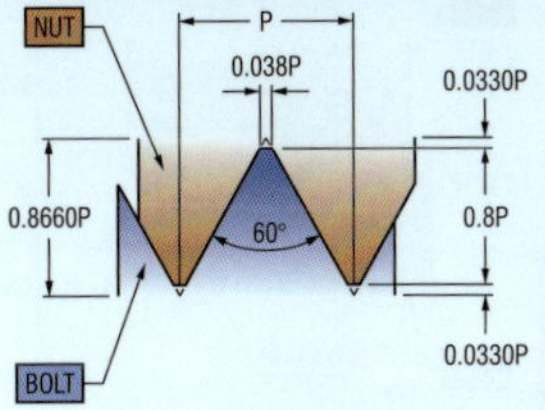

NPS (NATIONAL PIPE STRAIGHT THREAD)

Use : Pipe couplings where there is no demand for fluid-tightness.

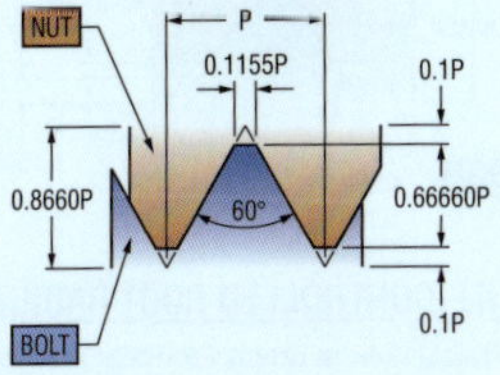

NPSL, NPSM, NPSH (NATIONAL PIPE STRAIGHT THREAD)

Use : Mechanical couplings of tubings normally not fluid-tight.

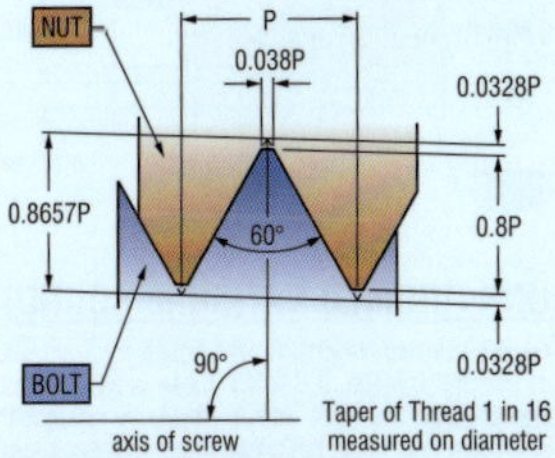

NPT (NATIONAL PIPE TAPER)

Use : Fittings and pipe couplings.

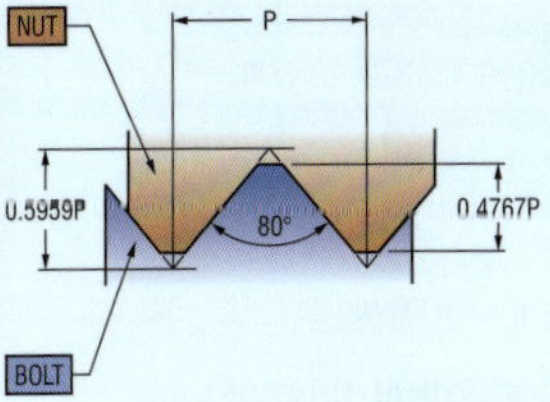

PG (PANZERROHR-GEWINDE)

Use : Steel conduit (electric).

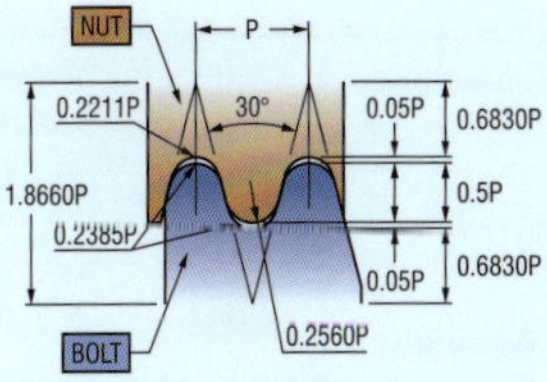

RD DIN 405 (ROUND DIN 405)

Use : Pipe couplings in fire protection and food industry.

COMMON THREAD FORMS USED IN THE ENGINEERING INDUSTRY

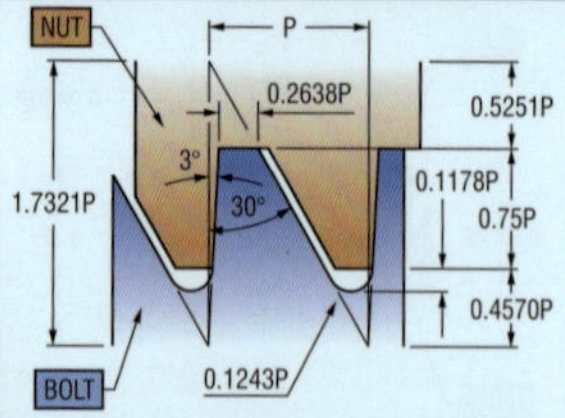

SAW-TOOTH THREAD 30° DIN 513

Use : Mechanical industry for motion transmission screws - one direction.

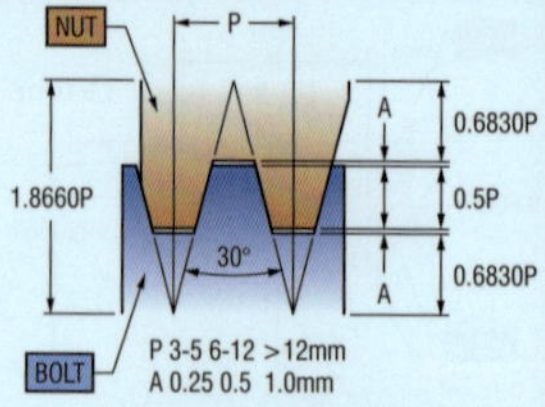

TR (TRAPEZ) DIN 103

Use : Mechanical industry for motion transmission screws.

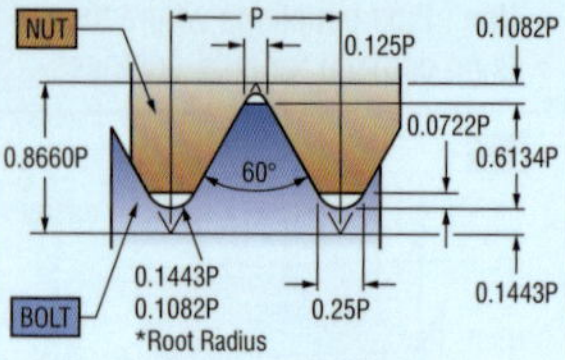

UNR (UNIFIED NATIONAL ROUND)

* UNR - Unified National Round (round root) applies to External Threads. It is a UN thread in all respects except that the root radius must be rounded. Inspected only when specified. Most manufacturers of externally threaded fasteners produce UNR rolled threads rather than plain UN.

Use : Mechanical industry.

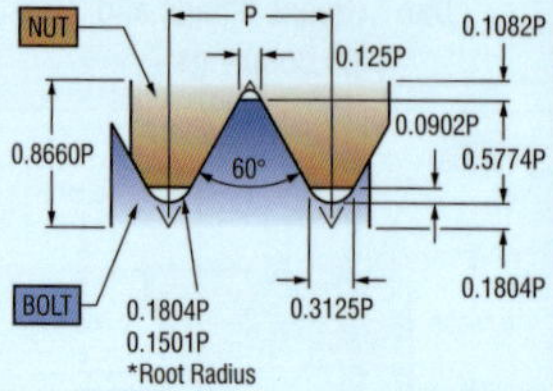

UNJ (CONTROLLED ROOT RADIUS)

* Applies to External Threads. Root Radius mandatory. Assembles only with UNJ Internal threads mandatory inspection required.

Use : High Tensile Fasteners in Aircraft & Space industry.

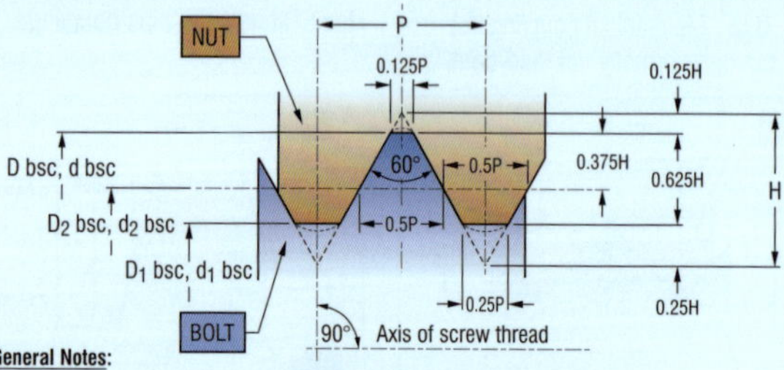

General Notes:

$H = \sqrt{3} / 2 \times P = 0.866025\ P$
$0.125H = 0.108253\ P$
$0.25H = 0.216506\ P$
$0.375H = 0.324760\ P$
$0.625H = 0.541266\ P$
Root rad (Bolt) = Optional

UNC (UNIFIED NATIONAL COARSE) & UNF (UNIFIED NATIONAL FINE)

according to ASME B1.1-2003 Basic Profile

Use : Mechanical industry.

MAIN TYPES OF GROUND THREAD TAPS

Straight Flute Tap (Short Machine)

A general purpose tap for hand and machine operation suitable for short production runs. It performs best in materials where the cutting action results in chips which break up readily.

Right-Hand Spiral Flute Tap

This tap is designed primarily for machine tapping of blind holes. Most suitable for Aluminum and Soft Steel which produce long stringy chips. The cutting action of the spiral flute curls the chips and draws them out of the hole, allowing greater depth of threading.

Interrupted Thread Tap

The interrupted thread ensures less friction and hence, less resistance, which is particularly important when threading materials which are resilient & difficult to machine, e.g. Stainless Steel, Bronze. This design also facilitates ease of delivery of coolant to the cutting edge.

Spiral / Gun Point Straight Flute Tap

The spiral point tap is used mainly for tapping through holes. The point helps to cut the chips with a shearing action directing them ahead of the tap, leaving the flutes clear for the free flow of coolant to the point.

Fluteless Tap without Oil Grooves

The fluteless tap is designed for machine tapping in Ductile materials. The fluteless design produces holes of increased strength with the absence of chips. A good lubricating fluid must be used with fluteless taps. Correct tapping drill sizes are equally important.

Fluteless Tap with Oil Grooves

The application of the fluteless tap with oil grooves is the same as the fluteless tap, with the oil grooves helping to deliver lubricant easier to the tap point.

Taper Pipe Tap

Pipe taps are designed primarily for threading pipe fittings and are suitable for hand and machine use.

Acme Tap

Acme taps have a 29° thread angle and is used to thread nuts for lead screws, jacks, presses and other mechanisms where heavy loads are encountered.

Machine Nut Tap

Nut taps are mainly used for tapping nuts, the reduced shank allows for several nuts to build up on the tap shank and eliminates the need to release the tap after each nut is threaded.

COLOR CODED HIGH-PERFORMANCE TAPS AND APPLICATIONS

High-Performance Machine Taps are produced from specially selected premium HSS-E Steel and are developed with unique geometries designed to provide accurately sized threads, superior tapping ability and longer tool life.

The taps come with color coded rings and a variety of surface coatings, facilitating ease of selection and ensuring optimum performance in specific material groups.

EXAMPLES OF HIGH-PERFORMANCE TAPS WITH COLOR CODED RINGS

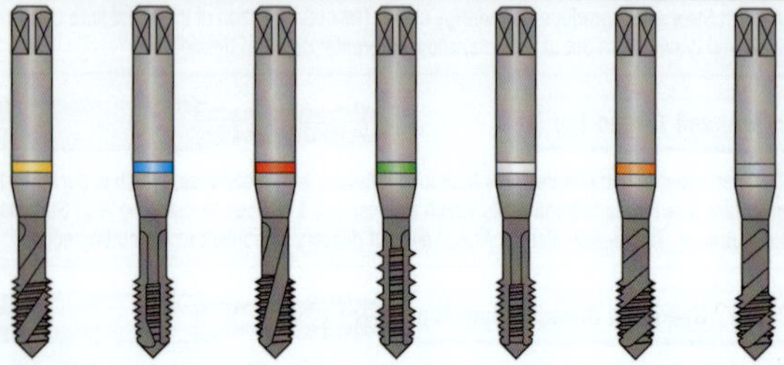

Most Tap Manufacturers adopt the following Color Coding System

Color Rings	Applications	Material Examples	Tapping Speeds ft/min. (Coated)
YELLOW	Soft Long Chipping Materials ≤ 50 Tons/inch²	Free-cutting Steels, Structural Steels, Aluminum, Copper, Thermoplastics	50 - 100 ft/min
BLUE	Stainless Steels ≤ 65 Tons/inch²	Free machining, Austenitic, Ferritic & Martensitic Stainless Steels	15 - 50 ft/min
RED	Hard, Short, Chipping Materials ≤ 75 Tons/inch²	Tool Steels, High-Speed Steels, Case Hardened Steels, Heat Treatable Steels	5 - 15 ft/min
GREEN	Universal Applications Steels ≤ 65 Tons/inch²	Free-Cutting Steels, Heat Treatable Steels, S. G. Iron, Nitriding & Case Hardened Steels	50 - 100 ft/min
WHITE	Cast Irons ≤ 55 Tons/inch²	Grey Cast Iron, Malleable Cast Iron, S. G. Iron, Brass & Bronze	50 - 100 ft/min
ORANGE	Heat Resistant Alloys ≤ 75 Tons/inch²	Nickel-based Alloys, Cobalt-based Alloys, Super Alloys with Hardness < 350 HB	10 - 20 ft/min
SILVER	MULTI: Steels ≤ 75 Tons/inch²	Structural Steels, Carbon Steels, Alloy Steels	30 - 60 ft/min
	MULTI: Stainless Steels ≤ 65 Tons/inch²	Free Machining, Austenitic & Ferritic Stainless Steels	15 - 50 ft/min

NOTE: The above information is intended as a general guide only. For detailed and up to date application information, it is recommended to seek the advice of the Tap Manufacturer.

TAP SELECTION GUIDE FOR HOLE TYPES

Hints on Tapping

- Select the correct design of the tap to suit the component material and hole type.
- Select the correct drill size - thread percentage (refer page 35).
- Clamp component securely.
- Select correct cutting speed if machine tapping.
- Use appropriate cutting fluid.
- When tapping blind holes, make sure to eliminate swarf collecting at the base of the hole which might result in tap breakage.

Tap Type and Description	Hole Type	Swarf
 **Straight Flute Tap (Short Machine)** They are the most commonly used type of taps and suitable for most short chips materials like cast iron, mild steel, etc.	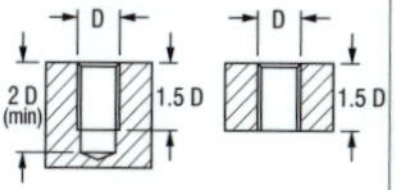	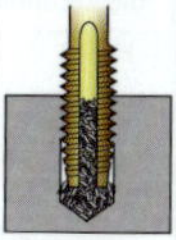
 **Right-Hand Spiral Flute Tap** The right hand spiral flute is designed to drive the swarf upwards and thus ensure smooth tapping particularly for blind holes.	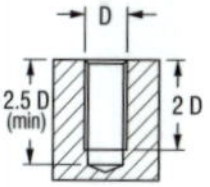	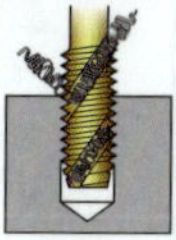
 **Interrupted Thread Tap** Used on difficult to machine materials such as aluminum, brass, bronze, etc. The interrupted thread ensures less friction and therefore less resistance.	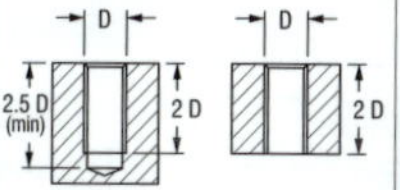	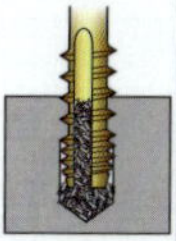
 **Spiral / Gun Point Straight Flute Tap** The spiral point is designed to drive the swarf forward and thus ensure smooth tapping. Not for use in blind holes.	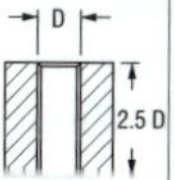	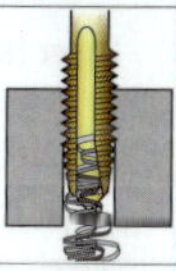
 **Fluteless Tap** The fluteless tap produces a strong thread by plastic deformation of the component material and therefore no swarf is produced.	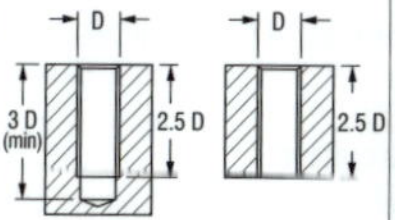	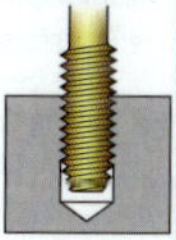
 **Machine Nut Tap** Specially designed to tap nuts which are allowed to accumulate on the long shank whose diameter is less than the minor diameter of the thread. Often used as an extension tap for deep down holes.	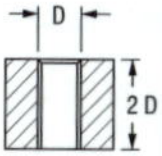	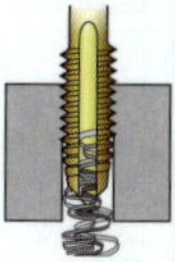

TAPS AND TAPPING

THREAD CUTTING TAP

THREAD FORMING TAP

A Tap is a cutting / forming tool manufactured from carbon steel, high speed steel, cobalt steel, carbide tipped and solid carbide and designed to produce threads within a predetermined-size drilled hole.

The chips (swarf) removed by a thread cutting tap are either deposited within the flutes, pushed forward in front of the tap or drawn up along the flute depending upon the style of the flute.

The thread forming (cold forming) fluteless tap forms a thread by plastic deformation of the component material and produces no swarf.

TAP GEOMETRY

Thread Cutting Taps

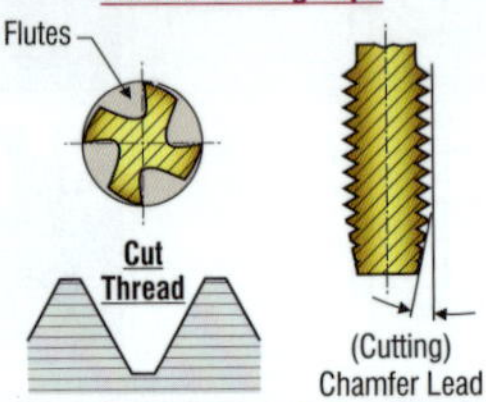

Fluteless Taps

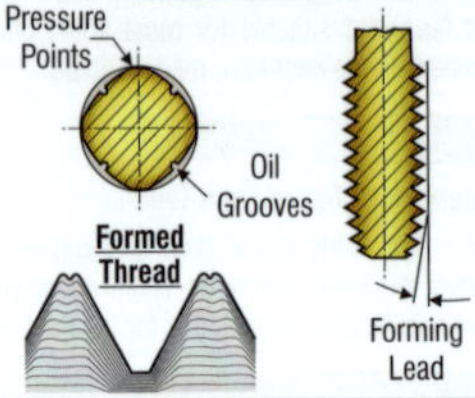

TAP CHAMFERS

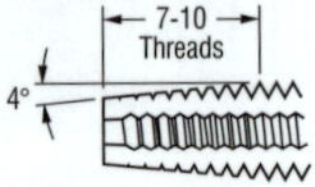

Taper (Starter Tap)

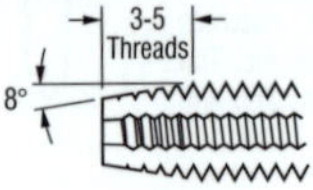

Inter (Second Tap)

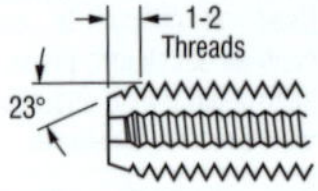

Bottoming (Third Tap)
For blind holes

Note: Chamfer length may vary between manufacturers.

THREAD PERCENTAGE AND TAP SELECTION

The success of any tapping operation depends upon a number of factors that will ultimately effect the quality of the thread produced in a finished component.

- Correct selection of tap flute geometry (based on through or blind hole)
- Drill size
- Cutting fluid
- Thread percentage

As a general rule, the tougher/harder the material, the less the percentage of thread is needed to create a hole strong enough. The following is a general guide for thread percentages.

Workpiece Material	Commercial Components	Deep Holes
Hard Steels, Cast Steels, Stainless Steels, Nickel, Monel, etc.	60% - 70%	55% - 65%
Cast Iron, Mild Steel, Free Cutting Steel, Aluminum, Brass, Bronze & Copper.	65% - 75%	60% - 70%

CALCULATION OF TAPPING DRILL SIZE

FOR THREAD CUTTING TAPS

TAPPING DRILL SIZE - INCH

$$\text{Tapping Drill Size} = \text{Basic Major Diameter of Thread} - \left(0.0130 \times \frac{\text{\% of Full Thread}}{\text{Threads per Inch}}\right)$$

Example : To find Tapping Drill Size for ¼ - 20 tap with 75% thread

$$\therefore \text{Tapping Drill Size} = 0.2500 - \left(0.0130 \times \frac{75}{20}\right) = \mathbf{0.2010\ inches\ (\#\ 7\ Drill)}$$

THREAD PERCENTAGE - INCH

$$\text{Thread Percentage} = \frac{(\text{Thread Major Diameter} - \text{Drill Diameter})}{0.0130} \times \text{Threads per Inch}$$

Example : To find Thread Percentage for ¼ - 20 UNC using 0.201" drill size

$$\therefore \text{Thread Percentage} = \frac{(0.250 - 0.201)}{0.0130} \times 20 = \mathbf{75.38\ \%}$$

TAPPING DRILL SIZE - METRIC

$$\text{Tapping Drill Size} = \text{Basic Major Diameter of Thread} - \frac{\text{\% of Full Thread} \times \text{Pitch (mm)}}{76.98}$$

Example : To find Tapping Drill Size for M6 x 1.0 tap with 70% thread

$$\therefore \text{Tapping Drill Size} = 6.0 - \frac{70 \times 1.0}{76.98} = \mathbf{5.1\ mm\ Drill}$$

THREAD PERCENTAGE - METRIC

$$\text{Thread Percentage} = \frac{(\text{Thread Major Diameter} - \text{Drill Diameter})}{0.0130 \times \text{Pitch (mm)}}$$

Example : To find Thread Percentage for M6 x 1.00 using 5.10 mm drill size

$$\therefore \text{Thread Percentage} = \frac{(6.0 - 5.10)}{0.0130 \times 1.00} = \mathbf{69.23\ \%}$$

FOR FLUTELESS TAPS

TAPPING DRILL SIZE - INCH

$$\text{Tapping Drill Size} = \text{Basic Major Diameter of Thread} - \left(0.0068 \times \frac{\text{\% of Full Thread}}{\text{Threads per Inch}}\right)$$

Example : To find Tapping Drill Size for ¼ - 20 tap with 75% thread

$$\therefore \text{Tapping Drill Size} = 0.2500 - \left(0.0068 \times \frac{75}{20}\right) = \mathbf{0.2245\ inches\ (\#\ 1\ Drill)}$$

TAPPING DRILL SIZE - METRIC

$$\text{Tapping Drill Size} = \text{Basic Major Diameter of Thread} - \frac{\text{\% of Full Thread} \times \text{Pitch (mm)}}{147.06}$$

Example : To find Tapping Drill Size for M6 x 1.0 tap with 70% thread

$$\therefore \text{Tapping Drill Size} = 6.0 - \frac{70 \times 1.0}{147.06} = \mathbf{5.5\ mm\ Drill}$$

NOTE: It is advisable to use a good cutting oil or lubricant rather than coolant for **Fluteless (Thread Forming)** taps. Correct tapping drill sizes are equally important.

TAPPING DRILL SIZE FOR THREAD CUTTING TAPS

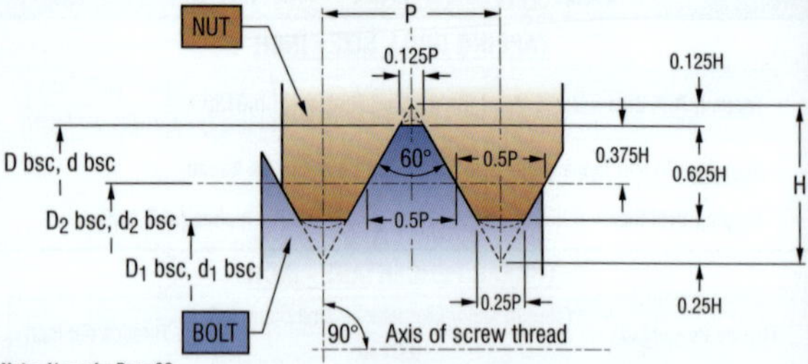

Note: Also refer Page 30

UNC (UNIFIED NATIONAL COARSE)

Screw Size UNC (No./inch)	(t.p.i.)	Basic Major Diameter (inch)	Basic Minor Diameter (inch)	*Tapping Drill Sizes For Mild Steel, Cast Iron, Aluminum, Brass, Bronze, Copper: Drill Size	Decimal	Thread %	Clearance Drill Size *(Normal Fit): Drill Size	Dec. Eq.
No. 1	64	0.0730	0.0538	# 53	0.0595	75	# 43	0.0890
No. 2	56	0.0860	0.0641	# 50	0.0700	75	# 38	0.1015
No. 3	48	0.0990	0.0734	# 47	0.0785	76	# 32	0.1160
No. 4	40	0.1120	0.0813	# 43	0.0890	71	# 30	0.1285
No. 5	40	0.1250	0.0943	# 38	0.1015	72	5/32"	0.1562
No. 6	32	0.1380	0.0997	# 36	0.1065	77	# 18	0.1695
No. 8	32	0.1640	0.1257	# 29	0.1360	69	# 9	0.1960
No.10	24	0.1900	0.1389	# 25	0.1495	75	# 2	0.2210
No.12	24	0.2160	0.1649	# 16	0.1770	72	D	0.2460
1/4"	20	0.2500	0.1887	# 7	0.2010	75	9/32"	0.2812
5/16"	18	0.3125	0.2443	F	0.2570	77	11/32"	0.3437
3/8"	16	0.3750	0.2983	O	0.3125	73	13/32"	0.4062
7/16"	14	0.4375	0.3499	U	0.3680	75	15/32"	0.4687
1/2"	13	0.5000	0.4056	27/64"	0.4219	78	9/16"	0.5625
9/16"	12	0.5625	0.4603	31/64"	0.4844	72	5/8"	0.6250
5/8"	11	0.6250	0.5135	17/32"	0.5312	80	11/16"	0.6875
3/4"	10	0.7500	0.6273	21/32"	0.6562	72	13/16"	0.8125
7/8"	9	0.8750	0.7387	49/64"	0.7656	76	15/16"	0.9375
1"	8	1.0000	0.8466	7/8"	0.8750	77	1 3/32"	1.0937
1 1/8"	7	1.1250	0.9497	63/64"	0.9860	76	1 7/32"	1.2187
1 1/4"	7	1.2500	1.0747	1 7/64"	1.1110	76	1 11/32"	1.3437
1 3/8"	6	1.3750	1.1705	1 7/32"	1.2130	72	1 1/2"	1.5000
1 1/2"	6	1.5000	1.2955	1 11/32"	1.3380	75	1 5/8"	1.6250
1 3/4"	5	1.7500	1.5046	1 9/16"	1.5550	75	1 7/8"	1.8750
2"	4 1/2	2.0000	1.7274	1 25/32"	1.7830	75	2 1/8"	2.1250
2 1/4"	4 1/2	2.2500	1.9774	2 1/32"	2.0330	75	2 3/8"	2.3750

For General Symbols refer page 26

* Refer page 63 for Close & Loose Fit Drill Sizes

TAPPING DRILL SIZE FOR THREAD CUTTING TAPS

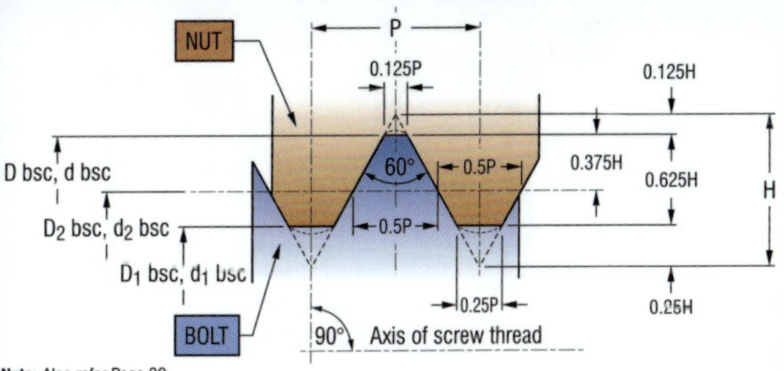

Note: Also refer Page 30

UNF (UNIFIED NATIONAL FINE)

Screw Size		Basic		✦Tapping Drill Sizes			Clearance Drill Size	
UNF		Major Diameter	Minor Diameter	For Mild Steel, Cast Iron, Aluminum, Brass, Bronze, Copper			*(Normal Fit)	
(No./inch)	(t.p.i.)	(inch)	(inch)	Drill Size	Decimal	Thread %	Drill Size	Dec. Eq.
No. 0	80	0.0600	0.0447	# 56	0.0469	77	# 48	0.0760
No. 1	72	0.0730	0.0560	# 53	0.0595	77	# 43	0.0890
No. 2	64	0.0860	0.0668	# 50	0.0700	77	# 38	0.1015
No. 3	56	0.0990	0.0771	# 45	0.0820	74	# 32	0.1160
No. 4	48	0.1120	0.0864	# 42	0.0935	69	# 30	0.1285
No. 5	44	0.1250	0.0971	# 37	0.1040	71	5/32"	0.1562
No. 6	40	0.1380	0.1073	# 33	0.1130	77	# 18	0.1695
No. 8	36	0.1640	0.1299	# 29	0.1360	78	# 9	0.1960
No.10	32	0.1900	0.1517	# 21	0.1590	76	# 2	0.2210
No.12	28	0.2160	0.1722	# 14	0.1820	74	1/4"	0.2500
1/4"	28	0.2500	0.2062	# 3	0.2130	80	9/32"	0.2812
5/16"	24	0.3125	0.2614	I	0.2720	75	11/32"	0.3437
3/8"	24	0.3750	0.3239	Q	0.3320	77	13/32"	0.4062
7/16"	20	0.4375	0.3762	25/64"	0.3906	72	15/32"	0.4687
1/2"	20	0.5000	0.4387	29/64"	0.4531	72	9/16"	0.5625
9/16"	18	0.5625	0.4943	33/64"	0.5156	65	5/8"	0.6250
5/8"	18	0.6250	0.5568	37/64"	0.5781	65	11/16"	0.6875
3/4"	16	0.7500	0.6733	11/16"	0.6875	77	13/16"	0.8125
7/8"	14	0.8750	0.7874	13/16"	0.8125	67	15/16"	0.9375
1"	12	1.0000	0.8978	15/16"	0.9375	58	1 3/32"	1.0937
1 1/8"	12	1.1250	1.0228	1 3/64"	1.0440	72	1 7/32"	1.2187
1 1/4"	12	1.2500	1.1478	1 11/64"	1.1690	72	1 11/32"	1.3437
1 3/8"	12	1.3750	1.2728	1 19/64"	1.2940	72	1 1/2"	1.5000
1 1/2"	12	1.5000	1.3978	1 27/64"	1.4190	72	1 5/8"	1.6250

✦ Recommended for average commercial work.

Drill Size and Decimal Equivalent refer pages 14-20

For General Symbols refer page 26

* Refer page 63 for Close & Loose Fit Drill Size

TAPPING DRILL SIZE FOR THREAD CUTTING TAPS

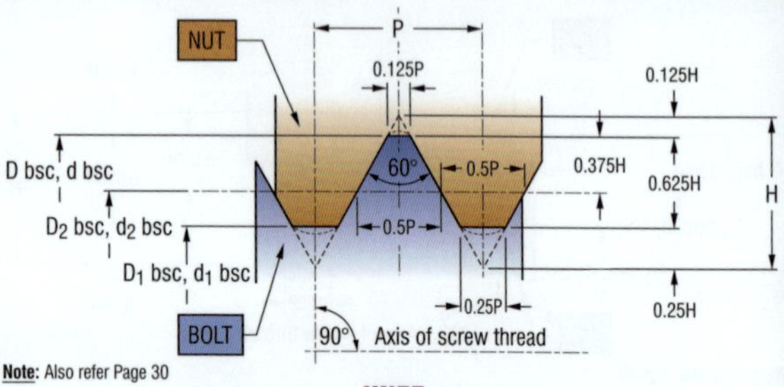

Note: Also refer Page 30

UNEF (UNIFIED NATIONAL EXTRA FINE)

Screw Size UNEF (No./inch)	(t.p.i.)	Basic Major Diameter (inch)	Basic Minor Diameter (inch)	*Tapping Drill Sizes For Mild Steel, Cast Iron, Aluminum, Brass, Bronze, Copper: Drill Size	Decimal	Thread %	Clearance Drill Size *(Normal Fit): Drill Size	Dec. Eq.
1/4"	32	0.2500	0.2128	7/32"	0.2205	78	9/32"	0.2812
5/16"	32	0.3125	0.2753	9/32"	0.2813	75	11/32"	0.3437
3/8"	32	0.3750	0.3378	11/32"	0.3437	75	13/32"	0.4062
7/16"	28	0.4375	0.3949	13/32"	0.4040	74	15/32"	0.4687
1/2"	28	0.5000	0.4574	15/32"	0.4646	70	9/16"	0.5625
9/16"	24	0.5625	0.5129	17/32"	0.5236	59	5/8"	0.6250
5/8"	24	0.6250	0.5754	19/32"	0.5807	60	11/16"	0.6875
11/16"	24	0.6875	0.6379	21/32"	0.6496	60	3/4"	0.7500
3/4"	20	0.7500	0.6905	45/64"	0.7031	70	13/16"	0.8125
13/16"	20	0.8125	0.7530	49/64"	0.7656	70	7/8"	0.8750
7/8"	20	0.8750	0.8155	53/64"	0.8268	70	15/16"	0.9375
15/16"	20	0.9375	0.8780	57/64"	0.8806	70	1"	1.0000
1"	20	1.0000	0.9405	61/64"	0.9531	71	1 3/32"	1.0937
1 1/16"	18	1.0625	0.9964	1 1/64"	1.0138	65	1 5/32"	1.1563
1 1/8"	18	1.1250	1.0589	1 5/64"	1.0728	65	1 7/32"	1.2187
1 3/16"	18	1.1875	1.1214	1 9/64"	1.1319	64	1 9/32"	1.2813
1 1/4"	18	1.2500	1.1839	1 13/64"	1.2008	65	1 11/32"	1.3437
1 5/16"	18	1.3125	1.2464	1 17/64"	1.2598	65	1 13/32"	1.4063
1 3/8"	18	1.3750	1.3089	1 21/64"	1.3189	65	1 1/2"	1.5000
1 7/16"	18	1.4375	1.3714	1 25/64"	1.3780	65	1 9/16"	1.5625
1 1/2"	18	1.5000	1.4339	1 29/64"	1.4531	65	1 5/8"	1.6250
1 9/16"	18	1.5625	1.4964	1 33/64"	1.5156	65	1 11/16"	1.6875
1 5/8"	18	1.6250	1.5589	1 37/64"	1.5748	65	1 3/4"	1.7500
1 11/16"	18	1.7500	1.6214	1 41/64"	1.6929	65	1 13/16"	1.8125

For General Symbols refer page 26

* Refer page 63 for Close & Loose Fit Drill Size

TAPPING DRILL SIZE FOR THREAD CUTTING TAPS

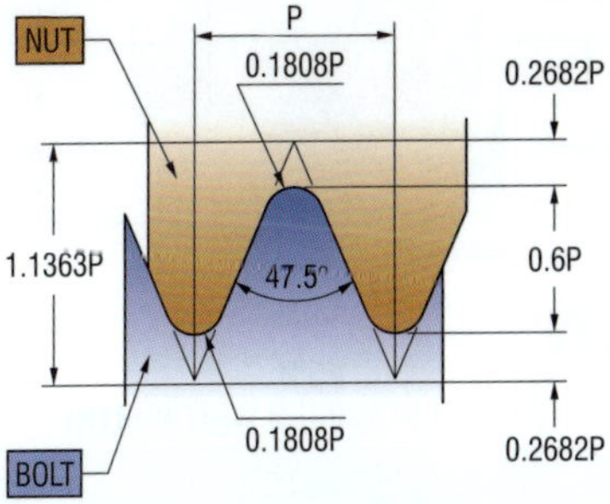

B.A. (BRITISH ASSOCIATION)

B.A. No.	Pitch (inch)	Basic Depth of Thread (inch)	Basic Major Diameter (inch)	Basic Effective Diameter (inch)	Basic Minor Diameter (inch)	Tapping Drill Size (inch)	Clearance Drill Size (inch)
0	**0.0394**	0.0236	0.2362	0.2126	0.1890	# 7	C
1	**0.0354**	0.0213	0.2087	0.1874	0.1661	# 16	# 3
2	**0.0319**	0.0191	0.1850	0.1659	0.1468	# 22	# 12
3	**0.0287**	0.0172	0.1614	0.1441	0.1268	# 29	# 19
4	**0.0260**	0.0156	0.1417	0.1262	0.1106	# 31	# 26
5	**0.0232**	0.0139	0.1260	0.1120	0.0980	# 37	# 30
6	**0.0209**	0.0125	0.1102	0.0976	0.0850	# 43	# 33
7	**0.0189**	0.0113	0.0984	0.0070	0.0756	# 47	# 38
8	**0.0169**	0.0102	0.0866	0.0764	0.0661	# 50	# 43
9	**0.0154**	0.0092	0.0748	0.0656	0.0563	# 53	# 48
10	**0.0138**	0.0083	0.0669	0.0587	0.0504	# 54	# 50
11	**0.0122**	0.0073	0.0591	0.0518	0.0445	# 56	# 52
12	**0.0110**	0.0066	0.0512	0.0445	0.0378	# 59	# 54
13	**0.0098**	0.0059	0.0472	0.0413	0.0354	# 62	# 55
14	**0.0091**	0.0054	0.0394	0.0339	0.0283	# 68	# 57
15	**0.0083**	0.0050	0.0354	0.0305	0.0256	# 71	# 61
16	**0.0075**	0.0045	0.0311	0.0266	0.0220	# 73	# 65

TAPPING DRILL SIZE FOR THREAD CUTTING TAPS

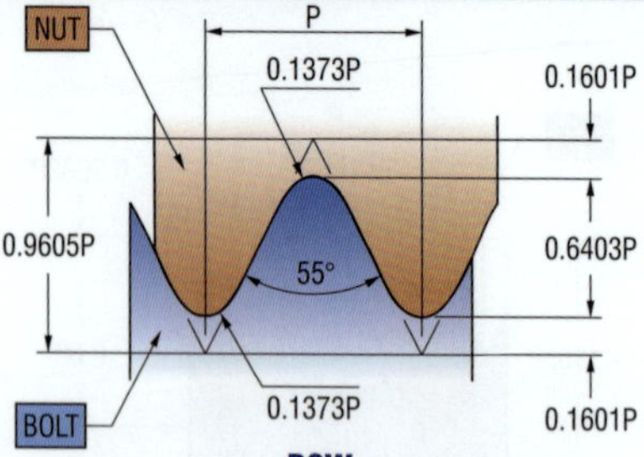

BSW (BRITISH STANDARD WHITWORTH)

Nom. Dia. BSW (inch)	Threads per Inch (t.p.i.)	Basic Depth of Thread (inch)	Basic Major Diameter (inch)	Basic Effective Diameter (inch)	Basic Minor Diameter (inch)	Tapping Drill Size (inch)	Clearance Drill Size (inch)
1/16"	60	0.0107	0.0625	0.0518	0.0411	# 56	# 51
3/32"	48	0.0133	0.0938	0.0805	0.0672	# 49	# 39
1/8"	40	0.0160	0.1250	0.1090	0.0930	# 39	# 30
5/32"	32	0.0200	0.1562	0.1362	0.1162	1/8"	# 20
3/16"	24	0.0267	0.1875	0.1608	0.1341	# 27	# 10
7/32"	24	0.0267	0.2188	0.1921	0.1654	# 16	# 1
1/4"	20	0.0320	0.2500	0.2180	0.1860	# 7	F
5/16"	18	0.0356	0.3125	0.2769	0.2413	F	P
3/8"	16	0.0400	0.3750	0.3350	0.2950	5/16"	W
7/16"	14	0.0457	0.4375	0.3918	0.3461	23/64"	29/64"
1/2"	12	0.0534	0.5000	0.4466	0.3932	13/32"	33/64"
9/16"	12	0.0534	0.5625	0.5091	0.4557	15/32"	37/64"
5/8"	11	0.0582	0.6250	0.5668	0.5086	17/32"	41/64"
11/16"	11	0.0582	0.6875	0.6293	0.5711	19/32"	45/64"
3/4"	10	0.0640	0.7500	0.6860	0.6220	41/64"	13/16"
13/16"	10	0.0640	0.8125	0.7485	0.6845	45/64"	53/64"
7/8"	9	0.0711	0.8750	0.8039	0.7328	3/4"	57/64"
15/16"	9	0.0711	0.9375	0.8664	0.7953	13/16"	61/64"
1"	8	0.0800	1.0000	0.9200	0.8400	7/8"	1 1/64"
1 1/8"	7	0.0915	1.1250	1.0335	0.9420	31/32"	1 5/32"
1 1/4"	7	0.0915	1.2500	1.1585	1.0670	1 3/32"	1 17/64"
1 3/8"	6	0.1067	1.3750	1.2683	1.1616	1 3/16"	1 13/32"
1 1/2"	6	0.1067	1.5000	1.3933	1.2866	1 5/16"	1 17/32"
1 5/8"	5	0.1281	1.6250	1.4969	1.3688	1 13/32"	1 41/64"
1 3/4"	5	0.1281	1.7500	1.6219	1.4938	1 17/32"	1 25/32"
1 7/8"	4 1/2	0.1423	1.8750	1.7327	1.5904	1 5/8"	1 57/64"
2"	4 1/2	0.1423	2.0000	1.8577	1.7154	1 3/4"	2 1/32"
2 1/4"	4	0.1601	2.2500	2.0899	1.9298	2"	2 5/16"
2 1/2"	4	0.1601	2.5000	2.3399	2.1798	2 1/4"	2 17/32"

TAPPING DRILL SIZE FOR THREAD CUTTING TAPS

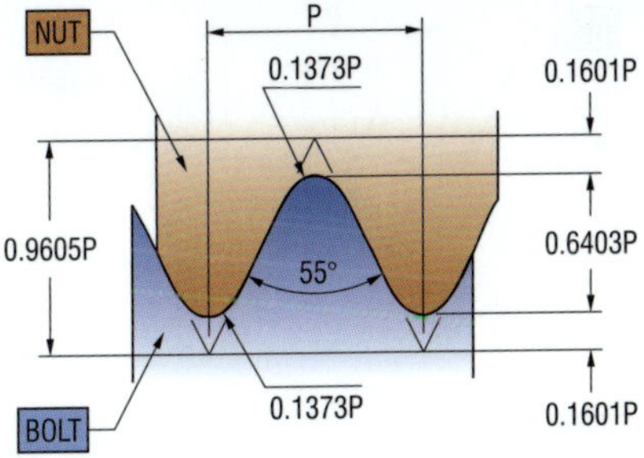

BSF (BRITISH STANDARD FINE)

Nom. Dia. BSF (inch)	Threads per Inch (t.p.i.)	Basic Depth Thread (inch)	Basic Major Diameter (inch)	Basic Effect Diameter (inch)	Basic Minor Diameter (inch)	Tapping Drill Size (inch)	Clearance Drill Size (inch)
3/16"	32	0.0200	0.1875	0.1675	0.1475	# 22	# 10
7/32"	28	0.0229	0.2188	0.1959	0.1730	# 15	# 1
1/4"	26	0.0246	0.2500	0.2254	0.2008	# 5	F
5/16"	22	0.0291	0.3125	0.2834	0.2543	H	P
3/8"	20	0.0320	0.3750	0.3430	0.3110	21/64"	W
7/16"	18	0.0356	0.4375	0.4019	0.3663	V	29/64"
1/2"	16	0.0400	0.5000	0.4600	0.4200	7/16"	33/64"
9/16"	16	0.0400	0.5625	0.5225	0.4825	1/2"	37/64"
5/8"	14	0.0457	0.6250	0.5793	0.5336	35/64"	41/64"
11/16"	14	0.0457	0.6875	0.6418	0.5961	39/64"	45/64"
3/4"	12	0.0534	0.7500	0.6966	0.6432	21/[illegible]"	49/64"
7/8"	11	0.0582	0.8750	0.8168	0.7586	25/32"	57/64"
1"	10	0.0640	1.0000	0.9360	0.8720	57/64"	1 1/64"
1 1/8"	9	0.0711	1.1250	1.0539	0.9828	1"	1 5/32"
1 1/4"	9	0.0711	1.2500	1.1789	1.1078	1 1/8"	1 17/64"
1 3/8"	8	0.0800	1.3750	1.2950	1.2150	1 1/4"	1 13/32"
1 1/2"	8	0.0800	1.5000	1.4200	1.3400	1 3/8"	1 17/32"
1 3/4"	7	0.0915	1.7500	1.6585	1.5670	1 5/8"	1 25/32"
2"	7	0.0915	2.0000	1.9085	1.8170	1 55/64"	2 1/16"
2 1/4"	6	0.1067	2.2500	2.1433	2.0366	2 3/32"	2 5/16"
2 1/2"	6	0.1067	2.5000	2.3933	2.2866	2 9/32"	2 17/32"

TAPPING DRILL SIZE FOR THREAD CUTTING TAPS

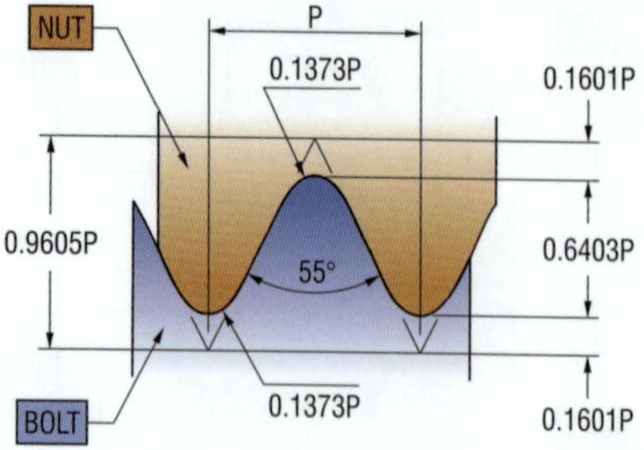

BSP (BRITISH STANDARD PIPE)

Nom. Dia. BSP (inch)	Threads per Inch (t.p.i.)	Basic Major Diameter (inch)	Basic Depth Thread (inch)	Basic Effective Diameter (inch)	Basic Minor Diameter (inch)	Tapping Drill (Bore) Size	
						Rp Series (Parallel) (BSPP) (inch)	G Series (Fastening) (BSPF) (inch)
1/16"	28	0.304	0.0229	0.2812	0.2583	G	H
1/8"	28	0.383	0.0229	0.3601	0.3372	R	S
1/4"	19	0.518	0.0337	0.4843	0.4506	29/64"	29/64"
3/8"	19	0.656	0.0337	0.6223	0.5886	19/32"	19/32"
1/2"	14	0.825	0.0457	0.7793	0.7336	47/64"	3/4"
5/8"	14	0.902	0.0457	0.8563	0.8106	13/16"	53/64"
3/4"	14	1.041	0.0457	0.9953	0.9496	61/64"	31/32"
7/8"	14	1.189	0.0457	1.1433	1.0976	1 3/32"	1 7/64"
1"	11	1.309	0.0582	1.2508	1.1926	1 3/16"	1 7/32"
1 1/4"	11	1.650	0.0582	1.5918	1.5336	1 17/32"	1 9/16"
1 1/2"	11	1.882	0.0582	1.8238	1.7656	1 49/64"	1 25/32"
1 3/4"	11	2.116	0.0582	2.0578	1.9996	1 63/64"	2"
2"	11	2.347	0.0582	2.2888	2.2306	2.227"	2.244"
2 1/4"	11	2.587	0.0582	2.5288	2.4706	2.467"	2.467"
2 1/2"	11	2.960	0.0582	2.9018	2.8436	2.840"	2.840"
2 3/4"	11	3.210	0.0582	3.1518	3.0936	3.090"	3.090"
3"	11	3.460	0.0582	3.4018	3.3436	3.340"	3.340"
3 1/2"	11	3.950	0.0582	3.8918	3.8336	3.830"	3.830"
4"	11	4.450	0.0582	4.3918	4.3336	4.330"	4.330"
5"	11	5.450	0.0582	5.3918	5.3336	5.330"	5.330"
6"	11	6.450	0.0582	6.3918	6.3336	6.330"	6.330"

TAPPING DRILL SIZE FOR THREAD CUTTING TAPS

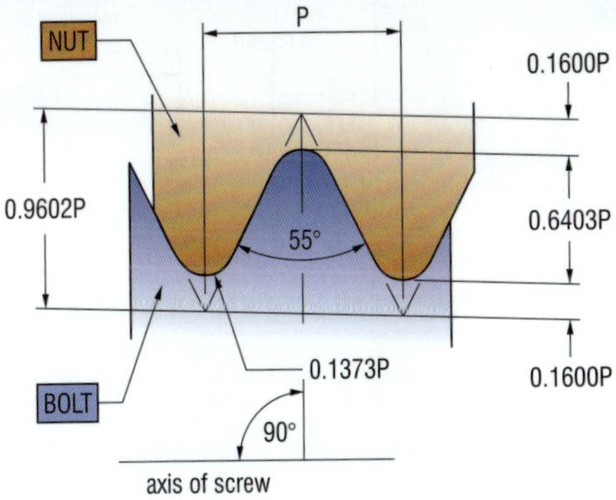

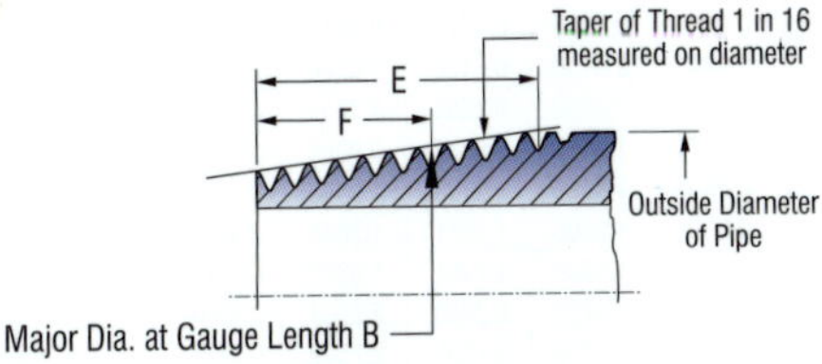

BSPT (Rc SERIES) (BRITISH STANDARD PIPE TAPER)

Nom. Size BSPT (inch)	Threads per Inch (t.p.i.)	Basic Depth of Thread (inch)	Diameter at Gauge Plane			Gauge Length (F) (inch)	Length of Useful Thread (E) (inch)	Tapping Drill Size (without reamer) (inch)
			Major Diameter (B) (inch)	Effective Diameter (inch)	Minor Diameter (inch)			
1/16"	28	0.0229	0.304	0.2812	0.2583	0.1563	0.2545	1/4"
1/8"	28	0.0229	0.383	0.3601	0.3372	0.1563	0.2545	21/64"
1/4"	19	0.0337	0.518	0.4843	0.4506	0.2367	0.3814	7/16"
3/8"	19	0.0337	0.656	0.6223	0.5886	0.2500	0.3947	37/64"
1/2"	14	0.0457	0.825	0.7793	0.7336	0.3214	0.5178	23/32"
3/4"	14	0.0457	1.041	0.9953	0.9496	0.3750	0.5714	15/16"
1"	11	0.0582	1.309	1.2508	1.1926	0.4091	0.6591	1 3/16"
1 1/4"	11	0.0582	1.650	1.5918	1.5336	0.5000	0.7500	1 1/2"
1 1/2"	11	0.0582	1.882	1.8238	1.7656	0.5000	0.7500	1 3/4"
2"	11	0.0582	2.347	2.2888	2.2306	0.6250	0.9204	2.204"
2 1/2"	11	0.0582	2.960	2.9018	2.8436	0.6875	1.0511	2.795"

TAPPING DRILL SIZE FOR THREAD CUTTING TAPS

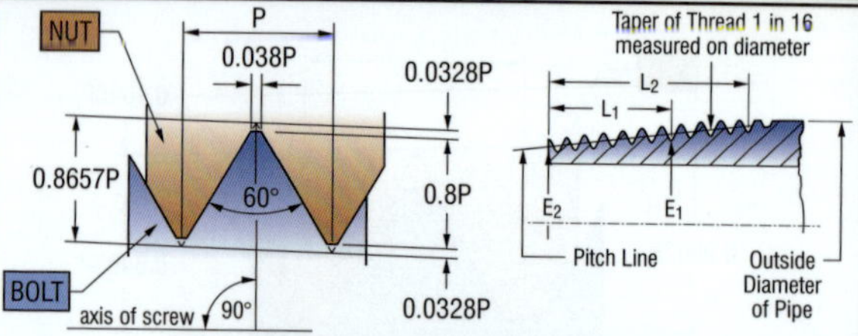

NPT (AMERICAN NATIONAL PIPE TAPER)

Nom. Size NPT (inch)	Threads per Inch (t.p.i.)	Basic Depth of Thread (inch)	Outside Diameter of Pipe (inch)	Eff. Dia. at Small End of Ext. Thd. (E_2) (inch)	Eff. Dia. at Large End of Ext. Thd. (E_1) (inch)	Gauge Length L_1 (inch)	Length of Useful Thread L_2 (inch)	Tapping Drill Size (w/o Taper Reaming) (inch)
1/16"	27	0.0296	0.3125	0.2712	0.2812	0.1598	0.2611	D
1/8"	27	0.0296	0.4050	0.3635	0.3736	0.1613	0.2639	R
1/4"	18	0.0444	0.5400	0.4774	0.4916	0.2275	0.4018	7/16"
3/8"	18	0.0444	0.6750	0.6120	0.6270	0.2398	0.4078	37/64"
1/2"	14	0.0571	0.8400	0.7584	0.7784	0.3199	0.5337	45/64"
3/4"	14	0.0571	1.0500	0.9677	0.9889	0.3391	0.5457	59/64"
1"	11 1/2"	0.0696	1.3150	1.2136	1.2386	0.3997	0.6828	1 5/32"
1 1/4"	11 1/2"	0.0696	1.6600	1.5571	1.5834	0.4197	0.7068	1 1/2"
1 1/2"	11 1/2"	0.0696	1.9000	1.7961	1.8223	0.4197	0.7235	1 47/64"
2"	11 1/2"	0.0696	2.3750	2.2690	2.2963	0.4354	0.7565	2 7/32"
2 1/2"	8	0.1000	2.8750	2.7195	2.7622	0.6825	1.1375	2 5/8"

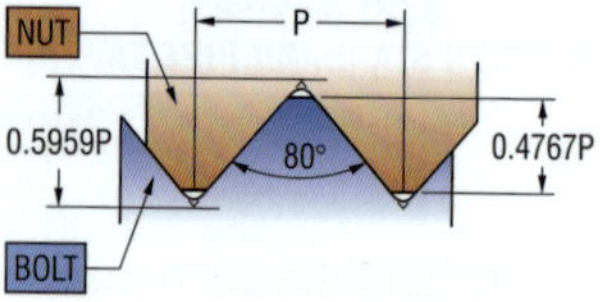

PG (PANZERROHR-GEWINDE)

Designation Pg	Threads per Inch (t.p.i.)	Basic Major Diameter (mm)	Basic Pitch Diameter (mm)	Basic Minor Diameter (mm)	Tapping Drill Size (mm)	Clearance Drill Size (mm)
Pg 7	20	12.5	11.89	11.28	11.30	12.60
Pg 9	18	15.2	14.53	13.86	13.90	15.30
Pg 11	18	18.6	17.93	17.26	17.30	18.70
Pg 13.5	18	20.4	19.73	19.06	19.10	20.50
Pg 16	18	22.5	21.83	21.16	21.20	22.75
Pg 21	16	28.3	27.54	26.78	26.80	28.75
Pg 29	16	37.0	36.24	35.48	35.50	37.25
Pg 36	16	47.0	46.24	45.48	45.50	47.25
Pg 42	16	54.0	53.24	52.48	52.50	54.25
Pg 48	16	59.3	58.54	57.78	57.80	59.55

TAPPING DRILL SIZE FOR THREAD CUTTING TAPS

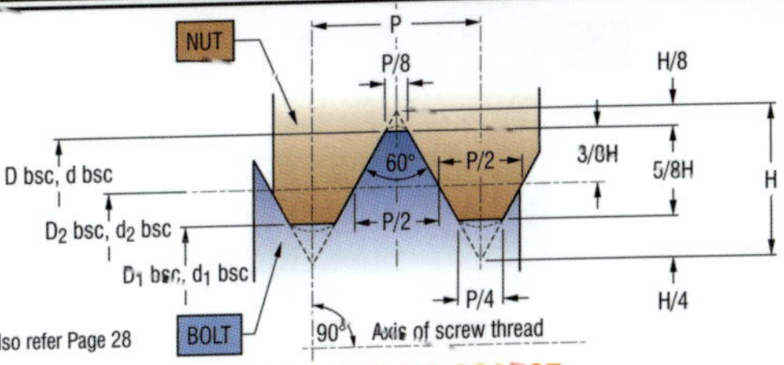

Note: Also refer Page 28

M (ISO) METRIC COARSE

Screw Size	Basic		✦Tapping Drill Sizes			Clearance Drill Size	
Nominal Dia. x Pitch	Major Diameter	Minor Diameter	For Mild Steel, Cast Iron, Aluminum, Brass, Bronze, Copper			*(Normal Fit)	
M (mm)	(mm)	(mm)	Drill (inch)	Drill (mm)	Thread %	Drill (inch)	Drill (mm)
M1.0 x 0.25	1.00	0.693	# 69	0.75	75	# 56	1.15
M1.1 x 0.25	1.10	0.793	# 66	0.85	75	# 55	1.25
M1.2 x 0.25	1.20	0.893	# 62	0.95	75	# 54	1.35
M1.4 x 0.30	1.40	1.032	# 56	1.10	75	# 53	1.55
M1.6 x 0.35	1.60	1.170	# 55	1.25	75	# 50	1.80
M1.8 x 0.35	1.80	1.370	# 53	1.45	75	# 47	2.00
M2.0 x 0.40	2.00	1.509	# 52	1.60	75	# 41	2.40
M2.2 x 0.45	2.20	1.648	# 50	1.75	77	# 37	2.60
M2.5 x 0.45	2.50	1.948	# 46	2.05	77	# 32	2.90
M3.0 x 0.50	3.00	2.387	# 39	2.50	77	# 29	3.40
M3.5 x 0.60	3.50	2.764	# 32	2.90	77	# 23	3.90
M4.0 x 0.70	4.00	3.141	# 30	3.30	77	# 15	4.50
M4.5 x 0.75	4.50	3.580	# 26	3.70	75	# 8	5.00
M5.0 x 0.80	5.00	4.019	# 19	4.20	77	# 2	5.50
M6.0 x 1.00	6.00	4.773	# 8	5.00	77	G	6.60
M7.0 x 1.00	7.00	5.773	B	6.00	77	N	7.60
M8.0 x 1.25	8.00	6.466	H	6.80	74	T	9.00
M9.0 x 1.25	9.00	7.466	N	7.80	75	X	10.00
M10.0 x 1.50	10.00	8.160	R	8.50	77	7/16"	11.00
M11.0 x 1.50	11.00	9.160	3/8"	9.50	75	15/32"	12.00
M12.0 x 1.75	12.00	9.853	13/32"	10.30	75	17/32"	13.50
M14.0 x 2.00	14.00	11.546	15/32"	12.00	73	39/64"	15.50
M16.0 x 2.00	16.00	13.546	35/64"	14.00	77	11/16"	17.50
M18.0 x 2.50	18.00	14.933	39/64"	15.50	77	49/64"	19.50
M20.0 x 2.50	20.00	16.933	11/16"	17.50	77	7/8"	22.00
M22.0 x 2.50	22.00	18.933	49/64"	19.50	77	15/16"	24.00
M24.0 x 3.00	24.00	20.319	53/64"	21.00	77	1 1/32"	26.00
M27.0 x 3.00	27.00	23.319	61/64"	24.00	77	1 9/64"	29.00
M30.0 x 3.50	30.00	25.706	1 3/64"	26.50	77	1 5/16"	33.00
M33.0 x 3.50	33.00	28.706	1 11/64"	29.50	77	1 13/32"	36.00
M36.0 x 4.00	36.00	31.093	1 17/64"	32.00	77	1 17/32"	39.00
M39.0 x 4.00	39.00	34.093	1 3/8"	35.00	75	1 21/32"	42.00
M42.0 x 4.50	42.00	36.479	1 31/64"	37.50	77	1 25/32"	45.00
M45.0 x 4.50	45.00	39.479	1 39/64"	40.50	77	1 7/8"	48.00
M48.0 x 5.00	48.00	41.866	1 45/64"	43.00	77	2 1/32"	52.00
M52.0 x 5.00	52.00	45.866	1 55/64"	47.00	77	2 7/32"	56.00
M56.0 x 5.50	56.00	49.252	2"	50.50	70	2 7/16"	62.00

Drill Size and Decimal Equivalent refer pages 14-20

✦ Recommended for average commercial work.

For General Symbols refer page 26

* Refer page 63 for Close & Loose Fit Drill Size

TAPPING DRILL SIZE FOR THREAD CUTTING TAPS

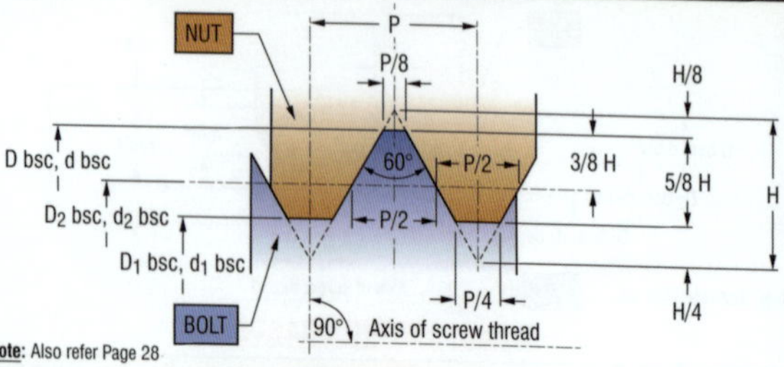

Note: Also refer Page 28

M (ISO) METRIC FINE

Screw Size Nominal Dia. x Pitch M (mm)	Basic Major Diameter (mm)	Basic Minor Diameter (mm)	+Tapping Drill Sizes For Mild Steel, Cast Iron, Aluminum, Brass, Bronze, Copper: Drill (inch)	Drill (mm)	Thread %	Clearance Drill Size *(Normal Fit): Drill (inch)	Drill (mm)
M3.0 x 0.35	3.00	2.570	# 37	2.65	77	# 29	3.40
M4.0 x 0.50	4.00	3.387	# 29	3.50	77	# 16	4.50
M5.0 x 0.50	5.00	4.387	# 16	4.50	77	# 3	5.50
M6.0 x 0.75	6.00	5.080	# 6	5.20	80	F	6.60
M7.0 x 0.75	7.00	6.080	C	6.20	77	M	7.60
M8.0 x 1.00	8.00	6.773	I	7.00	77	S	9.00
M10.0 x 1.25	10.00	8.466	R	8.70	80	27/64"	11.00
M12.0 x 1.25	12.00	10.466	27/64"	10.70	80	17/32"	13.50
M14.0 x 1.50	14.00	12.160	31/64"	12.50	77	39/64"	15.50
M16.0 x 1.50	16.00	14.160	37/64"	14.50	77	11/16"	17.50
M18.0 x 1.50	18.00	16.160	41/64"	16.50	77	25/32"	20.00
M20.0 x 1.50	20.00	18.160	47/64"	18.50	77	55/64"	22.00
M22.0 x 1.50	22.00	20.160	13/16"	20.50	77	15/16"	24.00
M24.0 x 2.00	24.00	21.546	55/64"	22.00	77	1 1/32"	26.00
M27.0 x 2.00	27.00	24.506	31/32"	25.00	77	1 9/64"	29.00
M30.0 x 2.00	30.00	27.546	1 3/32"	28.00	77	1 19/64"	33.00

(ISO) METRIC CONDUIT

Screw Size Nominal Dia. x Pitch M (mm)	Basic Major Diameter (mm)	Basic Minor Diameter (mm)	Tapping Drill (inch)	Tapping Drill (mm)	Thread %	Clearance Drill (inch)	Clearance Drill (mm)
M16.0 x 1.50	16.00	14.160	37/64"	14.50	77	11/16"	17.50
M20.0 x 1.50	20.00	18.160	47/64"	18.50	77	55/64"	22.00
M25.0 x 1.50	25.00	23.160	59/64"	23.50	77	1 1/16"	27.00
M32.0 x 1.50	32.00	30.160	1 3/16"	30.50	77	1 3/8"	35.00
M40.0 x 1.50	40.00	38.160	1 33/64"	38.50	77	1 11/16"	43.00
M50.0 x 1.50	50.00	48.160	1 29/32"	48.50	77	2 1/8"	54.00

For General Symbols refer page 26

* Refer page 63 for Close & Loose Fit Drill Size

TAPPING DRILL SIZE FOR THREAD CUTTING TAPS

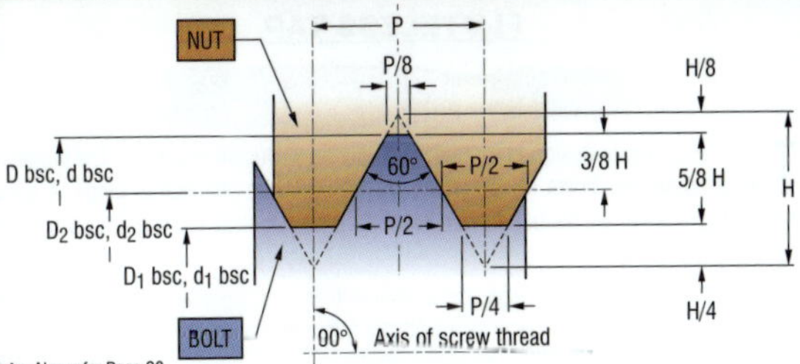

Note: Also refer Page 28

M (ISO) METRIC SPECIAL PITCHES

Screw Size Nominal Dia. x Pitch M (mm)	Basic Major Diameter (mm)	Basic Minor Diameter (mm)	+Tapping Drill Sizes For Mild Steel, Cast Iron, Aluminum, Brass, Bronze, Copper: Drill (inch)	Drill (mm)	Thread %	Clearance Drill Size *(Normal Fit): Drill (inch)	Drill (mm)
M1.7 x 0.35	1.70	1.270	# 55	1.35	77	# 50	1.85
M2.0 x 0.45	2.00	1.448	# 53	1.55	77	# 42	2.40
M2.3 x 0.40	2.30	1.809	# 49	1.90	77	# 40	2.50
M2.6 x 0.45	2.60	2.048	# 45	2.15	77	# 35	2.80
M3.0 x 0.60	3.00	2.264	# 42	2.40	77	# 30	3.40
M4.0 x 0.75	4.00	3.308	# 30	3.30	79	# 29	4.50
M5.0 x 0.90	5.00	3.896	# 20	4.10	77	# 3	5.50
M5.5 x 0.90	5.50	4.396	# 15	4.60	77	A	6.00
M8.0 x 0.75	8.00	7.080	9/32"	7.30	72	S	9.00
M9.0 x 1.00	9.00	7.773	5/16"	8.00	77	W	10.00
M10.0 x 0.75	10.00	9.080	23/64"	9.30	72	27/64"	11.00
M10.0 x 1.00	10.00	8.773	11/32"	9.00	77	27/64"	11.00
M12.0 x 1.00	12.00	10.773	7/16"	11.00	77	17/32"	13.50
M12.0 x 1.50	12.00	10.160	13/32"	10.50	77	17/32"	13.50
M14.0 x 1.25	14.00	12.466	1/2"	12.70	80	39/64"	15.50
M16.0 x 1.00	16.00	14.773	19/32"	15.00	77	11/16"	17.50
M20.0 x 1.00	20.00	18.773	3/4"	19.00	77	7/8"	22.00
M20.0 x 2.00	20.00	17.546	45/64"	18.00	77	7/8"	22.00
M24.0 x 1.00	24.00	22.773	57/64"	23.00	77	1 1/64"	26.00
M24.0 x 1.50	24.00	22.160	7/8"	22.50	77	1 1/64"	26.00
M25.0 x 1.50	25.00	23.160	59/64"	23.50	77	1 1/16"	27.00
M32.0 x 1.50	32.00	30.160	1 13/64"	30.50	77	1 3/8"	35.00

(ISO) METRIC SPARKPLUG

Screw Size	Major Diameter (mm)	Minor Diameter (mm)	Drill (inch)	Drill (mm)	Thread %	Clearance Drill (inch)	Clearance Drill (mm)
M10.0 x 1.00	10.00	8.773	S	9.00	77	27/64"	11.00
M12.0 x 1.25	12.00	10.466	27/64"	10.80	74	17/32"	13.50
M14.0 x 1.25	14.00	12.466	1/2"	12.70	80	39/64"	15.50
M18.0 x 1.50	18.00	16.160	41/64"	16.50	77	49/64"	19.50

+ Recommended for average commercial work. Drill Size and Decimal Equivalent refer pages 14-20

For General Symbols refer page 26

* Refer page 63 for Close & Loose Fit Drill Size

TAPPING DRILL SIZE FOR FLUTELESS (THREAD FORMING) TAPS

FLUTELESS TAP

The fluteless tap is designed for machine tapping in Ductile materials. The fluteless design produces holes of increased strength with the absence of chips. A good lubricating fluid must be used with fluteless taps. Correct tapping drill sizes are equally important.

NOTE : When using **Fluteless Taps**, it is advisable to use a straight Semi-Synthetic Oil with an extreme high pressure additive as a lubricant rather than a coolant.

UNC (UNIFIED NATIONAL COARSE)

Screw Size UNC (No./inch)	(t.p.i.)	+Tapping Drill Sizes For Mild Steel, Aluminum, Brass, Bronze, Copper — Drill (inch)	Drill (mm)	Thread %
No. 1	64	# 51	1.70	60
No. 2	56	# 47	2.00	65
No. 3	48	# 43	2.30	70
No. 4	40	# 39	2.50	75
No. 5	40	# 33	2.90	70
No. 6	32	1/8"	3.20	60
No. 8	32	# 25	3.80	70
No.10	24	# 18	4.30	70
No.12	24	# 9	5.00	78
1/4"	20	# 1	5.80	65
5/16"	18	L	7.30	60
3/8"	16	11/32"	8.80	75

UNF (UNIFIED NATIONAL FINE)

Screw Size UNF (No./inch)	(t.p.i.)	+Tapping Drill Sizes For Mild Steel, Aluminum, Brass, Bronze, Copper — Drill (in)	Drill (mm)	Thread %
No. 1	72	# 51	1.70	67
No. 2	64	# 48	1.90	90
No. 3	56	# 43	2.30	86
No. 4	48	# 37	2.60	60
No. 5	44	# 33	2.90	75
No. 6	40	1/8"	3.20	76
No. 8	36	# 24	3.90	61
No.10	32	# 16	4.50	63
No.12	28	# 7	5.10	60
1/4"	28	A	5.90	65
5/16"	24	M	7.50	63
3/8"	24	T	9.00	65

METRIC COARSE

Screw Size M (mm)	Pitch	+Tapping Drill Sizes For Mild Steel, Aluminum, Brass, Bronze, Copper — Drill (inch)	Drill (mm)	Thread %
M2.0	0.40	# 50	1.80	72
M3.0	0.50	# 36	2.75	72
M3.5	0.60	# 30	3.20	72
M4.0	0.70	# 27	3.60	84
M4.5	0.75	# 20	4.10	78
M5.0	0.80	# 14	4.60	73
M6.0	1.00	7/32"	5.50	73
M7.0	1.00	F	6.50	73
M8.0	1.25	M	7.40	70
M9.0	1.25	Q	8.40	70
M10.0	1.50	U	9.30	68
M11.0	1.50	Y	10.30	68
M12.0	1.75	7/16"	11.20	67
M14.0	2.00	33/64"	13.00	76
M16.0	2.00	19/32"	15.00	73
M18.0	2.50	43/64"	16.75	73
M20.0	2.50	3/4"	18.75	73

METRIC FINE & SPECIAL PITCHES

Screw Size M (mm)	Pitch	+Tapping Drill Sizes For Mild Steel, Aluminum, Brass, Bronze, Copper — Drill (inch)	Drill (mm)	Thread %
M5.0	0.50	# 12	4.80	65
M6.0	0.75	# 2	5.60	75
M7.0	0.75	G	6.60	75
M8.0	1.00	M	7.50	73
M9.0	1.00	Q	8.50	73
M10.0	1.00	V	9.50	73
M10.0	1.25	U	9.40	70
M11.0	1.00	Z	10.50	73
M12.0	1.00	29/64"	11.50	73
M12.0	1.25	29/64"	11.40	70
M12.0	1.50	7/16"	11.30	68
M14.0	1.00	17/32"	13.50	73
M14.0	1.25	17/32"	13.40	70
M14.0	1.50	33/64"	13.20	78
M16.0	1.00	39/64"	15.50	74
M16.0	1.50	19/32"	15.25	74
M18.0	1.50	11/16"	17.25	74

+ Recommended for average commercial work.

REMOVAL OF BROKEN TAPS

Removing a Broken Tap

Broken Taps that get lodged into a workpiece can be removed without damaging the component. Many methods are employed, the most common being:

- Using a Tap Extractor, e.g., WALTON®.
- Electrical Discharge Machining (EDM) / (Spark Eroding).
- Using a Tap Disintegrator.
- Welding a Nut onto the protruding part of the Broken Tap (if practicable).
- * Drilling out the Broken Tap using a Solid Carbide Spade Point Drill.

* Drilling Out a Broken Tap using a Solid Carbide Spade Point (Tap Removal) Drill

* Solid Carbide Spade Point Drill

This drill is specially designed for drilling out broken taps and other materials with hardness ≤ 72 Rc. Recommended drill speed : 1500 - 3000 rpm. No coolant necessary.

Broken Tap lodged in the Workpiece

PROCEDURE

1. Using a Ball Nose Carbide Rotary Burr, grind a semi-circular indent into the core of the broken tap. This provides a center-point (seat) for locating the Solid Carbide Spade Point (Tap Removal) Drill bit and prevents it from wandering.This operation is very critical to the success of subsequent drilling.

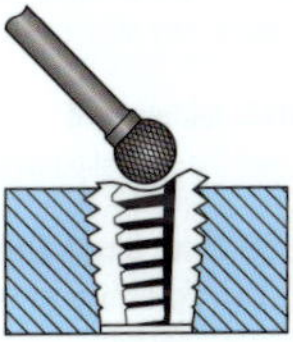

2. Using a suitable diameter Solid Carbide Spade Point (Tap Removal) Drill at a speed of 1500 to 3000 rpm and a steady and continuous manual feed, drill out the core of the broken tap. Clean out the swarf and any broken chips of the tap with compressed air at regular intervals during the drilling operation. This process should only be carried out using a rigid setup with the workpiece solidly clamped in a vice & preferably be done on a Milling/Drilling machine. No cutting fluid is required.

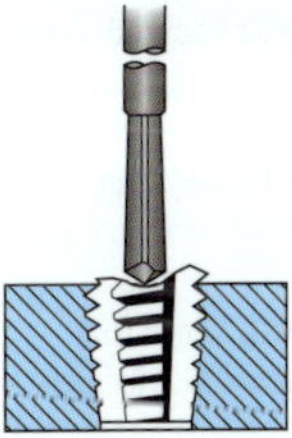

3. Using a Scriber or sharp instrument, remove the remaining pieces of the tap. Clean out the tapped hole with compressed air and retap the hole again.

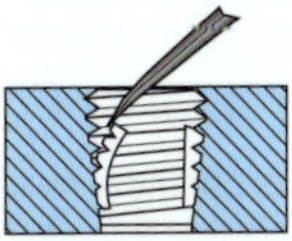

COMMON DRILL BITS

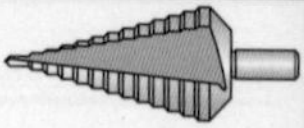

HSS Step Drill
Used for drilling/enlarging holes in thin materials

HSS Center Drill
Used for producing center holes that requires machining between centers

HSS Double Ended Panel Drill
Economical for use in thin metal sheets because of double end

HSS Spotting Drill
Used for accurate hole locating prior to using a jobber drill

HSS Stub Drill
Short length provides increased rigidity in machine drilling

Solid Carbide Straight Flute Drill
140° Drill point angle suitable for hardened and difficult to machine materials

Solid Carbide Low Helix Drill
Used for high speed applications on cast iron, non ferrous materials, plastics, etc.

Solid Carbide Slow Spiral Drill
Used to drill hardened and gummy materials

HSS Jobber Drill
Used for general purpose drilling

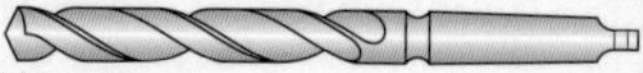

HSS Taper Shank Drill
Used for general purpose drilling in a rigid machine setup

HSS Reduced Shank Drill
Used for general purpose large hole drilling when chuck capacity is limited

HSS Step Drill
Combination drill and counterbore in one operation

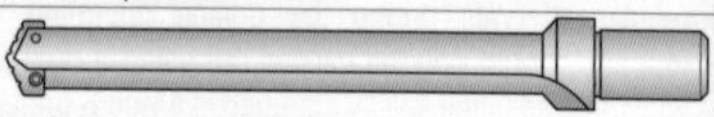

HSS Spade Drill
Economical for use in large hole production applications

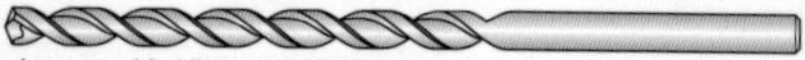

HSS Long Series Drill
Used for drilling straight holes upto 10-12 times drill diameter

Carbide Tipped Masonry Drill without/with SDS Shank (Special Direct System)
Used for drilling cement/concrete
suitable for spring loaded drill chucks

DRILL POINT GEOMETRY FOR VARIOUS MATERIALS

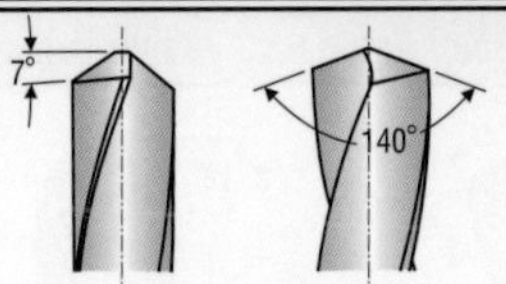

Tough Steels

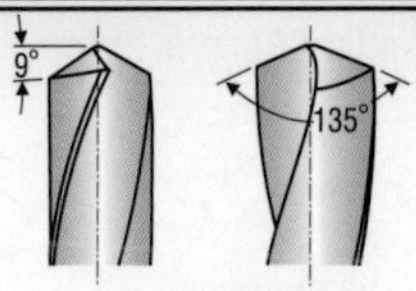

Stainless Steels

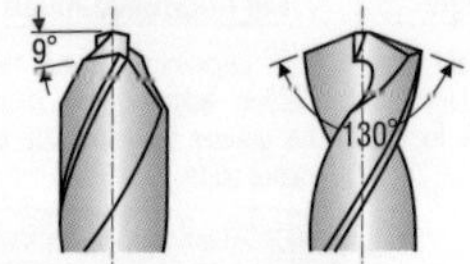

Deep Hole Drilling

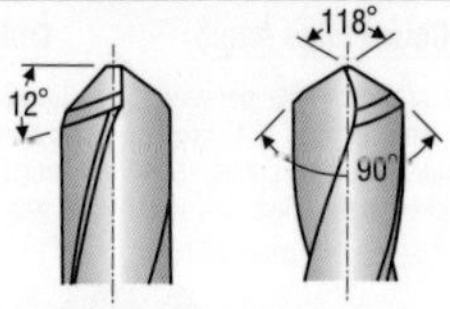

Cast Iron (Double-Angle Point)

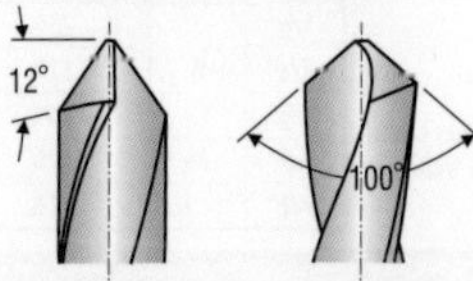

Copper & Some Copper Alloys

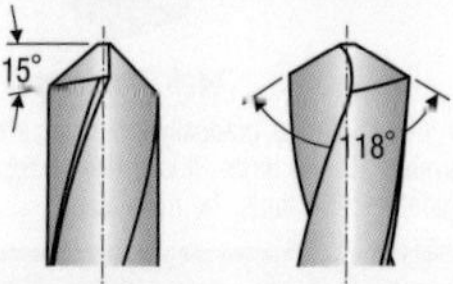

Aluminum, Brass & Soft Bronze

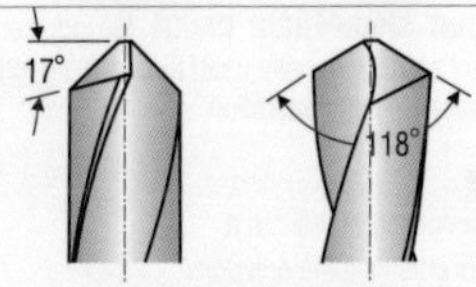

Bakelite, Plastics & Molded Materials

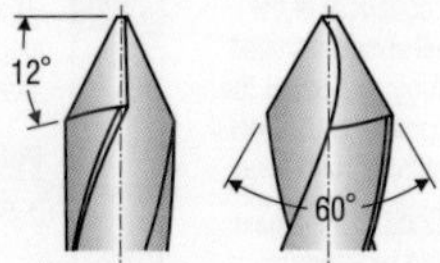

Wood, Hard Rubber & Fiber

Drill Point Styles

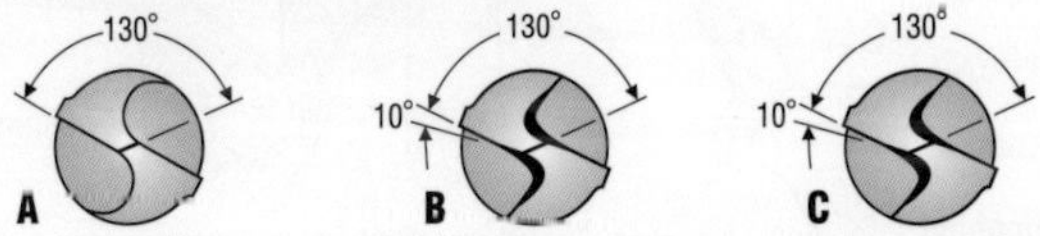

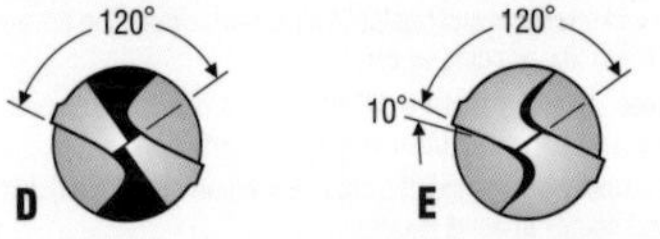

A - Standard Point
B - Point Relieved
C - Point Thinned
D - Split Point
E - Heavy Duty Point

DRILL POINT SHARPENING

Standard General Purpose 118° Point Angle HSS Jobber Drills

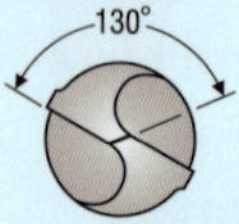

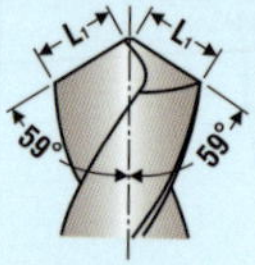

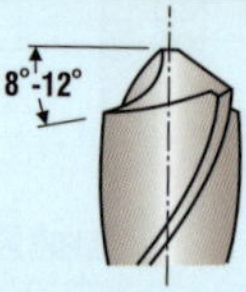

Chisel Edge Angle

The correct angle between the line across the dead center and the cutting lip must be maintained

Drill Point Angle

It is vitally important that both lips should be identical in length and angle to the drill axis.

Lip Clearance Angle

The clearance behind the cutting edges (lips) must be greater than the rate of penetration.

Web Thinning

The web thickness gradually increases after the drill has been reground several times. It is therefore recommended that the original web thickness be maintained.

RECOMMENDED WEB THICKNESS		
Drill Diameter		Web Thickness % of Diameter
(inch)	(mm)	
1/8"	3	20%
1/4"	6	17%
1/2"	13	14%
1"	25	12%
> 1"	> 25	11%

Hints for Hand Grinding HSS Twist Drills

The periphery of the wheel should be used for roughing-out of the drill point if much metal must be ground away.

The side of the wheel should be used for finishing.

Grinding Wheel Recommendations

38A60KVBE
or
SGB60JVH

The **DRILL POINT SHARPENING GAUGE** (on adjacent page) is the tool most frequently used to check the drill point during the sharpening operation.

Points to Note

- Avoid sudden cooling of drill point.
- Avoid overheating the drill point.
- Avoid excessive grinding pressures
- Keep the drill point constantly sharp.
- Use copious supply of coolant.

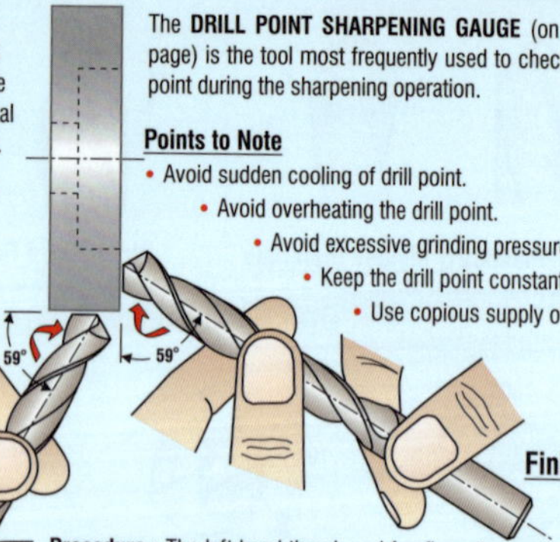

Finishing

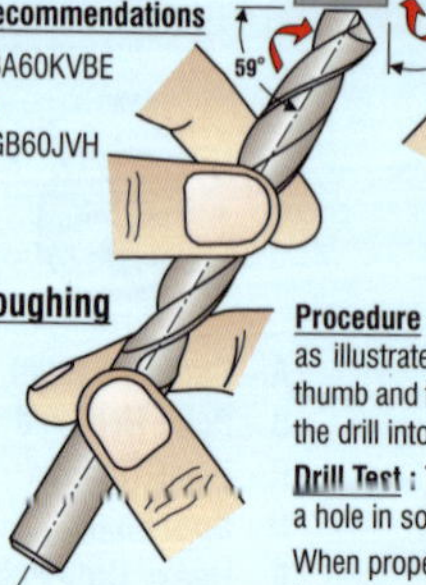

Roughing

Procedure : The left hand thumb and forefinger is used as a pivot as illustrated and the back of the drill is held with the right hand thumb and forefinger and rotated in a clockwise direction advancing the drill into the grinding wheel.

Drill Test : To test whether a drill has been properly sharpened, drill a hole in soft metal and observe the chip formation.

When properly sharpened, the chips will come out of the flutes in curled spirals of equal lengths.

DRILL POINT SHARPENING GAUGE AND EFFECTS OF INCORRECT GRINDING

Drill Point Sharpening Gauge

This Gauge is available from **www.EngineersBlackBook.com** with either **INCH** or **METRIC** Markings

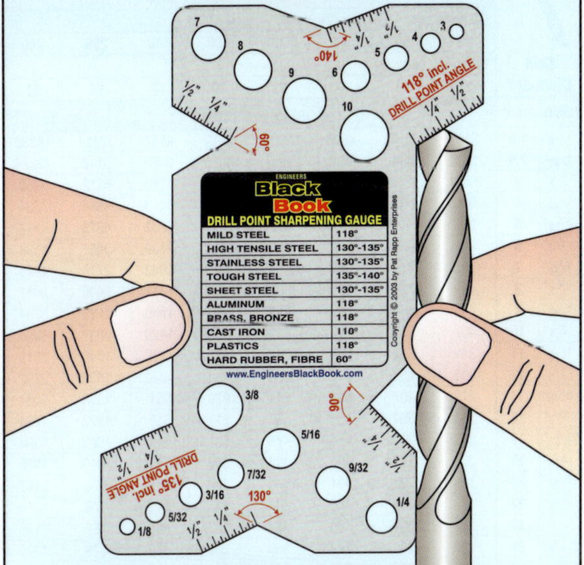

Effects of Incorrect Grinding of the Drill Point

Incorrect grinding causes the drill bit to run out of true, producing oversized and often, out-of-round holes. This puts severe strain on the drill bit causing rapid wear and premature drill failure.

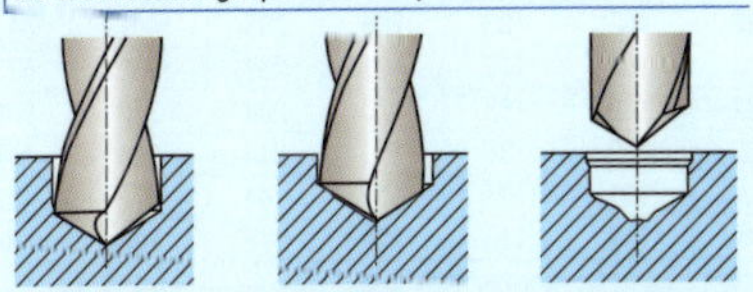

Lips of **EQUAL** lengths but **UNEQUAL** angles.

Lips of **UNEQUAL** lengths but **EQUAL** angles.

Lips of **UNEQUAL** lengths and **UNEQUAL** angles.

Drilling Thin Metal Sections

The point of the drill must be ground so that the outer corners are cutting before the points break through.

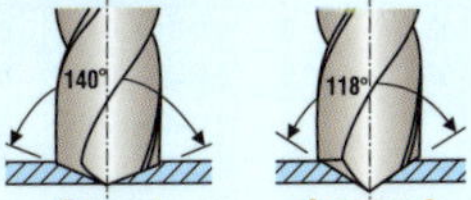

Correct

Incorrect

COMMON SPEEDS AND FEEDS FOR HIGH SPEED STEEL TWIST DRILLS

Drill Diameter		Work Material							
		Stainless Steel	Tool Steel	Cast Iron	Malleable Iron	Medium Carbon Steel	Free Cutting Steel	Aluminum Brass	Wood Plastic
		Cutting Speed : Feet per Minute							
		40	60	80	90	100	140	200	300
		Cutting Speed : Meters per Minute							
		12	18	24	27	30	43	61	91
(inch)	(mm)	Spindle Speed RPM							
1/16"	1.5	2445	3667	4890	5501	6112	8557	12224	18336
3/32"	2.5	1630	2445	3260	3667	4075	5705	8149	12224
1/8"	3	1222	1834	2445	2750	3056	4278	6112	9168
5/32"	4	978	1467	1956	2200	2445	3423	4890	7334
3/16"	5	815	1222	1630	1834	2037	2852	4075	6112
1/4"	6	611	917	1222	1375	1528	2139	3056	4584
9/32"	7	543	815	1087	1222	1358	1902	2716	4075
5/16"	8	489	733	978	1100	1222	1711	2445	3667
11/32"	9	445	667	889	1000	1111	1556	2223	3334
3/8"	10	407	611	815	917	1019	1426	2037	3056
7/16"	11	349	524	699	786	873	1222	1746	2619
1/2"	13	306	458	611	688	764	1070	1528	2292
9/16"	14	272	407	543	611	679	951	1358	2037
5/8"	16	244	367	489	550	611	856	1222	1834
23/32"	18	213	319	425	478	531	744	1063	1594
3/4"	19	204	306	407	458	509	713	1019	1528
25/32"	20	196	293	391	440	489	685	978	1467
7/8"	22	175	262	349	393	437	611	873	1310
1"	25	153	229	306	344	382	535	764	1146
1 1/8"	28	136	204	272	306	340	475	679	1019
1 1/4"	32	122	183	244	275	306	428	611	917
1 3/8"	35	111	167	222	250	278	389	556	833
1 1/2"	38	102	153	204	229	255	357	509	764
1 5/8"	41	94	141	188	212	235	329	470	705
1 3/4"	44	87	131	175	196	218	306	437	655
1 7/8"	47	81	122	163	183	204	285	407	611
2"	50	76	115	153	172	191	267	382	573
2 1/4"	57	68	102	136	153	170	238	340	509
2 1/2"	63	61	92	122	138	153	214	306	458
2 3/4"	70	56	83	111	125	139	194	278	417
3"	76	51	76	102	115	127	178	255	382

Recommended Feeds for Various Drill Diameters			
Diameter of Drill		Feed per Revolution	
(inch)	(mm)	(inch)	(mm)
1/8"	3	0.001" - 0.003"	0.02 - 0.10
1/4"	6	0.002" - 0.00[illegible]"	0.05 - 0.15
1/2"	13	0.004" - 0.010"	0.10 - 0.25
< 1"	< 25	0.007" - 0.015"	0.20 - 0.40
> 1"	> 25	0.015" - 0.025"	0.40 - 0.65

Note : The information on this page is intended as a general guide only. Figures will vary with grades of material used.

It is best to start with a moderate speed and feed, increasing either one, or both, after observing the cutting action and condition of the drill bit.

NOTE: Also refer pages 60 & 61 for detailed Speeds/Feeds Conversion table.

SPEEDS AND FEEDS FOR STANDARD SOLID CARBIDE DRILLS

Types of Solid Carbide Drills

Straight Flute Drill

Strong construction with 140° drill point angle suitable for hardened & difficult to machine materials. Hole depth should not exceed 2 x dia.

Slow Spiral Drill - 15°-20°

Designed to drill hardened and gummy materials and High-temperature Alloys. Also recommended for Stainless Steel, Tool Steel & Titanium Nickel Alloys.

Twist Drill - 25°-30°

2 & 3 Flute general purpose drill ideal for low-tensile & non-ferrous materials such as high Si Aluminum, Bronze, Masonite & Resin/ Fiberglass Laminates.

Material	Cutting Speed ft/min	Solid Carbide Drill Diameter					
		rpm in./rev	1/16"	1/8"	1/4"	3/8"	1/2"
Aluminum/ Al Alloys	200-450	rpm	16004	8002	4007	2670	2004
		feed	0.001"	0.003"	0.007"	0.009"	0.012"
Brass	200-450	rpm	16004	8002	4007	2670	2004
		feed	0.001"	0.002"	0.003"	0.003"	0.004"
Bronze/Copper	150-300	rpm	12720	6360	3180	1900	1590
		feed	0.001"	0.002"	0.003"	0.003"	0.004"
Cast Iron (Soft)	130-200	rpm	10002	5001	2505	1669	1252
		feed	0.002"	0.003"	0.005"	0.008"	0.010"
Cast Iron (Hard)	50-130	rpm	6360	3180	1590	960	795
		feed	0.001"	0.002"	0.003"	0.005"	0.008"
Plastics/ Composites	200-600	rpm	16004	8002	4007	2670	2004
		feed	0.003"	0.006"	0.008"	0.012"	0.016"
Mild Steel	60-130	rpm	4880	2440	1220	730	610
		feed	0.001"	0.002"	0.004"	0.005"	0.007"
Carbon Steel	100-180	rpm	6360	3180	1590	960	795
		feed	0.001"	0.002"	0.003"	0.004"	0.006"
Alloy Steel/ High Tensile	40-60	rpm	3000	1500	751	501	376
		feed	0.001"	0.002"	0.003"	0.003"	0.003"
Hardened/Tool Steel	20-40	rpm	2120	1060	530	320	265
		feed	0.0005"	0.0007"	0.001"	0.001"	0.001"
Stainless Steel/ 300 Series	30-90	rpm	4240	2120	1060	640	530
		feed	0.001"	0.003"	0.005"	0.006"	0.008"
Stainless Steel/ 400 Series	40-120	rpm	4880	2440	1220	730	610
		feed	0.001"	0.001"	0.002"	0.002"	0.003"

NOTE : To enhance Hole Quality and Tool Life, the use of copious quantity of coolant is recommended, particularly directed at the drill point to prevent chipping or flaking of the Carbide cutting edge.

LOCATING HOLE COORDINATES FOR EQUALLY SPACED HOLES

Multiply Values Shown by Diameter of Pitch Circle

(INCH or METRIC Units)

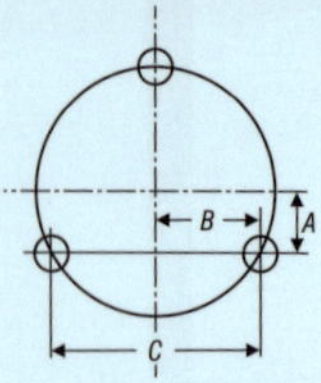

3-Hole

A = 0.25000
B = 0.43301
C = 0.86603

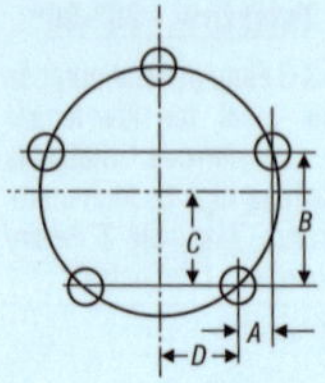

5-Hole

A = 0.18164
B = 0.55902
C = 0.40451
D = 0.29389

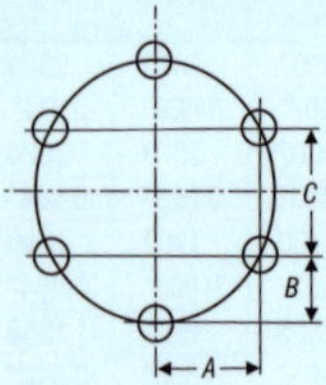

6-Hole

A = 0.43301
B = 0.25000
C = 0.50000

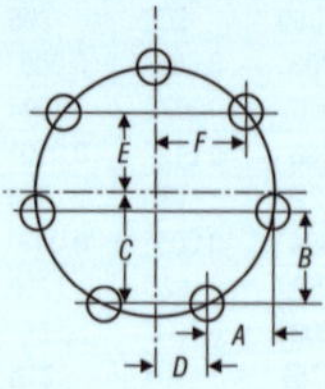

7-Hole

A = 0.27052
B = 0.33922
C = 0.45049
D = 0.21694
E = 0.31175
F = 0.39092

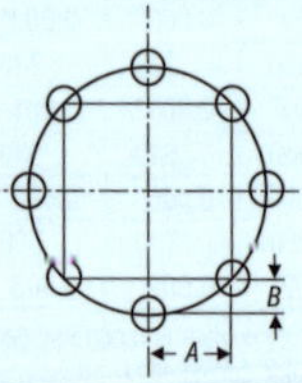

8-Hole

A = 0.35355
B = 0.14645

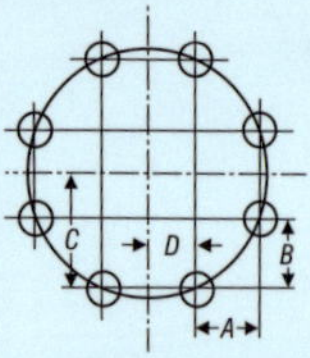

8-Hole

A = 0.27059
B = 0.27059
C = 0.46194
D = 0.19134

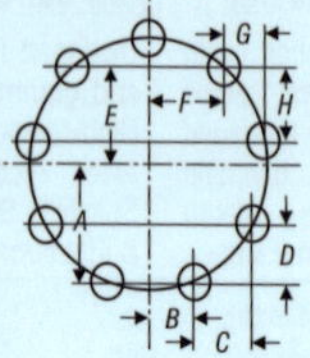

9-Hole

A = 0.46985
B = 0.17101
C = 0.26201
D = 0.21985
E = 0.38302
F = 0.32139
G = 0.17101
H = 0.29620

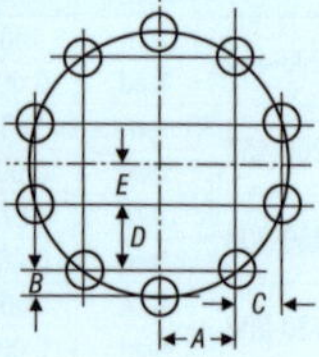

10-Hole

A = 0.29389
B = 0.09549
C = 0.18164
D = 0.25000
E = 0.15451

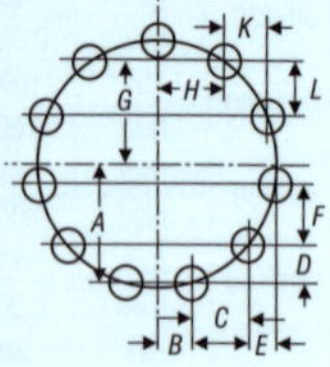

11-Hole

A = 0.47975
B = 0.14087
C = 0.23700
D = 0.15231
E = 0.11704
F = 0.25627
G = 0.42063
H = 0.27032
K = 0.18449
L = 0.21291

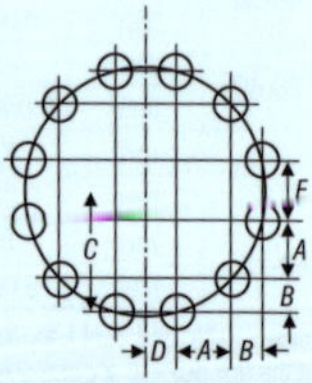

12-Hole

A = 0.22415
B = 0.12941
C = 0.48296
D = 0.12941
E = 0.25882

INCH TAPER PINS
ACCORDING TO ASME B18.8.2

Taper 1:48
Include Angle 1° 11' 38"
Taper per inch on diameter 0.020833

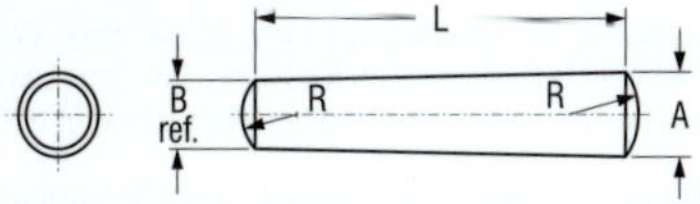

Small End Diameter B = Basic Large End Diameter – (L x 0.02083)

INCH Taper Pins (according to ASME B18.8.2) — Dimensions in inches

Pin Size Number	Basic Pin Diameter Large End	Major Diameter (Large End) A				End Crown Radius R	
		Commercial Class		Precision Class			
		max	min	max	min	max	min
7/0	0.0625	0.0638	0.0618	0.0635	0.0625	0.072	0.052
6/0	0.0780	0.0793	0.0773	0.0790	0.0780	0.088	0.068
5/0	0.0940	0.0953	0.0933	0.0950	0.0940	0.104	0.084
4/0	0.1090	0.1103	0.1083	0.1100	0.1090	0.119	0.099
3/0	0.1250	0.1263	0.1243	0.1260	0.1250	0.135	0.115
2/0	0.1410	0.1423	0.1403	0.1420	0.1410	0.151	0.131
0	0.1560	0.1573	0.1553	0.1570	0.1560	0.166	0.146
1	0.1720	0.1733	0.1713	0.1730	0.1720	0.182	0.162
2	0.1930	0.1943	0.1923	0.1940	0.1930	0.203	0.183
3	0.2190	0.2203	0.2183	0.2200	0.2190	0.229	0.229
4	0.2500	0.2513	0.2493	0.2510	0.2500	0.260	0.260
5	0.2890	0.2903	0.2883	0.2900	0.2890	0.299	0.279
6	0.3410	0.3423	0.3403	0.3420	0.3410	0.361	0.331
7	0.4090	0.4103	0.4983	0.4100	0.4090	0.419	0.399
8	0.4920	0.4933	0.4913	0.4930	0.4920	0.502	0.482
9	0.5010	0.5923	0.5903	0.5920	0.5910	0.601	0.581
10	0.7060	0.7073	0.7053	0.7070	0.7060	0.716	0.696
11	0.8600	0.8613	0.8593	–	–	0.870	0.850
12	1.0320	1.0333	1.0313	–	–	1.042	1.022
13	1.2410	1.2423	1.2403	–	–	1.251	1.231
14	1.5230	1.5243	1.5223	–	–	1.533	1.513

TYPES AND APPLICATIONS OF HIGH SPEED STEEL END MILLS

Standard 30° Spiral

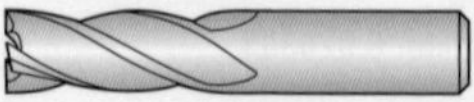

Suitable for both Finishing and Roughing operations on Mild Steel, Heat Treatable and Structural Steels with tensile strengths ≤ 65 Tons/inch². This cutter can also be used for short chipping Non-Ferrous metals. Usually comes with a variety of surface coatings.

Round Knuckle (Sinusoidal) Teeth Roughing

Suitable for maximum Roughing where rapid metal removal is the prime requisite. The cutting action produces short chips with excellent chip removal. As this cutter is mainly designed for rapid metal removal, additional finishing operations are necessary.

Flat Knuckle Teeth Roughing

Suitable for Medium Roughing operations where rapid metal removal is the prime requisite. This cutter produces a finish that is generally better than the Round Knuckle Teeth cutter, yet requires additional finishing operations.

Fine Pitch Knuckle Teeth Roughing

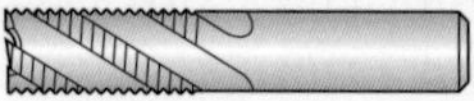

Suitable for Medium Roughing applications on high tensile material where rapid metal removal plus surface finish are the prime requisites. It produces a finish that is better than both of the above roughing cutters and generally no additional finishing is required.

40° Quick Spiral

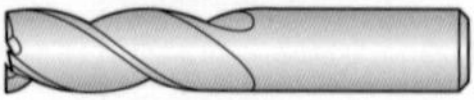

Suitable for Roughing and Finishing Soft, Tough and Long-Chipping materials. This cutter produces an excellent surface finish with no additional finishing required. Comes in a variety of surface coatings.

45° - 55° Extremely Quick Spiral

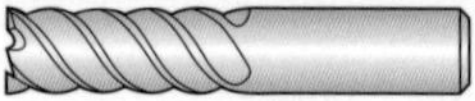

Suitable for Finishing High Tensile materials up to 75 Tons/inch² as well as Titanium, Nodular Cast Iron and difficult to machine materials. Comes in a variety of surface coatings.

REAMERS*

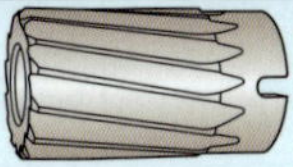

Shell Reamer
Mounted on arbor and mainly for reaming large holes upto approximately 4" diameter

Pipe Reamer
Used primarily for reaming a component prior to using a taper pipe tap

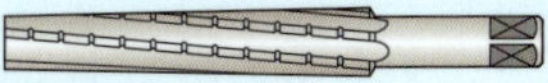

Morse Taper Roughing Reamer
Used as a roughing reamer for morse taper holes

Morse Taper Finishing Reamer
Used as a finishing reamer for morse taper holes

Repairman's Taper Reamer
Used to remove burrs in pipes, conduit and for enlarging holes in sheet metal

Spiral Flute Hand Reamer
Used for through-holes applications and moves swarf forward

Chucking Reamer
Used in a machine for reaming holes to a tight tolerance

Expansion Reamer
Allows for minor adjustments to size and can be reground several times

Taper Pin Reamer
Used for reaming holes into which a standard taper pin will fit

Adjustable Reamer
Allows for reamer size to be adjusted

Inline Chucking Reamer
Similar use as Chucking reamer but with pilot guide for alignment purpose

Bridge Reamer
Used to enlargen and align holes mainly in steel constructions for rivets

Taper Shank Machine Reamer
Similar to a Chucking reamer and also used in a machine

Taper Pin Drill / Reamer

Used as a combination drill and reamer for holes into which a standard taper pin will fit

***Note:** RH flutes move the swarf upwards whilst LH flutes move the swarf forward

CONVERSION TABLE SURFACE / CUTTING SPEED TO R.P.M

		Surface / Cutting Speed : Feet per Minute								
		10	16	26	33	49	66	82	98	115
		Surface / Cutting Speed : Meters per Minute								
Diameter		3	5	8	10	15	20	25	30	35
(inch)	(mm)	Spindle Speed (RPM)								
	2	477	796	1273	1592	2387	3183	3979	4775	5570
	3	318	531	849	1061	1592	2122	2653	3183	3714
1/8"		301	501	802	1003	1504	2005	2506	3008	3509
	4	239	398	637	796	1194	1592	1989	2387	2785
3/16"		200	334	535	668	1002	1337	1671	2005	2339
	5	191	318	509	637	955	1273	1592	1910	2228
	6	159	265	424	531	796	1061	1326	1592	1857
1/4"		150	251	401	501	752	1003	1253	1504	1754
5/16"		120	200	321	401	601	802	1002	1203	1403
	8	119	199	318	398	597	796	995	1194	1393
3/8"		100	167	267	334	501	668	835	1003	1170
	10	95	159	255	318	477	637	796	955	1114
7/16"		86	143	229	286	430	573	716	859	1003
	12	80	133	212	265	398	531	663	796	928
1/2"		75	125	201	251	376	501	627	752	877
	14	68	114	182	227	341	455	568	682	796
9/16"		67	111	178	223	334	446	557	668	780
5/8"		60	100	160	201	301	401	501	602	702
	16	60	99	159	199	298	398	497	597	696
	18	53	88	141	177	265	354	442	531	619
3/4"		50	84	134	167	251	334	418	501	585
	20	48	80	127	159	239	318	398	477	557
	22	43	72	116	145	217	289	362	434	506
7/8"		43	72	115	143	215	286	358	430	501
	24	40	66	106	133	199	265	332	398	464
	25	38	64	102	127	191	255	318	382	446
1"		38	63	100	125	188	251	313	376	439
	30	32	53	85	106	159	212	265	318	371
1 1/4"		30	50	80	100	150	201	251	301	351
1 1/2"		25	42	67	84	125	167	209	251	292
	40	24	40	64	80	119	159	199	239	279
1 3/4"		21	36	57	72	107	143	179	215	251
	50	19	32	51	64	95	127	159	191	223
2"		19	31	50	63	94	125	157	188	219
	75	13	21	34	42	64	85	106	127	149
3"		13	21	33	42	63	84	104	125	146
	100	10	16	25	32	48	64	80	95	111
4"		9	16	25	31	47	63	78	94	110
	200	5	8	13	16	24	32	40	48	56
8"		5	8	13	16	24	31	39	47	55
	300	3	5	8	11	16	21	27	32	37
12"		3	5	8	10	16	21	26	31	37

NOTE . R.P.M. for Surface / Cutting Speeds not given, can be obtained by simple addition or subtraction. Example :

180 ft/min = 262 – 82 = 2004 R.P.M. – 626 R.P.M. = 1378 R.P.M. (for 1/2" dia.)
120 m/min = 100 + 20 = 3183 R.P.M. + 636 R.P.M. = 3819 R.P.M. (for 10mm dia.)

CONVERSION TABLE
SURFACE / CUTTING SPEED TO R.P.M

Diameter		Surface / Cutting Speed : Feet per Minute								
		131	148	164	197	230	262	295	328	656
		Surface / Cutting Speed : Meters per Minute								
		40	45	50	60	70	80	90	100	200
(inch)	(mm)	Spindle Speed (RPM)								
	2	6366	7162	7950	9549	11141	12732	14324	15910	31831
	3	4244	4775	5305	6366	7427	8488	9549	10610	21221
1/8"		4010	4511	5013	6015	7018	8020	9023	10026	20051
	4	3183	3581	3979	4775	5570	6366	7162	7958	15916
3/16"		2673	3007	3341	4010	4678	5346	6015	6683	13366
	5	2546	2865	3183	3820	4456	5093	5730	6366	12732
	6	2122	2387	2653	3183	3714	4244	4775	5305	10610
1/4"		2005	2256	2506	3008	3509	4010	4511	5013	10026
5/16"		1604	1804	2005	2406	2807	3208	3609	4010	8020
	8	1592	1790	1989	2387	2785	3183	3581	3979	7958
3/8"		1337	1504	1671	2005	2339	2673	3008	3342	6684
	10	1273	1432	1592	1910	2228	2546	2865	3183	6366
7/16"		1146	1289	1432	1719	2005	2291	2578	2864	5729
	12	1061	1194	1326	1592	1857	2122	2387	2653	5305
1/2"		1003	1128	1253	1504	1754	2005	2256	2506	5013
	14	909	1023	1137	1364	1592	1819	2046	2274	4547
9/16"		891	1003	1114	1337	1559	1782	2005	2228	4456
5/8"		802	902	1003	1203	1404	1604	1805	2005	4010
	16	796	895	995	1194	1393	1592	1790	1989	3979
	18	707	796	884	1061	1238	1415	1592	1768	3537
3/4"		668	752	835	1003	1170	1337	1504	1671	3342
	20	637	716	796	955	1114	1273	1432	1592	3183
	22	579	651	723	868	1013	1157	1302	1447	2894
7/8"		573	644	716	859	1003	1146	1289	1432	2864
	24	531	597	663	796	928	1061	1194	1326	2653
	25	509	573	637	764	891	1019	1146	1273	2546
1"		501	564	627	752	877	1003	1128	1253	2506
	30	424	477	531	637	743	849	955	1061	2122
1 1/4"		401	451	501	602	702	802	902	1003	2005
1 1/2"		334	376	418	501	585	668	752	835	1671
	40	318	358	398	477	557	637	716	796	1592
1 3/4"		286	322	358	430	501	573	644	716	1432
	50	255	286	318	382	446	509	573	637	1273
2"		251	282	313	376	439	501	564	627	1253
	75	170	191	212	255	297	340	382	424	849
3"		167	188	209	251	292	334	376	418	835
	100	127	143	159	191	223	255	286	318	637
4"		125	141	157	188	220	251	282	314	627
	200	64	72	80	95	111	127	143	159	318
8"		63	71	78	94	110	125	141	157	314
	300	42	48	53	64	74	85	95	106	212
12"		42	47	52	63	73	84	94	104	209

To Calculate RPM in ft/min

$$RPM = \frac{SFM \times 3.82}{D}$$

Where SFM = surface speed (ft/min), D = diameter

To Calculate RPM in m/min

$$RPM = \frac{1000 \times Vc}{\pi \times D}$$

Where Vc = surface speed (m/min), D = diameter

SURFACE SPEED TO RPM

DIMENSIONS OF COUNTERBORE AND COUNTERSINK FOR SOCKET HEAD CAP SCREWS

Counterbore

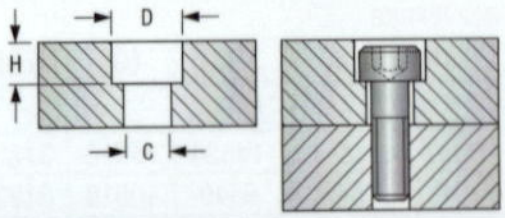

Countersink

*INCH α = 82°
*METRIC α = 90°

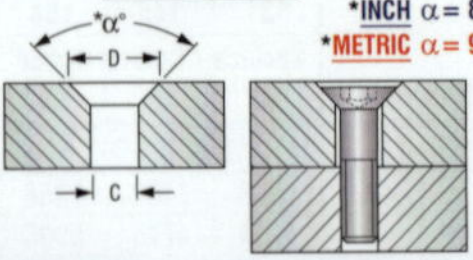

Nominal Thread Size	Counterbore: Clearance For Thread 'C'	Counterbore: Clearance For Head 'D'	Counterbore: Depth 'H'	Countersink: Clearance For Thread 'C'	Countersink: Diameter of Countersink 'D'
INCH					Dimensions in inches
No. 4	# 29	7/32"	0.125"	# 29	0.255"
No. 5	# 23	1/4"	0.140"	# 23	0.281"
No. 6	# 18	9/32"	0.160"	# 18	0.307"
No. 8	# 10	5/16"	0.185"	# 10	0.359"
No. 10	# 2	3/8"	0.215"	# 2	0.411"
1/4"	9/32"	7/16"	0.280"	9/32"	0.531"
5/16"	11/32"	17/32"	0.350"	11/32"	0.656"
3/8"	13/32"	5/8"	0.415"	13/32"	0.781"
7/16"	15/32"	23/32"	0.475"	15/32"	0.844"
1/2"	17/32"	13/16"	0.540"	17/32"	0.938"
5/8"	21/32"	1"	0.675"	21/32"	1.188"
3/4"	25/32"	1 3/16"	0.810"	25/32"	1.438"
METRIC					Dimensions in mm
M3.0	3.40	6.00	3.40	3.40	6.70
M4.0	4.50	8.00	4.60	4.50	9.00
M5.0	5.50	10.00	5.70	5.50	11.20
M6.0	6.60	11.00	6.80	6.60	13.40
M8.0	9.00	15.00	9.00	9.00	17.90
M10.0	11.00	18.00	11.00	11.00	22.40
M12.0	14.00	20.00	13.00	14.00	26.90
M14.0	16.00	23.00	15.20	16.00	30.30
M16.0	18.00	26.00	17.50	18.00	33.60
M18.0	20.00	29.00	19.50	20.00	37.00
M20.0	22.00	32.00	21.50	22.00	40.30

Standard Cap Screw Counterbores

HSS Cap Screw Counterbores are tools used to machine a recess for socket head cap screws and shoulder screws. They are available in Straight Shank and Morse Taper Shank.

Cap Screw Size (inch)	Pilot Dia.	Cutting Dia.	Shank: Parallel	Shank: MTS
No. 5	0.141"	0.221"	3/16"	–
No. 6	0.150"	0.242"	7/32"	–
No. 8	11/64"	19/64"	1/4"	–
No. 10	13/64"	21/64"	5/16"	–
1/4"	9/32"	13/32"	3/8"	1
5/16"	11/32"	1/2"	7/16"	2
3/8"	13/32"	19/32"	1/2"	2
7/16"	15/32"	11/16"	1/2"	2
1/2"	17/32"	25/32"	1/2"	2
5/8"	21/32"	31/32"	3/4"	3
3/4"	13/16"	1 3/16"	1"	3

Cap Screw Size (mm)	Pilot Dia.	Cutting Dia.	Shank: Parallel	Shank: MTS
M3.0	3.40	6.00	7/32"	–
M4.0	4.50	8.00	5/16"	–
M5.0	5.50	10.00	5/16"	–
M6.0	6.60	11.00	3/8"	1
M8.0	9.00	15.00	1/2"	2
M10.0	11.00	18.00	1/2"	2
M12.0	14.00	20.00	5/8"	2
M14.0	15.00	24.00	3/4"	2
M16.0	17.00	26.00	1"	3
M20.0	21.00	26.00	1"	3
M24.0	26.00	40.00	–	3

CLEARANCE HOLES FOR BOLTS, SCREWS AND STUDS ACCORDING TO ASME B18.2.8-1999, R2005 & ISO 273

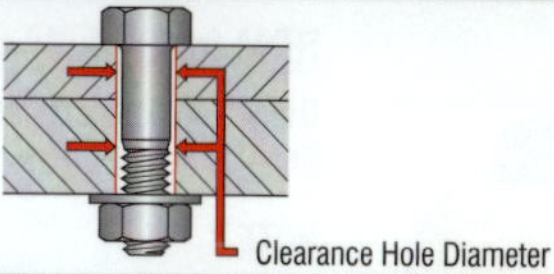

Clearance Holes for INCH Fasteners

Nominal Screw Size	Fit Classes – Nominal Drill Size		
	Normal	Close	Loose
# 0	# 48	# 51	3/32"
# 1	# 43	# 40	# 37
# 2	# 38	3/32"	# 32
# 3	# 32	# 36	# 30
# 4	# 30	# 31	# 27
# 5	5/32"	9/64"	11/64"
# 6	# 18	# 23	# 13
# 8	# 9	# 15	# 3
# 10	# 2	# 5	B
1/4"	9/32"	17/64"	19/64"
5/16"	11/32"	21/64"	23/64"
3/8"	13/32"	25/64"	27/64"
7/16"	15/32"	29/64"	31/64"
1/2"	9/16"	17/32"	39/64"
5/8"	11/16"	21/32"	47/64"
3/4"	13/16"	25/32"	29/32"
7/8"	15/16"	29/32"	1 1/32"
1"	1 3/32"	1 1/32"	1 5/32"
1 1/8"	1 7/32"	1 5/32"	1 5/16"
1 1/4"	1 11/32"	1 9/32"	1 7/16"
1 3/8"	1 1/2"	1 7/16"	1 39/64"
1 1/2"	1 5/8"	1 9/16"	1 47/64"

Clearance Holes for METRIC Fasteners

Nominal Screw Size	Fit Classes – Nominal Drill Size		
	Normal	Close	Loose
M1.6	1.80	1.70	2.00
M2.0	2.40	2.20	2.60
M2.5	2.90	2.70	3.10
M3.0	3.40	3.20	3.60
M4.0	4.50	4.30	4.80
M5.0	5.50	5.30	5.80
M6.0	6.60	6.40	7.00
M8.0	9.00	8.40	10.00
M10.0	11.00	10.50	12.00
M12.0	13.50	13.00	14.50
M14.0	15.50	15.00	16.50
M16.0	17.50	17.00	18.50
M20.0	22.00	21.00	24.00
M24.0	26.00	25.00	28.00
M30.0	33.00	31.00	35.00
M36.0	39.00	37.00	42.00
M42.0	45.00	43.00	48.00
M48.0	52.00	50.00	56.00
M56.0	62.00	58.00	66.00
M64.0	70.00	66.00	74.00
M72.0	78.00	74.00	82.00
M80.0	86.00	82.00	91.00

Common Countersinking Tools

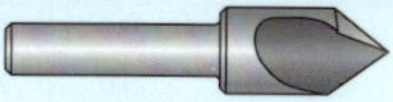

Single Flute Countersink

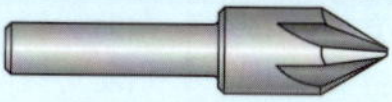

General Purpose Multi Flute Countersink

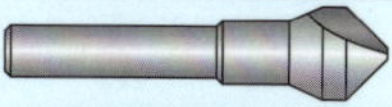

Single Flute Countersink

Through-Hole Countersink

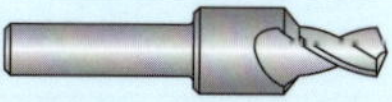

Combination Drill/Countersink

Through-Hole Countersink

Note: Included angle for Countersink Fasteners

INCH - Machine Screws 80°-82° (according to ASME B18.6.3 2002)	INCH - Self Tapping Screws 80°-82° (according to ASME B18.6.4 1998)
METRIC - Machine Screws 90°-92° (according to ASME B18.6.7M 1998)	INCH - Wood Screws 80°-82° (according to ASME B18.6.1 1981 - R 1997)

HOLE SAWS
RECOMMENDED CUTTING SPEEDS

RPM for High Speed Steel Hole Saws

Uses : Mild Steel, Cast Iron, Tool Steel, Stainless Steel, Brass, Aluminum, Wood, etc.

NOTE :

- Use Coolant when cutting Metals.
- Apply sufficient pressure to ensure that the Hole Saw cuts and does not rub.
- When cutting wood, finish off the operation from the reverse side to avoid splinters.

Hole Saw Diameter		Recommended Speed (rpm)					
(inch)	(mm)	Mild Steel	Cast Iron	Tool Steel Stainless	Brass	Aluminum	Wood
9/16"	14	580	400	300	790	900	3000
5/8"	16	550	365	275	730	825	3000
3/4"	19	460	300	230	600	690	3000
–	20	440	290	220	580	660	3000
13/16"	21	425	280	210	560	635	3000
7/8"	22	390	260	195	520	585	3000
1"	25	350	235	175	470	525	2700
1 1/8"	29	300	200	150	400	450	2700
–	30	285	190	145	380	425	2400
1 1/4"	32	275	180	140	380	410	2400
1 3/8"	35	250	165	125	330	375	2400
1 1/2"	38	230	150	115	300	345	2400
–	40	220	145	110	290	330	2100
1 5/8"	41	210	140	105	280	315	2100
1 3/4"	44	195	130	95	260	295	2100
1 7/8"	48	180	120	90	240	270	2100
2"	51	170	115	85	230	255	2000
2 1/4"	57	150	100	75	200	225	2000
2 3/8"	60	140	95	70	190	220	2000
2 1/2"	64	135	90	65	180	205	1800
2 3/4"	70	125	80	60	160	185	1800
3"	76	115	75	55	150	170	1500
3 1/4"	83	105	70	50	40	155	1500
3 1/2"	89	95	65	45	130	145	1200
4"	102	85	55	40	110	130	1000
5"	127	65	45	30	90	90	800
6"	152	55	35	25	75	75	800

HOLE SAWS
RECOMMENDED CUTTING SPEEDS

RPM for Carbide Tipped Hole Saws

Uses : Aluminum, Brass, Cast Iron, Mild Steel, Wood, Plastics, etc.

Hole Saw Diameter (inch)	Hole Saw Diameter (mm)	Aluminum	Brass	Cast Iron	Mild Steel	Wood & Plastics
9/16 - 1"	14-25	850-600	750-550	375-300	550-400	3000-2500
1 - 2"	25-50	600-300	550-250	300-125	400-200	2500-2000
2 - 3"	50-75	300-200	250-175	125-75	200-125	2000-1500
3 - 4"	75-100	200-150	175-125	75-50	125-100	1500-1000
4 - 8"	100-200	150-100	125-75	50-25	100-50	1000-500

RPM for Diamond Edged Hole Saws

Uses : Asbestos / Substitutes, Ceramics, Marble, Stone, etc.

Hole Saw Diameter (inch)	Hole Saw Diameter (mm)	Asbestos / Substitutes	Ceramics / Stone	Reinfo-Plastics, Fibreglass
3/4 - 1"	19-25	700-350	500-250	1000-500
1 - 2"	25-50	350-175	250-100	500-250
2 - 3"	50-75	175-100	100-75	250-150
3 - 5"	75-125	100-50	75-40	150-75
5 - 6"	125-150	50-25	40-20	75-50

RPM for Carbide Grit Hole Saws

Uses : Asbestos / Substitutes, Ceramic Tiles, Reinforced Plastics, etc.

Hole Saw Diameter (inch)	Hole Saw Diameter (mm)	Asbestos / Substitutes	Ceramic Tiles	Reinfo-Plastics, Fibreglass
3/4 - 1"	19-25	700-350	500-250	1000-500
1 - 2"	25-50	350-175	250-100	500-250
2 - 3"	50-75	175-100	100-75	250-150
3 - 5"	75-125	100-50	75-40	150-75
5 - 6"	125-150	50-25	40-20	75-50

NOTE : The above information is intended as a general guide only. Speeds will vary with grades of materials used. It is best to start with a slow speed an increase the speed after observing the cutting action and condition of the Hole Saw.

Coolant should be used, especially when cutting metals and ceramics. Always use enough pressure to ensure that the Hole Saw cuts and does not rub the material being cut.

HACKSAW AND POWERSAW BLADE SELECTION AND APPLICATIONS

Hacksaw Blades

Actual Tooth Size

Hacksaw Blade Types:

- **Standard** (All hard / Flexible) – High Speed Steel,
- **Bi-metal** (Flexible) – High Speed Steel,
- **Cobalt** (Flexible) – Cobalt High Speed Steel
- **Low-Alloy** (Flexible) – Carbon Steel

All hard blades are breakable but tend to cut straighter than flexible blades.
Flexible blades are shatter-proof and are designed for cutting odd shapes.

General Recommendations for tooth sizes:

18 teeth per inch : For cutting solid sections of Steel and Non-Ferrous Metals over 1/2" (13mm).
24 teeth per inch : For cutting solid sections of Steel between 1/8"(3 mm) and 1/2"(13mm), as well as Medium Tubing and Medium Angle-iron.
32 teeth per inch : For cutting sections less than 1/8" (3mm) such as Thin Wall Tubing, Sheet Steel and Light Angle-iron.

Powersaw Blades

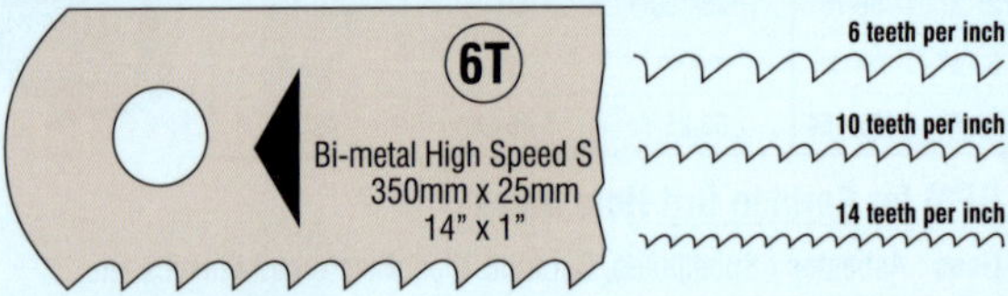

Actual Tooth Size

Powersaw Blade Types:

- **Standard** – All-Hard High Speed Steel
- **Bi-metal** – High Speed Steel

These blades are designed for use in a Power Hacksaw Machine and work in the direction of the down stroke.

General Recommendations for tooth sizes:

6 teeth per inch : For cutting large solid sections of soft materials and steels over 2" (50mm).
10 teeth per inch : For cutting medium solid sections of soft materials and all types of other materials over 1"(25mm).
14 teeth per inch : For cutting thin section of all types of materials such as pipe, tubing, light angle iron and solid metal sections under 1/2"(13mm).
NOTE : Coolant must be used to keep the blade cool at all times.

Carbide-Grit Rod Saw Blade

This saw blade is manufactured using a 1/8" (3 mm) diameter, high tensile, flexible cable onto which carbide particle grit is bonded. It fits into a standard hacksaw frame and is specifically designed for cutting odd shapes in Ceramic Tiles, Reinforced Plastics & Composites.

BANDSAW BLADES TYPES AND APPLICATIONS

Bandsaw Blades

Bi-Metal Blade : This blade is made of a thin strip of HSS welded on to a spring steel backing. The result is a durable blade that can withstand heat generated when cutting steels.
Uses : Mild / Carbon / Alloy / Tool and Stainless steels.

Carbide-Tipped Blade : This blade is made of Solid Carbide Tips welded on to a high-strength spring steel alloy backing. The very hard cutting edges of the Carbide tips ensure extended durability.
Uses : Alloy steels, Stainless steels, Nickel and Titanium-based alloys.

Carbide-Grit Blade : This blade is made of Tungsten carbide particle grit bonded on to a spring steel backing. This blade performs well on non-metals.
Uses : Fiberglass and Composite materials.

Diamond-Edge Blade : This blade is made of Industrial diamond grit bonded to a spring steel alloy backing. This blade performs well on abrasive non-metals.
Uses : Marble, Granite and Ceramics.

Carbon Flex-Back Blade : This blade is made of one-piece Carbon steel that is hardened and tempered and is especially designed for cutting curves.
Uses : Wood, Plastics and Non-ferrous materials.

Carbon Hard-Back Blade : This blade is made of one-piece Carbon steel that is hardened and tempered. The stiffer backing material ensures straighter cutting and resists swaging.
Uses : Wood, Plastics and Non-ferrous materials.

Standard Tooth Forms

Standard Tooth Set

Alternate

Alternate sequence - Every tooth set in an alternating sequence.

Wavy

Groups of teeth are set to each side in a controlled pattern

Raker

Three tooth sequence - Left, Right, Straight Uniform set angle

Vari-Raker

Multi-tooth sequence depending on tooth pitch. Varying set angles

Maximum Blade Width Selection Chart for Radius Cutting

Maximum Blade Width (inch)	1/8"	3/16"	1/4"	3/8"	1/2"	5/8"	3/4"	1"
Minimum Cutting Radius (inch)	1/8"	5/16"	5/8"	1 7/16"	2 1/2"	3 3/4"	5 1/16"	7 1/4"

SHARPENING OF HIGH SPEED STEEL TOOL BITS

Hints for Sharpening HSS Tool Bits

Proper sharpening of HSS Tool Bits results in increased production, better surface finish and accuracy of work. The tool bit should be roughed out using an aluminum oxide wheel of 46-60 grit followed by a wheel of finer grit to secure a keen edge.

Steels Used for HSS Tool Bits

M2 - Tungsten Molybdenum grade HSS, which has very good toughness and wear resistance properties and recommended for machining low alloy & free machining steels and most non-ferrous materials.

M42 - Tungsten-Molybdenum-Cobalt grade HSS, possessing excellent toughness and greater resistance to abrasion with a high "red hardness". Tool Bits made from this material can withstand higher cutting speeds and are suited to machining tough steels.

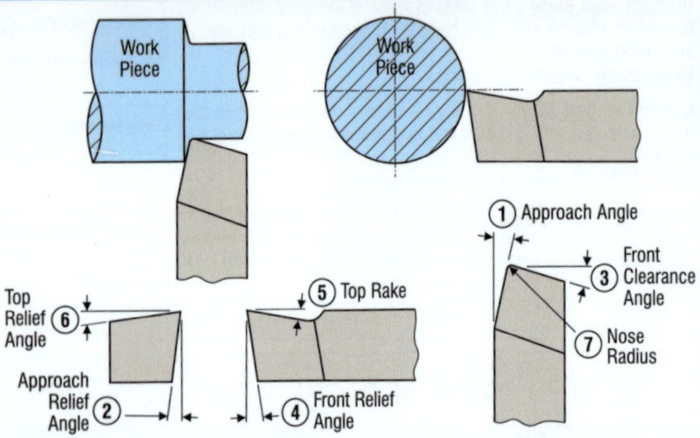

NOTE: A coolant must be used to minimise temperature build-up. An overheated tool bit should never be plunged into water as this will produce hair-line cracks, resulting in chipping of the cutting edge.

Numerical Order for Grinding HSS Toolbits

Material to be Machined	① Approach ∠	② Approach Relief ∠	③ Front Cl. ∠	④ Front Relief ∠	⑤ Top Rake ∠	⑥ Top Relief ∠	⑦ Nose Radius
Aluminum	Normally 15°	12°	Normally 15°	8°	35°	15°	To suit work profile.
Brass-Bronze		10°		8°	0°	5°	
Cast Iron		10°		8°	5°	12°	
Copper		14°		12°	16°	20°	
Fiber		15°		12°	0°	0°	
Monel		15°		13°	8°	14°	
Nickel		15°		13°	8°	14°	
Mild Steel	Varied to suit work profile.	12°	Varied to suit work profile.	8°	16°	14°	Larger radii extends tool life.
Med. Carbon Steel		10°		8°	12°	14°	
High Tensile Steel		10°		8°	8°	12°	
Stainless Steel		10°		8°	10°	15°	
Thermoplastics		20°		15°	20°	20°	
Molded Materials		12°		10°	10°	14°	

COOLANTS / LUBRICANTS / CUTTING FLUIDS USED IN MACHINING

It is advisable to always use a Coolant when cutting / machining metals

Coolants:

- Dissipate heat from both the tool & workpiece, permitting faster cutting speeds.
- Act as a lubricant between the tool and the workpiece, preventing heat build-up.
- Significantly extend the life of the cutting tool by reducing its wear.
- Remove swarf away from the cutting edge.
- Act as a short-term rust preventative.

The selection of the most appropriate coolant is important to ensure efficiency of the cutting tool. The following is a general guide for day-to-day workshop machining requirements. For long-run production jobs, it is advisable to seek the assistance of Coolant Manufacturers.

Material	General Cutting Fluid Recommendations for Drilling, Milling, Reaming, Tapping and Turning.
Aluminum	A mixture of Kerosene and Lard Oil is very effective as a coolant and lubricant in most machining operations. Soluble Oil is used as an alternative. In Rough Machining applications, where excessive heat is produced, a Soluble Oil mixture containing at least 90 per cent water may be used to dissipate the heat. Sulphurised Oils should be avoided for machining Aluminum as this may result in discoloration of the metal.
Brass	Brass is normally machined Dry. However, Soluble/Lard Oil is commonly used for Tapping.
Bronze	Soluble Oil/Mineral Oil is effective in machining Bronze.
Cast Iron	Cast Iron is usually machined Dry or with an Air blast. However, Sulphurized Oil/ Mineral Lard Oil is used for Tapping. To avoid excessive metal dust, Soluble Oil or low viscosity Mineral Oil may be used.
Cast Steel	Soluble Oil is very effective as a Cutting fluid.
Copper	Soluble Oil is very effective for machining copper. However Lard Oil is preferable for Tapping.
Magnesium	Magnesium is machined Dry as coolants containing water could react with Magnesium chips forming hydrogen gas that could pose a fire danger. When a cutting fluid is needed, Lard Oil may be used.
Malleable Iron	Malleable Iron is generally machined Dry. However, Lard Oil is preferred for Tapping.
Nickel	A Synthetic Oil with high pressure additives is recommended.
Plastics	Plastics are usually machined Dry. However, a medium Sulphurised Mineral Oil may be used.
Stainless Steel	A Semi-Synthetic cutting fluid with high pressure additives is recommended. For Tapping, a Lard Oil / Mineral Oil mixture is effective.
Alloy Steel	A light Soluble Oil is recommended. However, for Tapping applications, a mixture of Sulphur-based Oil with Mineral Oil is effective.
Mild Steel	General purpose Soluble Oil is recommended.
Tool Steel	A Semi-synthetic Oil with pressure additives is recommended. For Tapping, a mixture of Lard Oil and Mineral Oil is effective.
Titanium	A High-Soluble Synthetic Oil is recommended.
Zinc Alloys	A mixture of Kerosene and Lard Oil is effective.

NOTE : When using **Fluteless Taps**, it is advisable to use a straight Semi-Synthetic Oil with an extreme high pressure additive as a lubricant rather than a coolant.

TRIGONOMETRY
FORMULAE FOR SOLUTION OF TRIANGLES

The Pythagoras Theorem

In a Right-angle Triangle, the square of the hypotenuse is equal to the sum of the squares of the other two sides.

c = Hypotenuse; a & b = the other two sides.

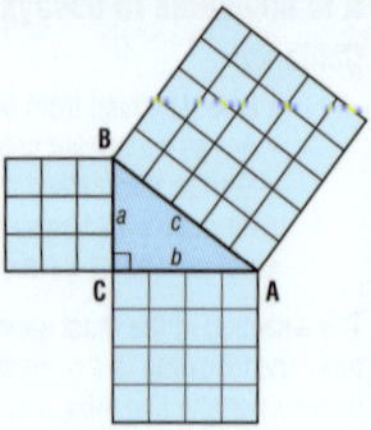

$$\mathbf{c^2 = a^2 + b^2}$$

$$\therefore \mathbf{c} = \sqrt{a^2 + b^2} \qquad \mathbf{a} = \sqrt{c^2 - b^2} \qquad \mathbf{b} = \sqrt{c^2 - a^2}$$

Example: *To find c when a = 6 inches and b = 8 inches.*

$$c^2 = a^2 + b^2$$

$$\therefore c = \sqrt{a^2 + b^2} = \sqrt{6^2 + 8^2} = \sqrt{36+64} = \sqrt{100} = 10 \text{ inches.}$$

The 3-4-5 Rule

To construct a right-angle triangle without measuring instruments, using the 3-4-5 rule.

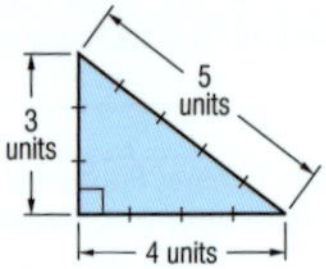

- Choose any suitable unit of measure (length).
- Construct a triangle whose sides measure 3 units, 4 units, 5 units.
- The angle opposite the 5 units is a right angle.

Formulae for the solution of OBLIQUE ANGLE TRIANGLES

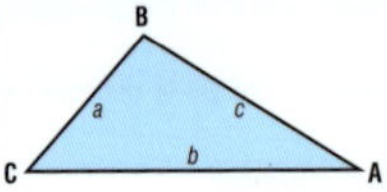

The Sine Rule : $\dfrac{a}{\text{Sin A}} = \dfrac{b}{\text{Sin B}} = \dfrac{c}{\text{Sin C}}$

The Cosine Rule :

$$a^2 = b^2 + c^2 - 2bc \ \text{Cos A}$$

$$b^2 = a^2 + c^2 - 2ac \ \text{Cos B}$$

$$c^2 = a^2 + b^2 - 2ab \ \text{Cos C}$$

Formulae for the solution of RIGHT ANGLE TRIANGLES

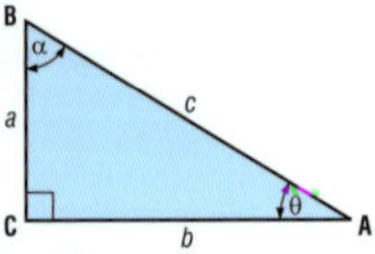

Sin θ $= \dfrac{\text{opposite}}{\text{hypotenuse}} = \dfrac{a}{c}$

Cos θ $= \dfrac{\text{adjacent}}{\text{hypotenuse}} = \dfrac{b}{c}$

Tan θ $= \dfrac{\text{opposite}}{\text{adjacent}} = \dfrac{a}{b}$

Cotan θ $= \dfrac{\text{adjacent}}{\text{opposite}} = \dfrac{b}{a}$

Cosec θ $= \dfrac{\text{hypotenuse}}{\text{opposite}} = \dfrac{c}{a}$

Sec θ $= \dfrac{\text{hypotenuse}}{\text{adjacent}} = \dfrac{c}{b}$

TRIGONOMETRY
FORMULAE FOR SOLUTION OF TRIANGLES

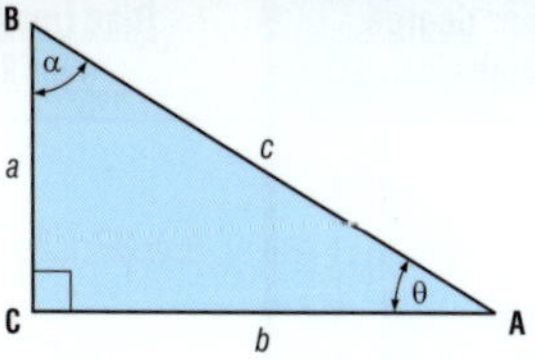

$$c^2 = a^2 + b^2$$

Parts Given	Parts to be Found				
	∠θ	∠α	side a	side b	side c
a & c	$\sin\theta = \frac{a}{c}$	$\cos\alpha = \frac{a}{c}$	given	$b = \sqrt{c^2 - a^2}$	given
a & b	$\tan\theta = \frac{a}{b}$	$\tan\alpha = \frac{b}{a}$	given	given	$c = \sqrt{a^2 + b^2}$
b & c	$\cos\theta = \frac{b}{c}$	$\sin\alpha = \frac{b}{c}$	$a = \sqrt{c^2 - b^2}$	given	given
∠θ & a	given	$\angle\alpha = 90° - \angle\theta$	given	$b = \frac{a}{\tan\theta}$	$c = \frac{a}{\sin\theta}$
∠θ & b	given	$\angle\alpha = 90° - \angle\theta$	$a = \tan\theta \times b$	given	$c = \frac{b}{\cos\theta}$
∠θ & c	given	$\angle\alpha = 90° - \angle\theta$	$a = \sin\theta \times c$	$b = \cos\theta \times c$	given
∠α & a	$\angle\theta = 90° - \angle\alpha$	given	given	$b = \tan\alpha \times a$	$c = \frac{a}{\cos\alpha}$
∠α & b	$\angle\theta = 90° - \angle\alpha$	given	$a = \frac{b}{\tan\alpha}$	given	$c = \frac{b}{\sin\alpha}$
∠α & c	$\angle\theta = 90° - \angle\alpha$	given	$a = \cos\alpha \times c$	$b = \sin\alpha \times c$	given

TRIG FUNCTIONS FOR COMMON ANGLES

$\sin 0° = 0$	$\cos 0° = 1$	$\tan 0° = 0$
$\sin 30° = \sin\frac{\pi}{6} = 0.5$	$\cos 30° = \cos\frac{\pi}{6} = 0.86602$	$\tan 30° = \tan\frac{\pi}{6} = 0.57735$
$\sin 45° = \sin\frac{\pi}{4} = 0.70710$	$\cos 45° = \cos\frac{\pi}{4} = 0.70710$	$\tan 45° = \tan\frac{\pi}{4} = 1$
$\sin 60° = \sin\frac{\pi}{3} = 0.86602$	$\cos 60° = \cos\frac{\pi}{3} = 0.5$	$\tan 60° = \tan\frac{\pi}{3} = 1.73205$
$\sin 90° = \sin\frac{\pi}{2} = 1$	$\cos 90° = \cos\frac{\pi}{2} = 0$	$\tan 90° = \tan\frac{\pi}{2} = $ infinity

RISE PER DEGREE - A QUICK REFERENCE

Rise (inches) per degree over 1 foot

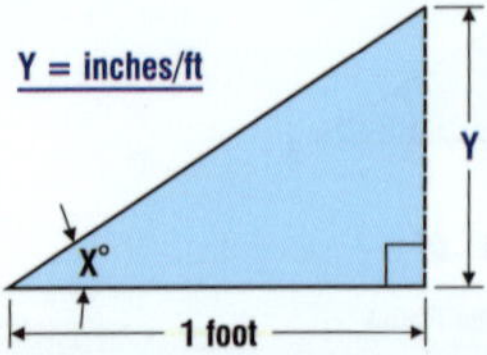

Rise (mm) per degree over 1000 mm

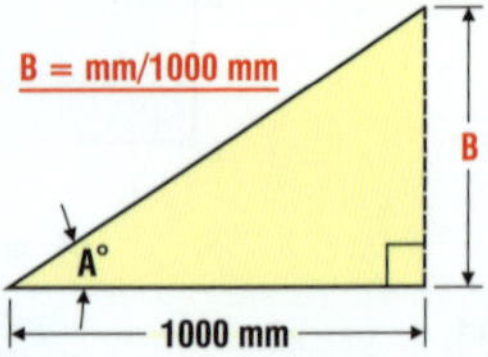

X°	Y (inch)	X°	Y (inch)	A°	B (mm)	A°	B (mm)
1°	0.21	24°	5.34	1°	17.46	24°	445.23
2°	0.42	25°	5.60	2°	34.92	25°	466.31
3°	0.63	26°	5.85	3°	52.41	26°	487.73
4°	0.84	27°	6.11	4°	69.93	27°	509.53
5°	1.05	28°	6.38	5°	87.49	28°	531.71
6°	1.26	29°	6.65	6°	105.10	29°	554.31
7°	1.47	30°	6.93	7°	122.78	30°	577.35
8°	1.69	31°	7.21	8°	140.54	31°	600.86
9°	1.90	32°	7.50	9°	158.38	32°	624.87
10°	2.12	33°	7.79	10°	176.33	33°	649.41
11°	2.33	34°	8.09	11°	194.38	34°	674.51
12°	2.55	35°	8.40	12°	212.56	35°	700.21
13°	2.77	36°	8.72	13°	230.87	36°	726.54
14°	2.99	37°	9.04	14°	249.33	37°	753.55
15°	3.22	38°	9.38	15°	267.95	38°	781.29
16°	3.44	39°	9.72	16°	286.75	39°	809.78
17°	3.67	40°	10.07	17°	305.73	40°	839.10
18°	3.90	41°	10.43	18°	324.92	41°	869.29
19°	4.13	42°	10.81	19°	344.33	42°	900.40
20°	4.37	43°	11.19	20°	363.97	43°	932.52
21°	4.61	44°	11.59	21°	383.86	44°	965.69
22°	4.85	45°	12.00	22°	404.03	45°	1000.00
23°	5.09			23°	424.47		

ANGLES IN DECIMALS, PERCENTAGES AND RADIANS

Angles are normally expressed in degrees, minutes and seconds, however, they can also be expressed in decimals, percentages or radians.

Conversion to Decimals

To convert degrees, minutes and seconds to decimals - first divide the number of seconds by 60. Then add the answer to the minutes and divide the result by 60 to get the decimals.

Example: To convert 20º 30' 15" to decimals.

1. Divide 15" by 60 = 0.25'
2. Add 0.25' to 30' = 30.25'
3. Now divide 30.25' by 60 = 0.504166
4. Finally, add 0.504166 to 20° = 20.504166

∴ **20° 30' 15"** expressed in decimals = **20.504166**

MINUTES TO DECIMALS OF A DEGREE

Minutes	Decimal	Minutes	Decimal	Minutes	Decimal	Minutes	Decimal	Minutes	Decimal
1	0.0167	13	0.2167	25	0.4167	37	0.6167	49	0.8167
2	0.0333	14	0.2333	26	0.4333	38	0.6333	50	0.8333
3	0.0500	15	0.2500	27	0.4500	39	0.6500	51	0.8500
4	0.0667	16	0.2667	28	0.4667	40	0.6667	52	0.8667
5	0.0833	17	0.2833	29	0.4833	41	0.6833	53	0.8833
6	0.1000	18	0.3000	30	0.5000	42	0.7000	54	0.9000
7	0.1167	19	0.3167	31	0.5167	43	0.7167	55	0.9167
8	0.1333	20	0.3333	32	0.5333	44	0.7333	56	0.9333
9	0.1500	21	0.3500	33	0.5500	45	0.7500	57	0.9500
10	0.1667	22	0.3667	34	0.5667	46	0.7667	58	0.9667
11	0.1833	23	0.3833	35	0.5833	47	0.7833	59	0.9833
12	0.2000	24	0.4000	36	0.6000	48	0.8000	60	1.0000

SECONDS TO DECIMALS OF A DEGREE

Seconds	Decimal	Seconds	Decimal	Seconds	Decimal	Seconds	Decimal	Seconds	Decimal
1	0.0003	13	0.0036	25	0.0069	37	0.0103	49	0.0136
2	0.0006	14	0.0039	26	0.0072	38	0.0106	50	0.0139
3	0.0008	15	0.0042	27	0.0075	39	0.0108	51	0.0142
4	0.0011	16	0.0044	28	0.0078	40	0.0111	52	0.0144
5	0.0014	17	0.0047	29	0.0081	41	0.0114	53	0.0147
6	0.0017	18	0.0050	30	0.0083	42	0.0117	54	0.0150
7	0.0019	19	0.0053	31	0.0086	43	0.0119	55	0.0153
8	0.0022	20	0.0056	32	0.0089	44	0.0122	56	0.0156
9	0.0025	21	0.0058	33	0.0092	45	0.0125	57	0.0158
10	0.0028	22	0.0061	34	0.0094	46	0.0128	58	0.0161
11	0.0031	23	0.0064	35	0.0097	47	0.0131	59	0.0164
12	0.0033	24	0.0067	36	0.0100	48	0.0133	60	0.0167

ANGLES IN DECIMALS, PERCENTAGES AND RADIANS

A radian represents the proportionate length of an arc resulting from an angle within the circle. The circumference of a circle is 2π r, therefore, a full circle (360°) can be seen as corresponding to an angle of 2 π radians.

CONVERSION TO RADIANS

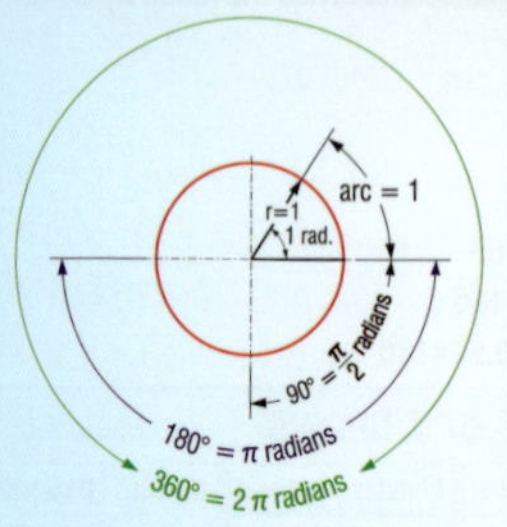

360° = 2 π rad	= 6.2831853071794 rad
180° = π rad	= 3.1415926535897 rad
90° = π ÷ 2 rad	= 1.5707963267948 rad
1°	= 0.01745329252 rad
1'	= 0.000290882 rad
1"	= 0.0000048481 rad
1 rad	= 57° 17' 44.806"
1 rad	= 57.2957795131°

To Convert Degrees, Minutes and Seconds to Radians

- First convert the degrees, minutes and seconds to decimal.

  Example: Convert 20° 30' 15" to radians:

  20° 30' 15" = 20.504166° (as worked out in "conversions to decimals" on page 73)

- Then divide the result by the decimal equivalent of 1 radian in degrees (57.2957795131).

  Example: 20.504166 ÷ 57.2957795131 = 0.357865207 rad

To Convert Radians to Degrees

- Multiply the number of radians by the number of degrees in 1 radian. eg. 57.2957795131°

  Example: Convert 2 radians to degrees:

  2 x 57.2957795131 = 114.5915590262°

To Convert Decimal Degrees to Degrees, Minutes and Seconds

Example: Convert using the result above:

- 114.5915590262° = (114°)
- Balance 0.5915590262 x 60 = 35.493541572' = (35')
- Balance 0.493541572 x 60 = 29.61249432" = (29")

Therefore 114.5915590262° = 114°, 35' and 29".

DEGREES TO RADIANS AND MINUTES TO RADIANS CONVERSION TABLE

DEGREES TO RADIANS

Degrees	Radians	Degrees	Radians	Degrees	Radians	Degrees	Radians
0.01	0.0001745	10	0.1745329	37	0.6457718	64	1.1170108
0.02	0.0003491	11	0.1919862	38	0.6632251	65	1.1344640
0.03	0.0005236	12	0.2094395	39	0.6806784	66	1.1519173
0.04	0.0006981	13	0.2268928	40	0.6981317	67	1.1693706
0.05	0.0008727	14	0.2443461	41	0.7155850	68	1.1868239
0.06	0.0010472	15	0.2617994	42	0.7330383	69	1.2042772
0.07	0.0012217	16	0.2792527	43	0.7504916	70	1.2217305
0.08	0.0013963	17	0.2967060	44	0.7679449	71	1.2391838
0.09	0.0015708	18	0.3141593	45	0.7853982	72	1.2566371
0.10	0.0017453	19	0.3316126	46	0.8028515	73	1.2740904
0.20	0.0034907	20	0.3490659	47	0.8203048	74	1.2915437
0.30	0.0052360	21	0.3665192	48	0.8377581	75	1.3089970
0.40	0.0069813	22	0.3839724	49	0.8552114	76	1.3264503
0.50	0.0087266	23	0.4014257	50	0.8726647	77	1.3439036
0.60	0.0104720	24	0.4188790	51	0.8901179	78	1.3613569
0.70	0.0122173	25	0.4363323	52	0.9075712	79	1.3788101
0.80	0.0139626	26	0.4537856	53	0.9250245	80	1.3962634
0.90	0.0157080	27	0.4712389	54	0.9424778	81	1.4137167
1	0.0174533	28	0.4886922	55	0.9599311	82	1.4311700
2	0.0349066	29	0.5061455	56	0.9773844	83	1.4486233
3	0.0523599	30	0.5235988	57	0.9948377	84	1.4660766
4	0.0698132	31	0.5410521	58	1.0122910	85	1.4835299
5	0.0872665	32	0.5585054	59	1.0297443	86	1.5009832
6	0.1047198	33	0.5759587	60	1.0471976	87	1.5184365
7	0.1221731	34	0.5934120	61	1.0646509	88	1.5358898
8	0.1396263	35	0.6108653	62	1.0821042	89	1.5533431
9	0.1570796	36	0.6283185	63	1.0995575	90	1.5707964

MINUTES TO RADIANS

Minutes	Radians	Minutes	Radians	Minutes	Radians	Minutes	Radians
1	0.0002909	16	0.0046542	31	0.0090175	46	0.0133809
2	0.0005818	17	0.0049451	32	0.0093084	47	0.0136717
3	0.0008727	18	0.0052360	33	0.0095993	48	0.0139626
4	0.0011636	19	0.0055269	34	0.0098902	49	0.0142535
5	0.0014544	20	0.0058178	35	0.0101811	50	0.0145444
6	0.0017453	21	0.0061087	36	0.0104720	51	0.0148353
7	0.0020362	22	0.0063995	37	0.0107629	52	0.0151262
8	0.0023271	23	0.0066904	38	0.0110538	53	0.0154171
9	0.0026180	24	0.0069813	39	0.0113446	54	0.0157080
10	0.0029089	25	0.0072722	40	0.0116355	55	0.0159989
11	0.0031998	26	0.0075631	41	0.0119264	56	0.0162897
12	0.0034907	27	0.0078540	42	0.0122173	57	0.0165806
13	0.0037815	28	0.0081449	43	0.0125082	58	0.0168715
14	0.0040724	29	0.0084358	44	0.0127991	59	0.0171624
15	0.0043633	30	0.0087266	45	0.0130900	60	0.0174533

SECONDS TO RADIANS CONVERSION TABLE

SECONDS TO RADIANS

Seconds	Radians	Seconds	Radians	Seconds	Radians	Seconds	Radians
1	0.0000048	16	0.0000776	31	0.0001503	46	0.0002230
2	0.0000097	17	0.0000824	32	0.0001551	47	0.0002279
3	0.0000145	18	0.0000873	33	0.0001600	48	0.0002327
4	0.0000194	19	0.0000921	34	0.0001648	49	0.0002376
5	0.0000242	20	0.0000970	35	0.0001697	50	0.0002424
6	0.0000291	21	0.0001018	36	0.0001745	51	0.0002473
7	0.0000339	22	0.0001067	37	0.0001794	52	0.0002521
8	0.0000388	23	0.0001115	38	0.0001842	53	0.0002569
9	0.0000436	24	0.0001164	39	0.0001891	54	0.0002618
10	0.0000485	25	0.0001212	40	0.0001939	55	0.0002666
11	0.0000533	26	0.0001261	41	0.0001988	56	0.0002715
12	0.0000582	27	0.0001309	42	0.0002036	57	0.0002763
13	0.0000630	28	0.0001357	43	0.0002085	58	0.0002812
14	0.0000679	29	0.0001406	44	0.0002133	59	0.0002860
15	0.0000727	30	0.0001454	45	0.0002182	60	0.0002909

Angles expressed as a Percent

When measuring angles in percentage, a ratio between the length of the base of the angle and the amount of rise of the angle needs to be established. The proportional part of the base is expressed as a percentage.

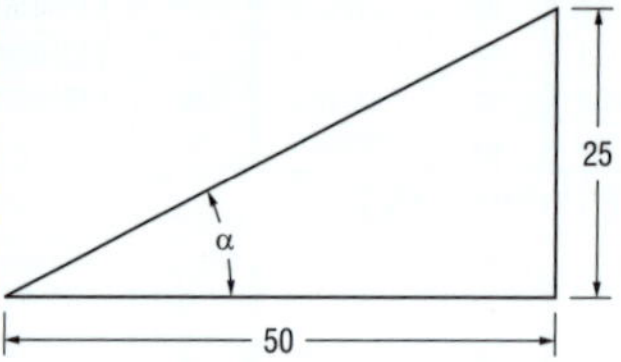

$\frac{25}{50}$ x 100 = 50% (1 in 2 taper)

Example: In the angle above, the base is 50 units and the rise is 25 units - therefore the percentage is 25 ÷ 50 x 100 = 50 %. Another way of expressing this angle would be to call it a 1 in 2 taper which means that for every 2 units of base length, the height increases by 1 unit. (Refer pages 78 & 79 for useful tapers and angle information).

To Convert Angles expressed as Percentage to Degrees, Minutes & Seconds

- Divide the percent of the angle by 100. The result is the tangent of the angle which can be looked up in any "tangent tables" (refer pages 219 & 220).

  Example: Convert an angle expressed as 50 % into degrees, minutes and seconds: 50 ÷ 100 = .500 = 26° 33' 54".

RADIANS AND DEGREES

Radians to Decimal Degrees

Radians	Decimal Degrees
0.0001	0.0057296
0.0002	0.0114592
0.0003	0.0171887
0.0004	0.0229183
0.0005	0.0286479
0.0006	0.0343775
0.0007	0.0401070
0.0008	0.0458366
0.0009	0.0515662
0.001	0.0572958
0.002	0.1145916
0.003	0.1718873
0.004	0.2291831
0.005	0.2864789
0.006	0.3437747
0.007	0.4010705
0.008	0.4583662
0.009	0.5156620
0.01	0.5729578
0.0174533	**1°**
0.02	1.1459156
0.03	1.7188734
0.04	2.2918312
0.05	2.8647890
0.06	3.4377468
0.07	4.0107046
0.08	4.5836624
0.09	5.1566202
0.1	5.7295780
0.2	11.4591559
0.3	17.1887339
0.4	22.9183118
0.5	28.6478898
0.6	34.3774677
0.7	40.1070457
0.8	45.8366236
0.9	51.5662016
1	57.2957795
1.57079633	**90°**
2	114.5915590
3	171.8873385
3.1415926	**180°**
4	229.1831181
4.7123889	**270°**
5	286.4788976
6	343.7746771
6.283185	**360°**

Radians to Degrees, Minutes & Seconds

Radians	Degrees	Minutes	Seconds
0.0001	0	0	21
0.0002	0	0	41
0.0003	0	1	2
0.0004	0	1	23
0.0005	0	1	43
0.0006	0	2	4
0.0007	0	2	24
0.0008	0	2	45
0.0009	0	3	6
0.001	0	3	26
0.002	0	6	53
0.003	0	10	19
0.004	0	13	45
0.005	0	17	11
0.006	0	20	38
0.007	0	24	4
0.008	0	27	30
0.009	0	30	56
0.01	0	34	23
0.0174533	1	0	0
0.02	1	8	45
0.03	1	43	8
0.04	1	17	31
0.05	2	51	53
0.06	3	26	16
0.07	4	0	39
0.08	4	35	2
0.09	5	9	24
0.1	5	43	46
0.2	11	27	33
0.3	17	11	19
0.4	22	55	6
0.5	28	38	52
0.6	34	22	39
0.7	40	6	25
0.8	45	50	12
0.9	51	33	58
1	57	17	45
1.57079633	90	0	0
2	114	35	30
3	171	53	14
3.1415926	180	0	0
4	229	10	59
4.7123889	270	0	0
5	286	28	44
6	343	46	29
6.283185	360	0	0

USEFUL TAPERS AND ANGLES USED IN THE WORKSHOP

Cone of	Taper per inch on diameter	INCLUDED ANGLE		
		Degrees	Minutes	Seconds
1 in 96	0.0104		35	47
1 in 60	0.0167		57	17
1 in 50	0.0200	1	8	46
1 in 48	0.0208	1	11	38
1 in 40	0.0250	1	25	56
1 in 30	0.0333	1	54	36
1 in 24	0.0417	2	23	12
1 in 20	0.0500	2	51	52
1 in 19	0.0526	3	0	54
1 in 18	0.0556	3	10	58
1 in 17	0.0588	3	22	9
1 in 16	0.0625	3	34	48
1 in 15	0.0667	3	49	6
1 in 14	0.0714	4	5	26
1 in 13	0.0769	4	24	18
1 in 12	0.0833	4	46	19
1 in 11	0.0909	5	12	18
1 in 10	0.1000	5	43	31
1 in 9 1/2	0.1053	6	1	32
1 in 9	0.1111	6	21	34
1 in 8 1/2	0.1176	6	43	58
1 in 8	0.1250	7	9	10
1 in 7 1/2	0.1333	7	37	41
1 in 7	0.1429	8	10	16
1 in 6 1/2	0.1538	8	47	51
1 in 6	0.1667	9	31	37
1 in 5 1/2	0.1818	10	23	20
1 in 5	0.2000	11	25	16
1 in 4	0.2500	14	15	0
1 in 3	0.3333	18	55	28
1 in 2	0.5000	28	4	20
7:24	0.2917	16	35	41

USEFUL TAPERS AND ANGLES USED IN THE WORKSHOP

Taper per foot on diameter	Taper per inch on diameter	INCLUDED ANGLE		
		Degrees	Minutes	Seconds
1/16"	0.0052		17	53
1/8"	0.0104		35	47
3/16"	0.0156		53	44
1/4"	0.0208	1	11	38
5/16"	0.0260	1	29	31
3/8"	0.0313	1	47	25
7/16"	0.0364	2	5	18
1/2"	0.0417	2	23	12
9/16"	0.0469	2	41	7
5/8"	0.0521	2	59	3
11/16"	0.0573	3	16	56
3/4"	0.0625	3	34	48
13/16"	0.0677	3	52	42
7/8"	0.0729	4	10	32
15/16"	0.0781	4	28	26
1"	0.0833	4	46	19
1 1/8"	0.0938	5	22	2
1 1/4"	0.1042	5	57	45
1 3/8"	0.1146	6	33	29
1 1/2"	0.1250	7	9	10
1 5/8"	0.1354	7	44	49
1 3/4"	0.1458	8	20	28
2"	0.1667	0	01	37
2 1/2"	0.2083	11	53	38
3"	0.2500	14	15	0
3 1/2"	0.2917	16	35	41
4"	0.3333	18	55	28
4 1/2"	0.3750	21	14	22
5"	0.4167	23	32	12
5 1/2"	0.4583	25	48	53
6"	0.5000	28	4	20

TAPER TURNING ON A STANDARD LATHE

1. Swiveling the Compound Slide

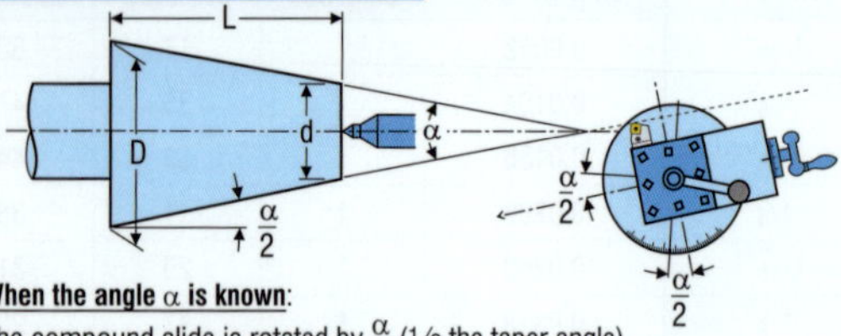

- **When the angle α is known**:
  The compound slide is rotated by $\frac{\alpha}{2}$ (1/2 the taper angle).
- **When the dimensions are given**:
  The angle can be calculated using the formula: $\tan\frac{\alpha}{2} = \frac{D-d}{2L}$

Example : When D = 8.0 inches, d = 5.63 inches, L = 5.0 inches

$\tan\frac{\alpha}{2} = \frac{8.0-5.63}{10}$ $\therefore$ $\tan\frac{\alpha}{2} = \frac{2.37}{10} = 0.23700"$

The result can now be looked up in the Table of Natural Tangents on pages 219 & 220.

Therefore $\frac{\alpha}{2} = 13°20'$ and $\alpha = 26°40'$

2. Offsetting the Tail Stock

- **When the length of taper = length of workpiece:** Offset $S = \frac{D-d}{2}$

The Tail Stock will have to be offset forward at a distance = S, noting that the tool must travel parallel to the axis of the machine.

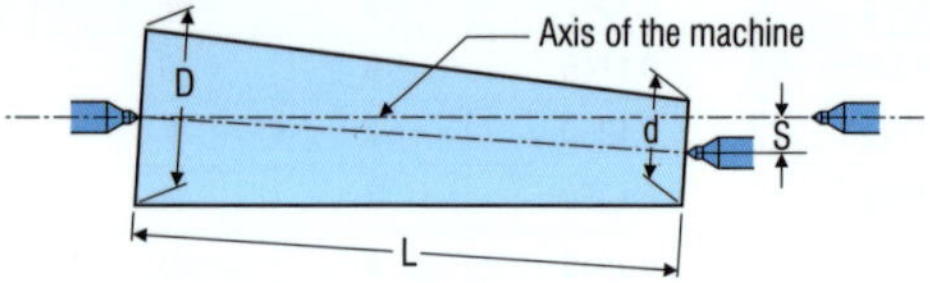

- **When the length of taper < length of workpiece:** Offset $S = \frac{D-d}{2 \times L} \times L_1$

The Tail Stock will have to be offset forward at a distance = S, noting that the tool must travel parallel to the axis of the machine.

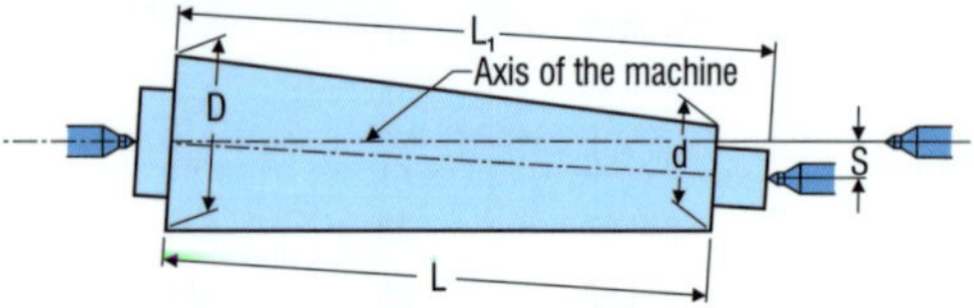

NOTE : Taper turning by offsetting the tail stock has its limitations and is applicable to tapers of less than 7°.

SINE BAR AND SINE PLATE SETTING

The Sine Bar is used for accurately setting up work for machining or for inspection purposes. It consists of a hardened Steel Bar with two pins mounted at an accurate pre-determined distance. A Sine Bar can be of any length. However, 5" and 10" are common Inch sizes and 100 mm,125 mm & 200 mm are common Metric sizes.

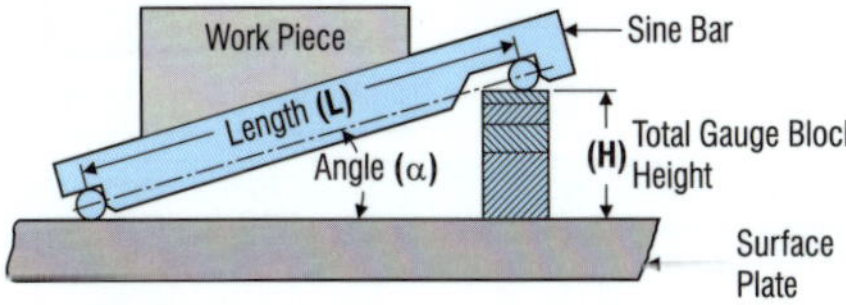

Gauge blocks are used for establishing the setting height **(H)** of the Sine Bar.

The setting height **(H)** can be determined by multiplying the Sine of the given angle by the length of the Sine Bar. (Refer pages 217 & 218 - Table of Natural Sines).

Height (H) $= \sin\alpha \times L$ where, H = Setting Height

Sine α $= \frac{H}{L}$ L = Length of Sine Bar

Examples using a 5 inch Sine Bar

To find H when given angle = 15°.

$H = \text{Sine } 15° \times L = 0.25882 \times 5$

$\therefore$ **H = 1.2941"** (total gauge block height)

Finding ∠ when given height (H) = 2.50 in.

$\text{Sine } \alpha = \frac{H}{L} = \frac{2.50}{5} = 0.5000$

$\therefore \alpha =$ **30°**

Examples using a 200 mm Sine Bar

To find H when given angle = 15°.

$H = \text{Sine } 15° \times L = 0.25882 \times 200$

$\therefore$ **H = 51.764 mm** (total gauge block height)

Finding ∠ when given height (H) = 100mm

$\text{Sine } \alpha = \frac{H}{L} = \frac{100}{200} = 0.5000$

$\therefore \alpha =$ **30°**

Rules for working out Inch Tapers (Also refer pages 78 & 79)

To Find	Given	Rule
Taper per inch	Taper per foot	Divide the taper per foot by 12.
Taper per foot	Taper per inch	Multiply the taper per inch by 12.
Taper per foot	End diameters and length of tapers in inches	Subtract small dia. from large dia., then divide by length of taper, & multiply quotient by 12.
Diameter at small end in inches	Large diameter, length of taper in inches, and taper per foot	Divide taper per foot by 12, multiply by length of taper, & subtract from large dia.
Diameter at large end in inches	Small diameter, length of taper in inches, and taper per foot	Divide taper per foot by 12, multiply by length of taper, & add result to small dia.
Distance between 2 given diameters in inches	Taper per foot and two diameters in inches	Subtract small dia. from large dia., then divide remainder by taper per foot & multiply quotient by 12.

DIMENSIONS OF MORSE TAPER SHANKS / SOCKETS (For Identification Purposes)

Morse Taper Shank

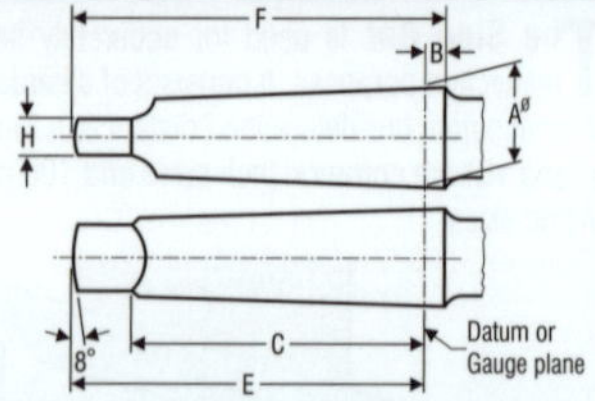

Morse Taper No.	A (inch)	B (inch)	C (inch)	E (inch)	F (inch)	H (inch)	Taper per inch on Dia.	Taper per foot on Dia.	Included Angle		
									Deg.	Min.	Sec.
0	0.3561	1/8	2	2 7/32	2 11/32	5/32	0.05205	0.62460	2	58	54
1	0.4750	1/8	2 1/8	2 7/16	2 9/16	13/64	0.04988	0.59858	2	51	27
2	0.7000	3/16	2 9/16	2 15/16	3 1/8	1/4	0.04995	0.59941	2	51	41
3	0.9380	3/16	3 3/16	3 11/16	3 7/8	5/16	0.05020	0.60235	2	52	31
4	1.2310	1/4	4 1/16	4 5/8	4 7/8	15/32	0.05194	0.62326	2	58	31
5	1.7480	1/4	5 3/16	5 7/8	6 1/8	5/8	0.05263	0.63151	3	0	52
6	2.4940	5/16	7 1/4	8 1/4	8 9/16	3/4	0.05214	0.62565	2	59	12
7	3.2700	3/8	10	1 1/4	11 5/8	1 1/8	0.05200	0.62400	2	56	44

Morse Taper Shank with Wedge Drive Slot

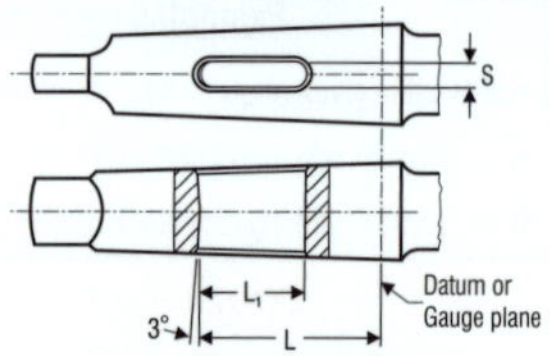

Morse Taper No.	L (inch)	L_1 (inch)	S (inch)
4	2 11/32	1 15/32	0.327
5	2 35/64	1 11/16	0.488

Morse Taper Socket

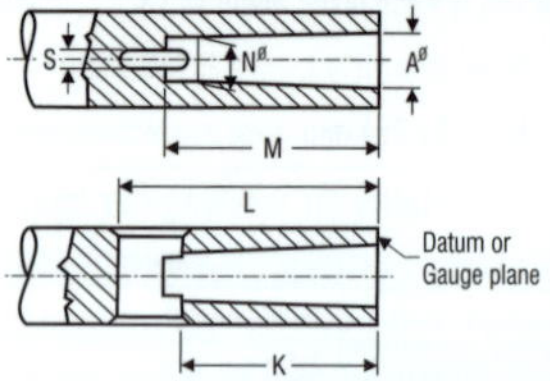

Morse Taper No.	A (inch)	L (inch)	M (inch)	N (inch)	K (inch)	S (inch)	Taper per inch. on Dia.	Included Angle		
								Deg.	Min.	Sec.
0	0.3561	2 1/2	2 1/16	0.260	1 15/16	11/64	0.05205	2	58	54
1	0.4750	2 13/16	2 3/16	0.378	2 1/16	7/32	0.04988	2	51	27
2	0.7000	3 3/8	2 21/32	0.588	2 1/2	17/64	0.04995	2	51	41
3	0.9368	4 1/4	3 5/16	0.797	3 1/16	21/64	0.05020	2	52	31
4	1.2310	5 1/8	4 3/16	1.044	3 7/8	31/64	0.05194	2	58	31
5	1.7480	6 7/16	5 5/16	1.502	4 15/16	21/32	0.05263	3	0	52
6	2.4940	8 3/4	7 13/32	2.150	7	25/32	0.05214	2	59	12
7	3.2700	12 1/8	10 5/32	2.786	9 1/2	1 5/32	0.05200	2	56	44

DIMENSIONS OF JACOBS AND DIN TAPERS FOR DRILL CHUCKS (For Identification Purposes)

Jacobs Taper

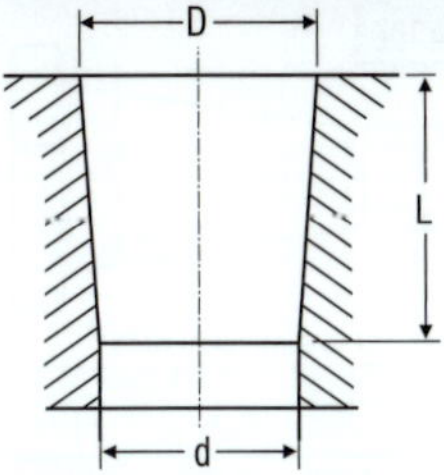

Jacobs Taper No.	D (inch)	d (inch)	L (inch)	Taper per inch on Dia.	Included Angle (Approx.)	
					Deg.	Min.
0	0.2500	0.2284	0.4375	0.04929	2	49
1	0.3840	0.3334	0.5652	0.07709	4	24
2 short	0.5488	0.4876	0.7500	0.08155	4	39
2	0.5590	0.4876	0.8750	0.08155	4	39
33	0.6240	0.5605	1.0000	0.06350	3	38
6	0.6760	0.6241	1.0000	0.05191	2	58
3	0.8110	0.7461	1.2187	0.05325	3	02
4	1.1240	1.0372	1.6562	0.05241	2	59
5	1.4130	1.3161	1.8750	0.05168	2	58

DIN Taper

DIN Taper No.	Morse Taper No.	D (inch)	d (inch)	L (inch)	Taper per inch on Dia.	Included Angle		
						Deg.	Min.	Sec.
B10	1	0.3974	0.3689	0.571	0.04988	2	51	27
B12	1	0.4750	0.4387	0.728	0.04988	2	51	27
B16	2	0.6194	0.5722	0.945	0.04995	2	51	41
B18	2	0.7000	0.6371	1.260	0.04995	2	51	41
B22	3	0.8580	0.7780	1.594	0.05020	2	52	31
B24	3	0.9380	0.8382	1.988	0.05020	2	52	31

NOTE : **DIN Tapered Sections are identical with the following Morse Tapers :**
B10 & B12 for Morse Taper No. 1
B16 & B18 for Morse Taper No. 2
B22 & B24 for Morse Taper No. 3

The length of each of these tapers is less than the overall length of the corresponding Morse Taper. Each taper may be regarded as corresponding approximately either to that part of the Morse Taper nearest the smaller end, e.g. B10 or to the part nearest to the large end, e.g. B12.

PRINCIPAL DIMENSIONS OF MILLING M/C ARBORS / SHANKS (For Identification Purposes)

ISO/NT DIN2080 Taper

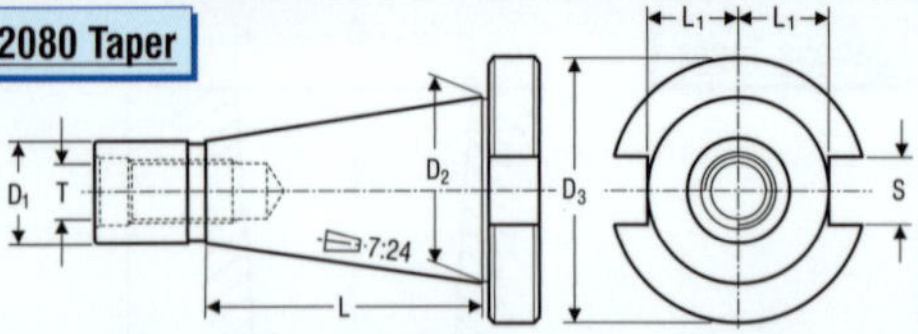

Dimensions in inches Taper 7:24 - also refer pages 78-79

ISO/NT	D_1	D_2	D_3	L	L_1	S	T (thread)	
							Inch	Metric
30	0.685	1 1/4"	1.968	1.905	0.638	0.634	1/2" UNC	M12
40	0.996	1 3/4"	2.480	2.574	0.886	0.634	5/8" UNC	M16
45	1.275	2 1/4"	3.150	3.260	1.142	0.760	3/4" UNC	M20
50	1.559	2 3/4"	3.838	4.008	1.390	1.012	1" UNC	M24
60	2.370	4 1/4"	6.142	6.370	2.362	1.012	1 1/4" UNC	M30

MAS 403 BT Taper

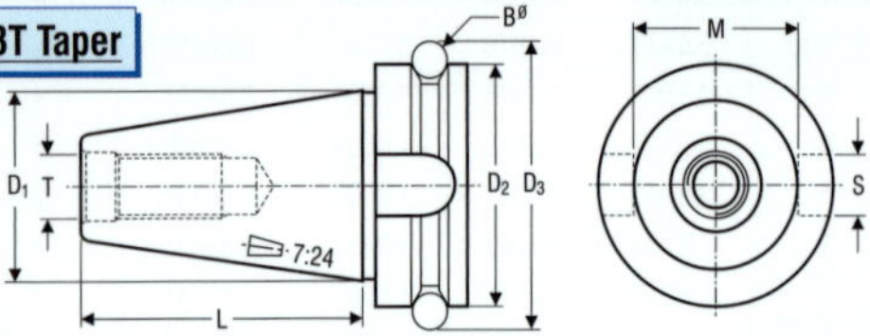

Dimensions in inches Taper 7:24 - also refer pages 78-79

MAS/BT	D_1	D_2	D_3	Bø	L	M	S	T (thread)
30	1 1/4"	1.811	2.210	0.3150	1.905	1.283	0.634	M12
35	1 1/2"	2.087	2.586	0.3937	2.220	1.543	0.634	M12
40	1 3/4"	2.480	2.979	0.3937	2.575	1.779	0.634	M16
45	2 1/4"	3.346	3.945	0.4724	3.260	2.291	0.760	M20
50	2 3/4"	3.937	4.486	0.5905	4.008	2.787	1.012	M24

HSK DIN 69893-1 Form A Taper

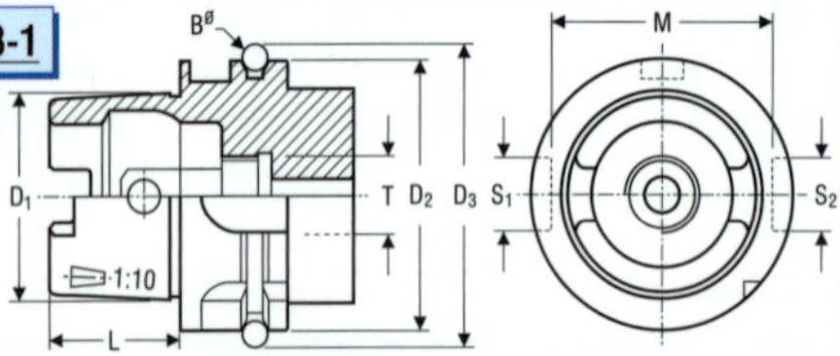

Dimensions in inches Cone 1:10 - also refer pages 78-79

HSK	D_1	D_2	D_3	Bø	L	M	S_1	S_2	T (thread)
32	0.945	1.260	1.457	0.1575	0.630	1.024	0.354	0.275	M10 x 1.0
40	1.181	1.575	1.772	0.1575	0.787	1.338	0.433	0.354	M12 x 1.0
50	1.496	1.968	2.335	0.2756	0.984	1.653	0.551	0.472	M16 x 1.0
63	1.890	2.480	2.846	0.2756	1.260	2.087	0.709	0.630	M18 x 1.0
80	2.362	3.150	3.496	0.2756	1.575	2.677	0.787	0.709	M20 x 1.5
100	2.953	3.937	4.319	0.2756	1.968	3.464	0.866	0.787	M24 x 1.5

PRINCIPAL DIMENSIONS OF MILLING M/C ARBORS / SHANKS (For Identification Purposes)

DIN 69871 Taper

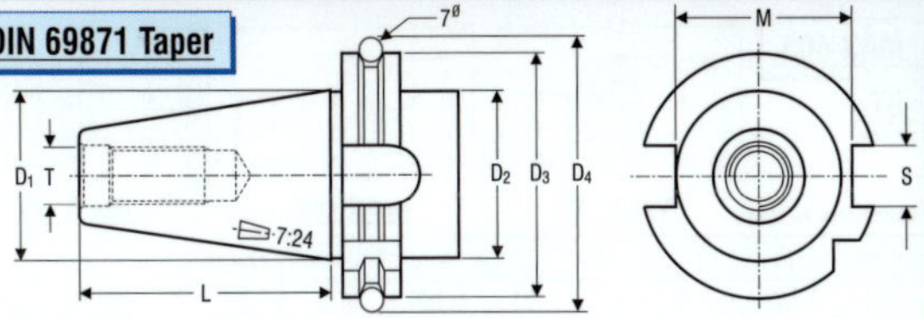

Dimensions in inches

Taper 7:24 - also refer pages 78-79

DIN	D_1	D_2	D_3	D_4	L	M	S	T (thread)
30	1 1/4"	1.772	1.968	2.335	1.882	1.394	0.634	M12
40	1 3/4"	1.968	2.502	2.846	2.693	1.882	0.634	M16
45	2 1/4"	2.480	3.250	3.596	3.256	2.378	0.760	M20
50	2 3/4"	3.149	3.838	4.222	4.004	2.882	1.012	M24

ANSI/CAT Taper

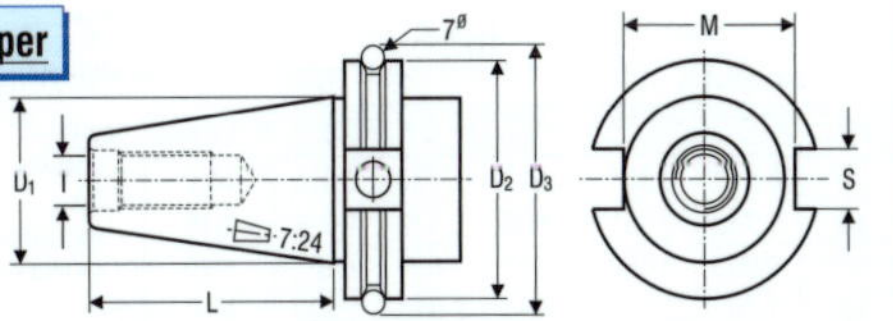

Dimensions in inches

Taper 7:24 - also refer pages 78-79

ANSI/CAT	D_1	D_2	D_3	L	M	S	T (thread)	
							Inch	Metric
40	1 3/4"	2.500	2.846	2.693	1.921	0.634	5/8" UNC	M16
45	2 1/4"	3.250	3.596	3.256	2.425	0.760	3/4" UNC	M20
50	2 3/4"	3.875	4.222	4.004	2.988	1.012	1" UNC	M24

R8 Taper

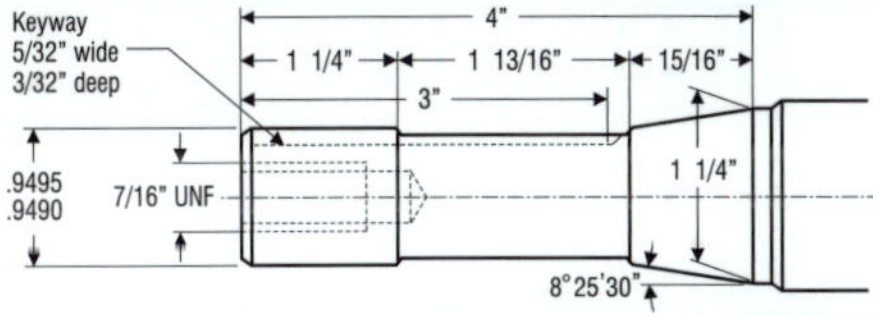

Dimensions in inches

ER Collets

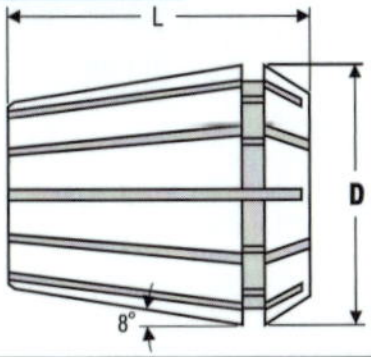

Type	Size Range (inches)	D	L	Number of Collets in Set
ER 8	0.020" - 0.107"	0.336"	17/32"	5
ER 11	0.020" - 0.276"	0.453"	23/32"	7
ER 16	0.020" - 0.394"	0.669"	1 3/32"	10
ER 20	0.040" - 0.512"	0.827"	1 1/4"	12
ER 25	0.040" - 0.630"	1.024"	1 11/32"	15
ER 32	0.080" - 0.787"	1.299"	1 9/16"	18
ER 40	0.120" - 1.024"	1.614"	1 13/16"	23
ER 50	0.197" - 1.338"	2.047"	2 3/8"	12

STANDARD PULL STUDS / RETENTION KNOBS

(For Identification Purposes)

MAS 403 BT

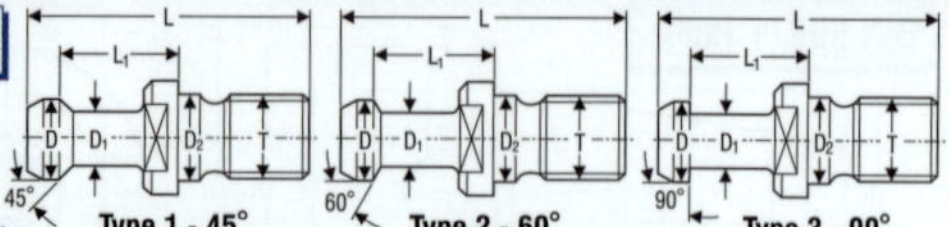

Type 1 - 45° Type 2 - 60° Type 3 - 90°

Dimensions in inches

Taper	D	D_1	D_2	L	L_1	T (thread)
30	0.433	0.276	0.650	1.693	0.709	M12
40	0.590	0.394	0.905	2.362	1.102	M16
50	0.906	0.669	1.496	3.346	1.378	M24

Specify:
- Taper No.
- Type
- With/without through coolant hole
- With/without O Ring

DIN 69872

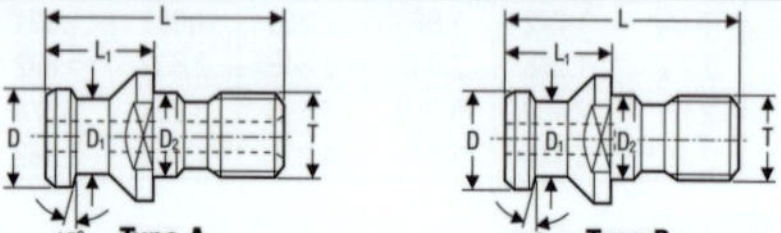

Type A (For tools with axial coolant supply) Type B (For tools with coolant supply via the flange)

Dimensions in inches

Taper	D	D_1	D_2	L	L_1	T (thread)
30	0.512	0.354	0.512	1.732	0.945	M12
40	0.748	0.551	0.669	2.126	1.024	M16
45	0.905	0.669	0.827	2.559	1.181	M20
50	1.102	0.827	0.984	2.913	1.339	M24

Specify:
- Taper No.
- Type
- With/without through coolant hole
- With/without O Ring

ISO 7388/2 (DIN 69871/A)

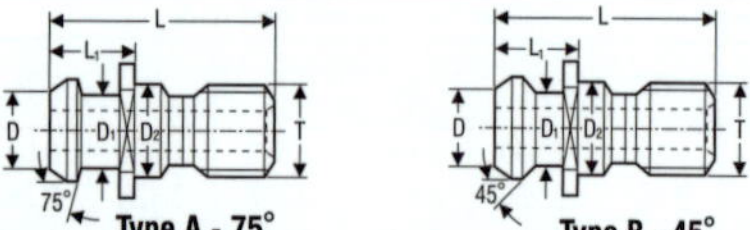

Type A - 75° Type B - 45°

Dimensions in inches

Taper	Type	D	D_1	D_2	L	L_1	T (thread)
40	A	0.748	0.551	0.669	2.126	1.024	M16
	B	0.744	0.508	0.669	1.752	0.646	M16
45	A	0.905	0.669	0.827	2.559	1.181	M20
	B	0.945	0.642	0.827	2.205	0.823	M20
50	A	1.102	0.827	0.984	2.913	1.338	M24
	B	1.142	0.772	0.984	2.579	1.004	M24

Specify:
- Taper No.
- Type
- With/without through coolant hole
- With/without O Ring

ANSI/CAT OTT

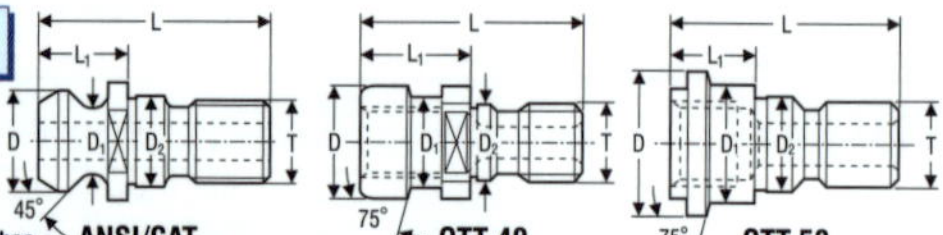

ANSI/CAT OTT 40 OTT 50

Dimensions in inches

Taper	D	D_1	D_2	L	L_1	T (thread)
ANSI/CAT 40	0.746	0.510	0.669	0.646	1.752	5/8" UNC
ANSI/CAT 45	0.947	0.642	0.827	0.825	2.205	3/4" UNC
ANSI/CAT 50	1.146	0.772	0.984	1.006	2.579	1" UNC
OTT 40	0.984	0.831	0.669	2.087	0.984	5/8" UNC
OTT 50	1.555	1.260	0.984	2.571	0.992	1" UNC

Specify:
- Taper No.
- With/without through coolant hole
- With/without O Ring

INCH SQUARE (PREFERRED) PARALLEL KEYS AND KEYWAYS (ANSI B17.1-1967, R1989)

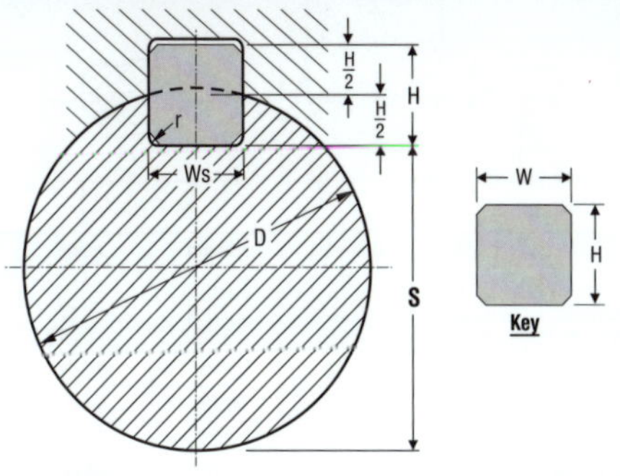

Class 1 - Side Fit

Dimensions in inches

Nominal Shaft Dia. (D)		Key		Keyway in Shaft		Keyway Radius Max. (r)
Over	Up to & including	Key Size (Square) (W x H)	Width / Thickness	Width (Ws)	Nominal Depth (H/2)	
5/16	7/16	3/32	0.093 0.091	0.095 0.093	3/64	1/32
7/16	9/16	1/8	0.125 0.123	0.127 0.125	1/16	1/32
9/16	7/8	3/16	0.187 0.185	0.189 0.187	3/32	1/32
7/8	1 1/4	1/4	0.250 0.248	0.252 0.250	1/8	1/32
1 1/4	1 3/8	5/16	0.312 0.310	0.314 0.312	5/32	1/32
1 3/8	1 3/4	3/8	0.375 0.373	0.377 0.375	3/16	1/32
1 3/4	2 1/4	1/2	0.500 0.498	0.502 0.500	1/4	1/32
2 1/4	2 3/4	5/8	0.625 0.623	0.628 0.625	5/16	1/16
2 3/4	3 1/4	3/4	0.750 0.747	0.753 0.750	3/8	1/16
3 1/4	3 3/4	7/8	0.875 0.872	0.878 0.875	7/16	1/16
3 3/4	4 1/2	1	1.000 0.997	1.003 1.000	1/2	1/16
4 1/2	5 1/2	1 1/4	1.250 1.247	1.254 1.250	5/8	1/8
5 1/2	6 1/2	1 1/2	1.500 1.497	1.504 1.500	3/4	1/8

Measurement from bottom of Keyseat	
Nominal Shaft Dia. (D)	(S)
1/2"	0.430
9/16"	0.493
5/8"	0.517
11/16"	0.581
3/4"	0.644
13/16"	0.708
7/8"	0.771
15/16"	0.796
1"	0.859
1 1/16"	0.923
1 1/8"	0.986
1 3/16"	1.049
1 1/4"	1.112
1 5/16"	1.137
1 3/8"	1.201
1 7/16"	1.225
1 1/2"	1.289
1 5/8"	1.416
1 3/4"	1.542
1 7/8"	1.591
2"	1.718
2 1/8"	1.845
2 1/4"	1.972
2 3/8"	2.021
2 1/2"	2.148
3"	2.577

INCH WOODRUFF KEYS

ACCORDING TO ANSI B17.2-1967 (R2003)

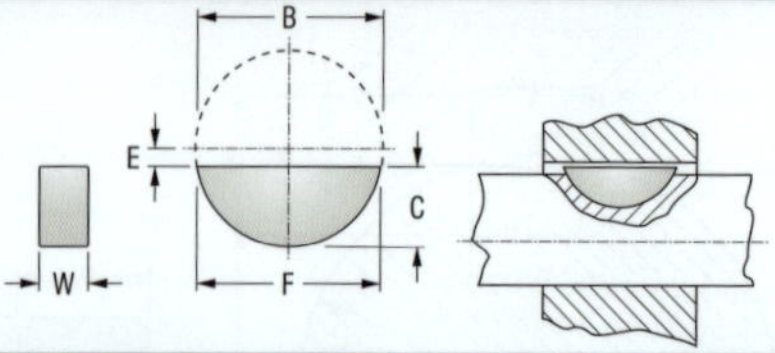

Key No.	Nominal Key Size	Width W +0.001 −0.000	Diameter B +0.000 −0.010	Height C +0.005 −0.000	Length F +0.000 −0.010	Distance below Center E (Nominal)
202	1/16 x 1/4	0.0625	0.250	0.104	0.248	1/64
202.5	1/16 x 5/16	0.0625	0.312	0.135	0.311	1/64
302.5	3/32 x 5/16	0.0937	0.312	0.135	0.311	1/64
203	1/16 x 3/8	0.6250	0.375	0.167	0.374	1/64
303	3/32 x 3/8	0.0937	0.375	0.167	0.374	1/64
403	1/8 x 3/8	0.1250	0.375	0.167	0.374	1/64
204	1/16 x 1/2	0.0625	0.500	0.198	0.491	3/64
304	3/32 x 1/2	0.0937	0.500	0.198	0.491	3/64
404	1/8 x 1/2	0.1250	0.500	0.198	0.491	3/64
305	3/32 x 5/8	0.0937	0.625	0.245	0.612	1/16
405	1/8 x 5/8	0.1250	0.625	0.245	0.612	1/16
505	5/32 x 5/8	0.1563	0.625	0.245	0.612	1/16
605	3/16 x 5/8	0.1875	0.625	0.245	0.612	1/16
406	1/8 x 3/4	0.1250	0.750	0.308	0.740	1/16
506	5/32 x 3/4	0.1563	0.750	0.308	0.740	1/16
606	3/16 x 3/4	0.1875	0.750	0.308	0.740	1/16
806	1/4 x 3/4	0.2500	0.750	0.308	0.740	1/16
507	5/32 x 7/8	0.1563	0.875	0.370	0.866	1/16
607	3/16 x 7/8	0.1875	0.875	0.370	0.866	1/16
707	7/32 x 7/8	0.2187	0.875	0.370	0.866	1/16
807	1/4 x 7/8	0.2500	0.875	0.370	0.866	1/16
608	3/16 x 1	0.1875	1.000	0.433	0.992	1/16
708	7/32 x 1	0.2187	1.000	0.433	0.992	1/16
808	1/4 x 1	0.2500	1.000	0.433	0.992	1/16
1008	5/16 x 1	0.3125	1.000	0.433	0.992	1/16
1208	3/8 x 1	0.3750	1.000	0.433	0.992	1/16
609	3/16 x 1 1/8	0.1875	1.125	0.479	1.114	5/64
709	7/32 x 1 1/8	0.2187	1.125	0.479	1.114	5/64
809	1/4 x 1 1/8	0.2500	1.125	0.479	1.114	5/64
1009	5/16 x 1 1/8	0.3125	1.125	0.479	1.114	5/64
610	3/16 x 1 1/4	0.1875	1.250	0.542	1.240	5/64
710	7/32 x 1 1/4	0.2187	1.250	0.542	1.240	5/64
810	1/4 x 1 1/4	0.2500	1.250	0.542	1.240	5/64
1010	5/16 x 1 1/4	0.3125	1.250	0.542	1.240	5/64

INCH WOODRUFF KEY-SEATS
ACCORDING TO ANSI B17.2-1967 (R2003)

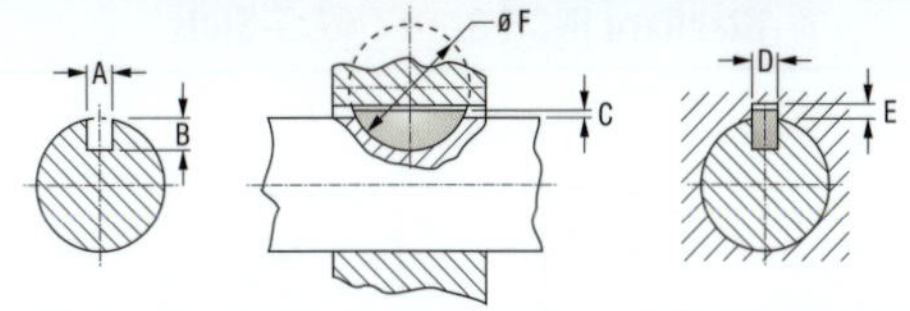

		Keyseat for Shaft					Key	Keyseat for Hub	
		Width		Depth	Diameter		Height	Width	Depth
		A		B	F		C	D	E
Key No.	Nominal Key Size	max	min	+0.005 – 0.000	max	min	±0.005	+0.002 – 0.000	+0.005 – 0.000
202	1/16 x 1/4	0.0630	0.0615	0.0728	0.268	0.25	0.0312	0.0635	0.0372
202.5	1/16 x 5/16	0.0630	0.0615	0.1038	0.330	0.312	0.0312	0.0635	0.0372
302.5	3/32 x 5/16	0.0943	0.0928	0.0882	0.330	0.312	0.0469	0.0948	0.0529
203	1/16 x 3/8	0.0630	0.0615	0.1358	0.3930	0.3750	0.0312	0.0635	0.0372
303	3/32 x 3/8	0.0943	0.0928	0.1202	0.3930	0.3750	0.0469	0.0948	0.0529
403	1/8 x 3/8	0.1255	0.1240	0.1045	0.3930	0.3750	0.0625	0.1260	0.0685
204	1/16 x 1/2	0.0630	0.0615	0.1668	0.5180	0.5000	0.0312	0.0635	0.0372
304	3/32 x 1/2	0.0943	0.0928	0.1511	0.5180	0.5000	0.0469	0.0948	0.0529
404	1/8 x 1/2	0.1255	0.1240	0.1355	0.5180	0.5000	0.0625	0.1260	0.0685
305	3/32 x 5/8	0.0943	0.0928	0.1981	0.6430	0.6250	0.0469	0.0948	0.0529
405	1/8 x 5/8	0.1255	0.1240	0.1825	0.6430	0.6250	0.0625	0.1260	0.0685
505	5/32 x 5/8	0.1568	0.1553	0.1669	0.6430	0.6250	0.0781	0.1573	0.0841
605	3/16 x 5/8	0.1880	0.1863	0.1513	0.6430	0.6250	0.0937	0.1885	0.0997
406	1/8 x 3/4	0.1255	0.1240	0.2455	0.7680	0.7500	0.0625	0.1260	0.0685
506	5/32 x 3/4	0.1568	0.1553	0.2299	0.7680	0.7500	0.0781	0.1573	0.0841
606	3/16 x 3/4	0.1880	0.1863	0.2143	0.7680	0.7500	0.0937	0.1885	0.0997
806	1/4 x 3/4	0.2505	0.2487	0.1830	0.7680	0.7500	0.1250	0.2510	0.1310
507	5/32 x 7/8	0.1568	0.1553	0.2919	0.8950	0.8750	0.0781	0.1573	0.0841
607	3/16 x 7/8	0.1880	0.1863	0.2763	0.8950	0.8750	0.0937	0.1885	0.0997
707	7/32 x 7/8	0.2193	0.2175	0.2607	0.8950	0.8750	0.1093	0.2198	0.1153
807	1/4 x 7/8	0.2505	0.2487	0.2450	0.8950	0.8750	0.1250	0.2510	0.1310
608	3/16 x 1	0.1880	0.1863	0.3393	1.0200	1.0000	0.0937	0.1885	0.0997
708	7/32 x 1	0.2193	0.2175	0.3237	1.0200	1.0000	0.1093	0.2198	0.1153
808	1/4 x 1	0.2505	0.2487	0.3080	1.0200	1.0000	0.1250	0.2510	0.1310
1008	5/16 x 1	0.3130	0.3111	0.2768	1.0200	1.0000	0.1562	0.3135	0.1622
1208	3/8 x 1	0.3755	0.3735	0.2455	1.0200	1.0000	0.1875	0.3760	0.1935
609	3/16 x 1 1/8	0.1880	0.1863	0.3853	1.1450	1.1250	0.0937	0.1885	0.0997
709	7/32 x 1 1/8	0.2193	0.2175	0.3697	1.1450	1.1250	0.1093	0.2198	0.1153
809	1/4 x 1 1/8	0.2505	0.2487	0.3540	1.1450	1.1250	0.1250	0.2510	0.1310
1009	5/16 x 1 1/8	0.3130	0.3111	0.3228	1.1450	1.1250	0.1562	0.3135	0.1622
610	3/16 x 1 1/4	0.1880	0.1863	0.4483	1.2730	1.2500	0.0937	0.1885	0.0997
710	7/32 x 1 1/4	0.2193	0.2175	0.4327	1.2730	1.2500	0.1093	0.2198	0.1153
810	1/4 x 1 1/4	0.2505	0.2487	0.4170	1.2730	1.2500	0.1250	0.2510	0.1310
1010	5/16 x 1 1/4	0.3130	0.3111	0.3858	1.2730	1.2500	0.1562	0.3135	0.1622

STANDARD T-SLOTS

Standard INCH & METRIC T-Slots

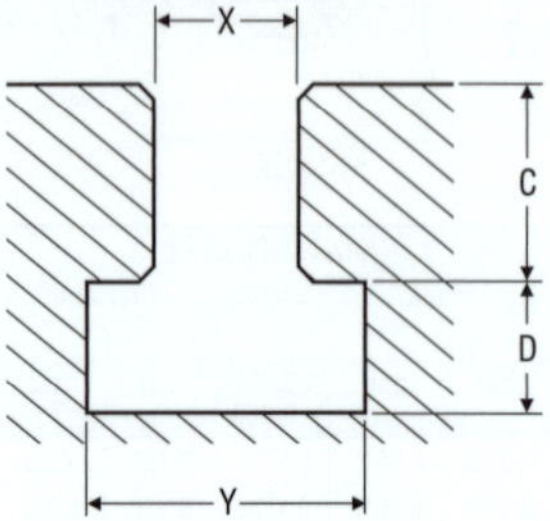

INCH T-SLOTS

Dimensions in inches

Nominal T-Bolt Size	'X'		'Y'		'C'		'D'	
	Min.	Max.	Min.	Max.	Min.	Max.	Min.	Max.
1/4"	0.282	0.292	0.500	0.562	0.125	0.375	0.203	0.234
5/16"	0.344	0.354	0.594	0.656	0.156	0.438	0.234	0.266
3/8"	0.438	0.448	0.719	0.781	0.219	0.562	0.297	0.328
1/2"	0.562	0.572	0.906	0.969	0.312	0.688	0.359	0.391
5/8"	0.688	0.698	1.188	1.250	0.438	0.875	0.453	0.484
3/4"	0.812	0.822	1.375	1.469	0.562	1.062	0.594	0.625
1"	1.062	1.072	1.750	1.844	0.750	1.250	0.781	0.828
1 1/4"	1.312	1.322	2.125	2.219	1.000	1.562	1.031	1.094
1 1/2"	1.562	1.572	2.562	2.656	1.250	1.938	1.281	1.344

METRIC T-SLOTS

Dimensions in mm

Nominal T-Bolt Size	'X'		'Y'		'C'		'D'	
	Min.	Max.	Min.	Max.	Min.	Max.	Min.	Max.
6 mm	8.00	8.15	14.50	16.00	7.00	11.00	7.00	8.00
8 mm	10.00	10.15	16.00	18.00	9.00	14.00	7.00	8.00
10 mm	12.00	12.18	19.00	21.00	11.00	17.00	8.00	9.00
12 mm	14.00	14.18	23.00	25.00	12.00	19.00	9.00	11.00
16 mm	18.00	18.21	30.00	32.00	16.00	24.00	12.00	14.00
20 mm	22.00	22.21	37.00	40.00	20.00	29.00	16.00	18.00
24 mm	28.00	28.21	46.00	50.00	26.00	36.00	20.00	22.00
30 mm	36.00	36.25	56.00	60.00	33.00	46.00	25.00	28.00
36 mm	42.00	42.25	68.00	72.00	39.00	53.00	32.00	35.00

STANDARD E-RING EXTERNAL CIRCLIPS (INCH)

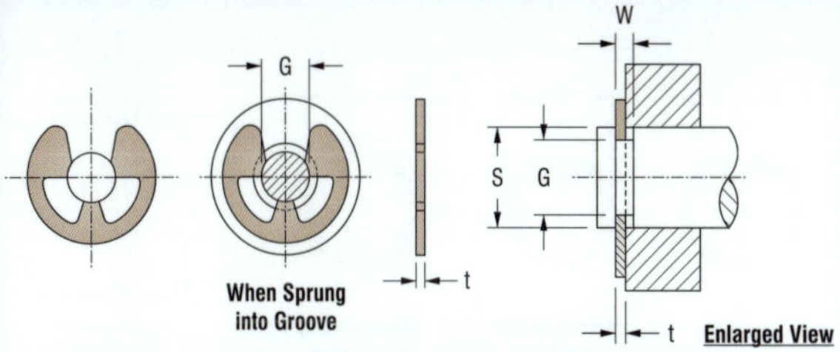

SHAFT		RING		GROOVE			
Diameter S		Thickness		Diameter		Width	
Fraction (inch)	Decimal (inch)	t (inch)	Tolerance (inch)	G (inch)	Tolerance (inch)	W (inch)	Tolerance (inch)
1/16"	0.062	0.010	±0.001	0.052	± 0.002	0.012	
1/16"	0.062	0.020	±0.002	0.052		0.023	
1/16"	0.062	0.010	±0.001	0.052		0.012	
3/32"	0.094	0.015		0.074		0.020	
3/32"	0.094	0.015		0.074		0.020	+0.002 −0.000
7/64"	0.110	0.015		0.079		0.020	
1/8"	0.125	0.015		0.095		0.020	
9/64"	0.140	0.015		0.102		0.020	
9/64"	0.140	0.015		0.110	+0.002 −0.000	0.020	
9/64"	0.140	0.025		0.105		0.029	
5/32"	0.156	0.025		0.116		0.029	
11/64"	0.172	0.025		0.127		0.029	
3/16"	0.188	0.025		0.125		0.029	
3/16"	0.188	0.025		0.147		0.029	
7/32"	0.219	0.025	±0.002	0.188		0.029	
1/4"	0.250	0.025		0.210		0.029	
5/16"	0.312	0.025		0.250		0.029	
3/8"	0.375	0.035		0.303		0.039	+0.003 −0.000
7/16"	0.438	0.035		0.343		0.039	
7/16"	0.438	0.035		0.380		0.039	
1/2"	0.500	0.042		0.396		0.046	
5/8"	0.625	0.042		0.485	+0.003 −0.000	0.046	
3/4"	0.744	0.050		0.625		0.056	
3/4"	0.750	0.050		0.580		0.056	
7/8"	0.875	0.050		0.675		0.056	
1"	0.984	0.050		0.835		0.056	
1 3/16"	1.188	0.062	±0.003	1.079		0.068	+0.004 −0.000
1 3/8"	1.375	0.062		1.230		0.068	

STANDARD INTERNAL RETAINING RINGS (INCH)

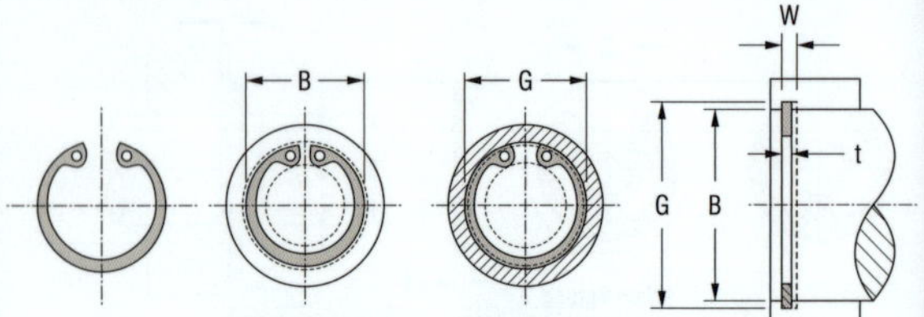

When Sprung into Bore

When Sprung into Groove

Enlarged View

BORE		RING		GROOVE			
Diameter B		Thickness		Diameter		Width	
Fraction (inch)	Decimal (inch)	t (inch)	Tolerance (inch)	G (inch)	Tolerance (inch)	W (inch)	Tolerance (inch)
1/4"	0.250	0.015	±0.002	0.268	±0.001	0.018	+0.002 −0.000
5/16"	0.312	0.015		0.330		0.018	
3/8"	0.375	0.025		0.397	±0.002	0.029	+0.003 −0.000
7/16"	0.438	0.025		0.461		0.029	
29/64"	0.453	0.025		0.477		0.029	
1/2"	0.500	0.035		0.530		0.039	
9/16"	0.562	0.035		0.596		0.039	
5/8"	0.625	0.035		0.665		0.039	
11/16"	0.688	0.035		0.732		0.039	
3/4"	0.750	0.035		0.796		0.039	
13/16"	0.812	0.042		0.862	±0.003	0.046	
7/8"	0.875	0.042		0.931		0.046	
15/16"	0.938	0.042		1.000		0.046	
1"	1.000	0.042		1.066		0.046	
1 1/16"	1.062	0.050		1.130	±0.004	0.056	
1 1/8"	1.125	0.050		1.197		0.056	
1 3/16"	1.188	0.050		1.262		0.056	+0.004 −0.000
1 1/4"	1.250	0.050		1.330		0.056	
1 5/16"	1.312	0.050		1.396		0.056	
1 3/8"	1.375	0.050		1.461		0.056	
1 7/16"	1.438	0.050		1.528		0.056	
1 1/2"	1.500	0.050		1.594		0.056	

STANDARD EXTERNAL RETAINING RINGS (INCH)

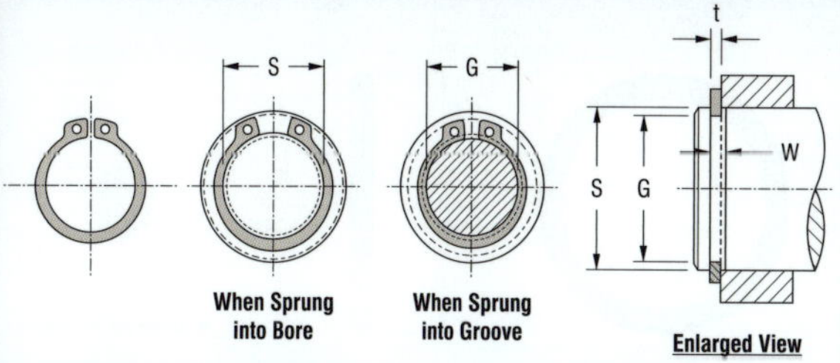

SHAFT		RING		GROOVE			
Diameter S		Thickness		Diameter		Width	
Fraction (inch)	Decimal (inch)	t (inch)	Tolerance (inch)	G (inch)	Tolerance (inch)	W (inch)	Tolerance (inch)
1/8"	0.125	0.010	±0.001	0.117	±0.0015	0.012	+0.002 −0.000
5/32"	0.156	0.010		0.146		0.012	
3/16"	0.188	0.015	±0.002	0.175		0.018	
7/32"	0.219	0.015		0.205		0.018	
15/64"	0.236	0.015		0.222		0.018	
1/4"	0.250	0.025		0.230	±0.002	0.029	+0.003 −0.000
9/32"	0.281	0.025		0.261		0.029	
5/16"	0.312	0.025		0.290		0.029	
11/32"	0.344	0.025		0.321		0.029	
3/8"	0.375	0.025		0.352		0.029	
13/32"	0.406	0.025		0.382		0.029	
7/16"	0.438	0.025		0.412		0.029	
15/32"	0.469	0.025		0.443		0.029	
1/2"	0.500	0.035		0.468		0.039	
9/16"	0.562	0.035		0.530		0.039	
19/32"	0.594	0.035		0.559	±0.003	0.039	
5/8"	0.625	0.035		0.588		0.039	
43/64"	0.672	0.035		0.631		0.039	
11/16"	0.688	0.042		0.646		0.046	
3/4"	0.750	0.042		0.704		0.046	
25/32"	0.781	0.042		0.733		0.046	
13/16"	0.812	0.042		0.762		0.046	
7/8"	0.875	0.042		0.821		0.046	
15/16"	0.938	0.042		0.882		0.046	
63/64"	0.984	0.042		0.926		0.046	
1"	1.000	0.042		0.940		0.046	

O-RINGS
ACCORDING TO SAE AS-568A

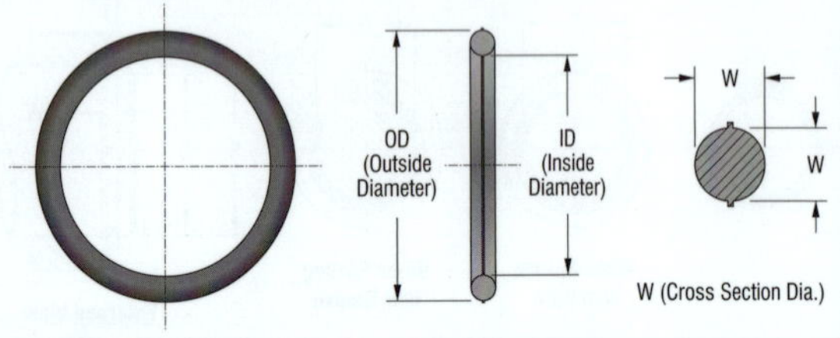

Size Ref. AS 568	Nom. Inside Dia. ID	Actual Inside Dia. ID	Actual Inside Dia. Tol.	Nom. Outside Dia. OD	Actual Cross Section Dia. W	Actual Cross Section Dia. Tol.
001	1/32	0.029		3/32	.040	
002	3/64	0.042	±.004	9/64	.050	
003	1/16	0.056		3/16	.060	
004	5/64	0.070		13/64	.070	
005	3/32	0.101		15/64	.070	
006	1/8	0.114		1/4	.070	
007	5/32	0.145		9/32	.070	
008	3/16	0.176		5/16	.070	
009	7/32	0.208	±.005	11/32	.070	
010	1/4	0.239		3/8	.070	
011	5/16	0.301		7/16	.070	
012	3/8	0.364		1/2	.070	
013	7/16	0.426		9/16	.070	±.003
014	1/2	0.489		5/8	.070	
015	9/16	0.551	±.007	11/16	.070	
016	5/8	0.614		3/4	.070	
017	11/16	0.676		13/16	.070	
018	3/4	0.739	±.009	7/8	.070	
019	13/16	0.801		15/16	.070	
020	7/8	0.864		1	.070	
021	15/16	0.926		1 1/16	.070	
022	1	0.989		1 1/8	.070	
023	1 1/16	1.051	±.010	1 3/16	.070	
024	1 1/8	1.114		1 1/4	.070	
025	1 3/16	1.176		1 5/16	.070	
026	1 1/4	1.239	±.011	1 3/8	.070	
027	1 5/16	1.301		1 7/16	.070	
028	1 3/8	1.364		1 1/2	.070	
029	1 1/2	1.489	±.013	1 5/8	.070	
030	1 5/8	1.614		1 3/4	.070	

Size Ref. AS 568	Nom. Inside Dia. ID	Actual Inside Dia. ID	Actual Inside Dia. Tol.	Nom. Outside Dia. OD	Actual Cross Section Dia. W	Actual Cross Section Dia. Tol.
031	1 3/4	1.739	±.015	1 7/8	.070	
032	1 7/8	1.864		2	.070	
033	2	1.989		2 1/8	.070	
034	2 1/8	2.114		2 1/4	.070	
035	2 1/4	2.239	±.018	2 3/8	.070	
036	2 3/8	2.364		2 1/2	.070	
037	2 1/2	2.489		2 5/8	.070	
038	2 5/8	2.614		2 3/4	.070	
039	2 3/4	2.739	±.020	2 7/8	.070	
040	2 7/8	2.864		3	.070	
041	3	2.989		3 1/8	.070	
042	3 1/4	3.239	±.024	3 3/8	.070	
043	3 1/2	3.489		3 5/8	.070	±.003
044	3 3/4	3.739	±.027	3 7/8	.070	
045	4	3.989		4 1/8	.070	
046	4 1/4	4.239		4 3/8	.070	
047	4 1/2	4.489	±.030	4 5/8	.070	
048	4 3/4	4.739		4 7/8	.070	
049	5	4.989	±.037	5 1/8	.070	
050	5 1/4	5.239		5 3/8	.070	
051 Thru 101 O-Ring Sizes Not Assigned						
102	1/16	0.049		1/4	.103	
103	3/32	0.081		9/32	.103	
104	1/8	0.112		5/16	.103	
105	5/32	0.143		11/32	.103	
106	3/16	0.174	±.005	3/8	.103	
107	7/32	0.206		13/32	.103	
108	1/4	0.237		7/16	.103	
109	5/16	0.299		1/2	.103	
110	3/8	0.362		9/16	.103	

Note: All sizes are based on 70 Durometer Nitrile compound

O-RINGS
ACCORDING TO SAE AS-568A

Size Ref. AS 568	Nom. Inside Dia. ID	Actual Inside Dia. ID	Actual Inside Dia. Tol.	Nom. Outside Dia. OD	Actual Cross Section Dia. W	Actual Cross Section Dia. Tol.
111	7/16	0.424		5/8	.103	
112	1/2	0.487	±.005	11/16	.103	
113	9/16	0.549		3/4	.103	
114	5/8	0.612		13/16	.103	
115	11/16	0.674	±.009	7/8	.103	
116	3/4	0.737		15/16	.103	
117	13/16	0.799		1	.103	
118	7/8	0.862		1 1/16	.103	
119	15/16	0.924	±.010	1 1/8	.103	
120	1	0.987		1 3/16	.103	
121	1 1/16	1.049		1 1/4	.103	
122	1 1/8	1.112		1 5/16	.103	
123	1 3/16	1.174		1 3/8	.103	
124	1 1/4	1.237		1 7/16	.103	
125	1 5/16	1.299	±.012	1 1/2	.103	
126	1 3/8	1.362		1 9/16	.103	
127	1 7/16	1.424		1 5/8	.103	
128	1 1/2	1.487		1 11/16	.103	
129	1 9/16	1.549		1 3/4	.103	
130	1 5/8	1.612		1 13/16	.103	
131	1 11/16	1.674	±.015	1 7/8	.103	
132	1 3/4	1.737		1 15/16	.103	±.003
133	1 13/16	1.799		2	.103	
134	1 7/8	1.862		2 1/16	.103	
135	1 15/16	1.925		2 1/8	.103	
136	2	1.987		2 3/16	.103	
137	2 1/16	2.050	±.017	2 1/4	.103	
138	2 1/8	2.112		2 5/16	.103	
139	2 3/16	2.175		2 3/8	.103	
140	2 1/4	2.237		2 7/16	.103	
141	2 5/16	2.300		2 1/2	.103	
142	2 3/8	2.362		2 9/16	.103	
143	2 7/16	2.425	±.020	2 5/8	.103	
144	2 1/2	2.487		2 11/16	.103	
145	2 9/16	2.550		2 3/4	.103	
146	2 5/8	2.612		2 13/16	.103	
147	2 11/16	2.675		2 7/8	.103	
148	2 3/4	2.737	±.022	2 15/16	.103	
149	2 13/16	2.800		3	.103	
150	2 7/8	2.862		3 1/16	.103	
151	3	2.987		3 3/16	.103	
152	3 1/4	3.237	±.024	3 11/16	.103	
153	3 1/2	3.487		3 15/16	.103	

Size Ref. AS 568	Nom. Inside Dia. ID	Actual Inside Dia. ID	Actual Inside Dia. Tol.	Nom. Outside Dia. OD	Actual Cross Section Dia. W	Actual Cross Section Dia. Tol.
154	3 3/4	3.737	±.028	3 15/16	.103	
155	4	3.987		4 3/16	.103	
156	4 1/4	4.237		4 7/16	.103	
157	4 1/2	4.487	±.030	4 11/16	.103	
158	4 3/4	4.737		4 15/16	.103	
159	5	4.987		5 3/16	.103	
160	5 1/4	5.237		5 7/16	.103	
161	5 1/2	5.487	±.035	5 11/16	.103	
162	5 3/4	5.737		5 15/16	.103	
163	6	5.987		6 3/16	.103	
164	6 1/4	6.237		6 7/16	.103	
165	6 1/2	6.487	±.040	6 11/16	.103	
166	6 3/4	6.737		6 15/16	.103	±.003
167	7	6.987		7 3/16	.103	
168	7 1/4	7.237		7 7/16	.103	
169	7 1/2	7.487	±.045	7 11/16	.103	
170	7 3/4	7.737		7 15/16	.103	
171	8	7.987		8 3/16	.103	
172	8 1/4	8.237		8 7/16	.103	
173	8 1/2	8.487	±.050	8 11/16	.103	
174	8 3/4	8.737		8 15/16	.103	
175	9	8.987		9 3/16	.103	
176	9 1/4	9.237		9 7/16	.103	
177	9 1/2	9.487	±.055	9 11/16	.103	
178	9 3/4	9.737		9 15/16	.103	
179 Thru 200 O-Ring Sizes Not Assigned						
201	3/16	0.171		7/16	.139	
202	1/4	0.234		1/2	.139	
203	5/16	0.296	±.005	9/16	.139	
204	3/8	0.359		5/8	.139	
205	7/16	0.421		11/16	.139	
206	1/2	0.484		3/4	.139	
207	9/16	0.546	±.007	13/16	.139	
208	5/8	0.609	±.009	7/8	.139	
209	11/16	0.671		15/16	.139	±.004
210	3/4	0.734		1	.139	
211	13/16	0.796		1 1/16	.139	
212	7/8	0.859	±.010	1 1/8	.139	
213	15/16	0.921		1 3/16	.139	
214	1	0.984		1 1/4	.139	
215	1 1/16	1.046		1 5/16	.139	
216	1 1/8	1.109	±.012	1 3/8	.139	
217	1 3/16	1.171		1 7/16	.139	

Note: All sizes are based on 70 Durometer Nitrile compound

SPUR GEAR CALCULATIONS

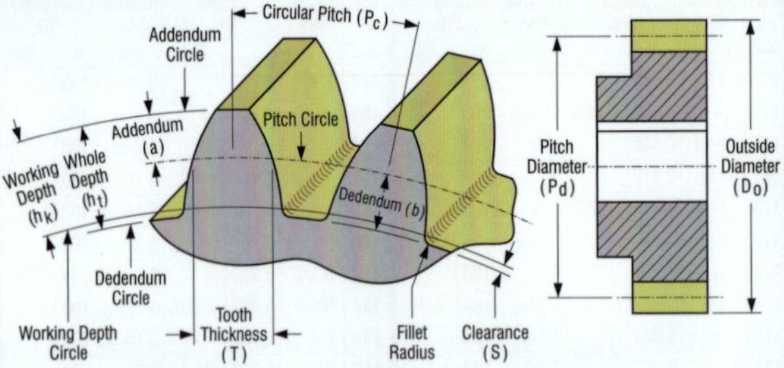

To Find	Formula
Diametral pitch P	$P = \frac{3.1416}{P_c} = \frac{N}{P_d} = \frac{25.4}{\text{Module}}$
Module (mm)	$\text{Module} = \frac{25.4}{P} = \frac{P_d\ (\text{mm})}{N}$
Circular pitch (inch) P_c	$P_c = \frac{3.1416}{P} = \frac{3.1416 \times P_d}{N}$
Circular pitch (mm) P_c	$P_{c(mm)} = 3.1416 \times \text{Module}$
Pitch diameter (inch) P_d	$P_d = \frac{N}{P} = \frac{N \times P_c}{3.1416}$
Pitch diameter (mm) P_d	$P_{d(mm)} = N \times \text{Module}$
Outside dia. (inch/mm) D_o	$D_o = \frac{N+2}{P} = P_d + 2a$
Number of teeth in the gear N	$N = P_d \times P = \frac{3.1416 \times P_d}{P_c}$
Center distance C	$C = \frac{N_1 + N_2}{2P}$
Length of rack L	$L = N \times P_c$

TOOTH PARTS - FULL DEPTH TEETH, METRIC MODULE AND STUB TOOTH GEARING

Tooth Parts - Full Depth Teeth

To Find		20° Pressure Angle	14½° Pressure Angle
Addendum	a	1.00 ÷ P	1.000 ÷ P
Minimum dedendum	b	1.25 ÷ P	1.157 ÷ P
Working depth	h_k	2.00 ÷ P	2.000 ÷ P
Minimum whole depth	h_t	2.25 ÷ P	2.157 ÷ P
Basic circular tooth thickness at pitch line	T	1.5708 ÷ P	1.5708 ÷ P
Minimum clearance	S	0.25 ÷ P	0.157 ÷ P

Tooth Parts Metric Module System - Pressure Angle = 20°

Dimensions in mm

Addendum	a	a = 1 x Module
Minimum dedendum	b	b = 1.25 x Module
Working depth	h_k	h_k = 2 x Module
Minimum whole depth	h_t	h_t = 2.25 x Module
Basic circular tooth thickness at pitch line	T	T = 1.5708 x Module
Minimum clearance	S	S = 0.25 x Module

Tooth Parts - Stub Tooth gearing - Pressure Angle = 20°

To Find		American	Fellows	Nuttal
Addendum	a	0.800 ÷ P	1.000 ÷ P_s	0.250 x P_c
Minimum dedendum	b	1.000 ÷ P	1.250 ÷ P_s	0.300 x P_c
Working depth	h_k	1.600 ÷ P	2.000 ÷ P_s	0.500 x P_c
Minimum whole depth	h_t	1.800 ÷ P	2.250 ÷ P_s	0.550 x P_c
Basic circular tooth thickness at pitch line	T	1.5708 ÷ P	1.5708 ÷ P_s	0.500 x P_c
Minimum clearance	S	0.200 ÷ P	0.250 ÷ P_s	0.050 x P_c

P_s = Stub diametral pitch, e.g. "**8**" in 6/8 DP

GRAPHICAL REPRESENTATION OF GEAR RACK AND PINION - DP* (ACTUAL SIZE)

RACK	PINION
 32 DP	 32 DP
 24 DP	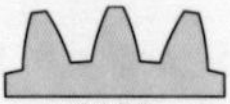 24 DP
 14 DP	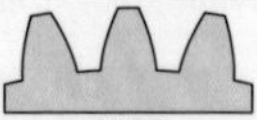 14 DP
 12 DP	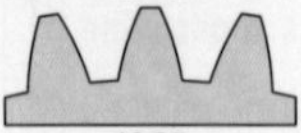 12 DP
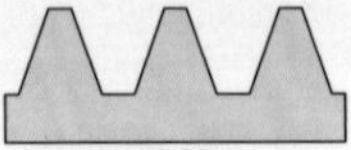 10 DP	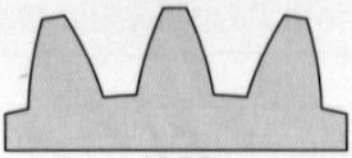 10 DP
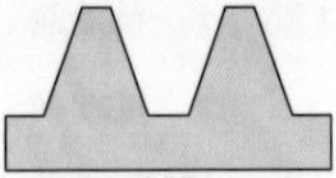 8 DP	 8 DP
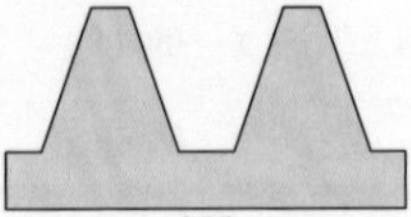 6 DP	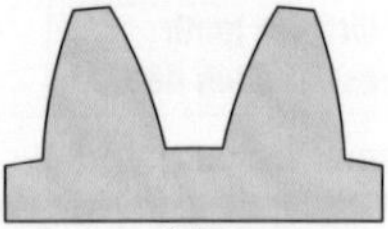 6 DP
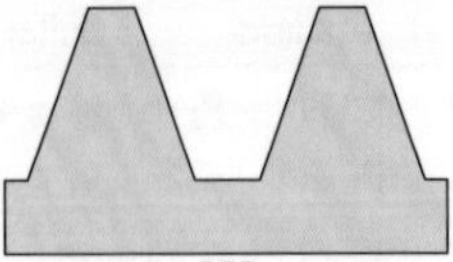 5 DP	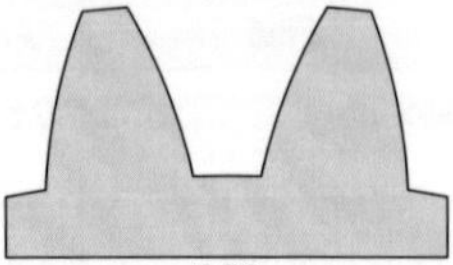 5 DP
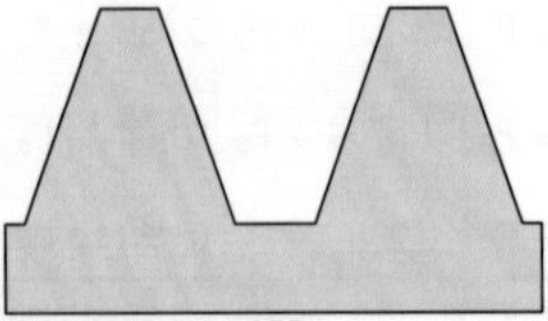 4 DP	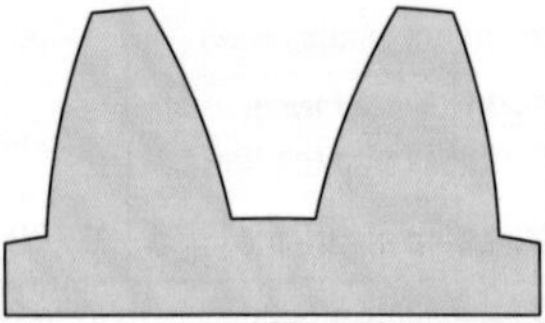 4 DP

* Diameteral Pitch - 20° Pressure Angle

GRAPHICAL REPRESENTATION OF GEAR RACK AND PINION - MODULE* (ACTUAL SIZE)

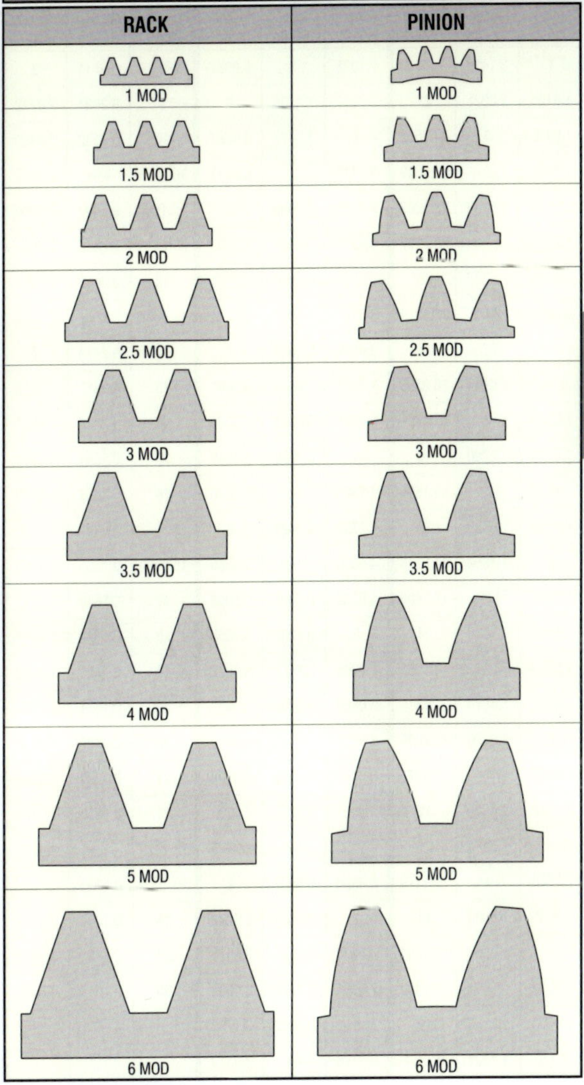

* Module - 20° Pressure Angle

SPUR GEARING
TABLE OF EQUIVALENT PITCHES

Diametral Pitch	Circular Pitch (in.)	Module mm.	Diametral Pitch	Circular Pitch (in.)	Module mm.	Diametral Pitch	Circular Pitch (in.)	Module mm.
1	3.1416	25.4000	2.7925	1 1/8	9.0957	8.4667	0.3711	**3**
1.0160	3.0921	**25**	2.8222	1.1132	**9**	**9**	0.3491	2.8222
1.0472	**3**	24.2552	**3**	1.0472	8.4667	**10**	0.3142	2.5400
1.1424	2 3/4	22.2339	3.1416	**1**	8.0851	10.0531	**5/16**	2.5266
1 1/4	2.5133	20.3202	3.1750	0.9895	**8**	10.1600	0.3092	2.5000
1.2566	2 1/2	20.2127	3.3510	15/16	7.5797	**11**	0.2856	2.3091
1.2700	2.4737	**20**	3 1/2	0.8976	7.2571	**12**	0.2618	2.1167
1.3963	2 1/4	18.1914	3.5904	**7/8**	7.0744	12.5664	**1/4**	2.0213
1.4111	2.2263	**18**	3.6286	0.8658	**7**	12.7000	0.2474	**2**
1 1/2	2.0944	16.9333	3.8666	13/16	6.5691	**13**	0.2417	1.9538
1.5708	**2**	16.1701	3.9078	0.8040	6.5000	**14**	0.2244	1.8143
1.5875	1.9790	**16**	**4**	0.7854	6.3500	**15**	0.2094	1.6933
1.6755	1 7/8	15.1595	4.1888	**3/4**	6.0638	**16**	0.1963	1.5875
1.6933	1.8553	**15**	4.2333	0.7421	**6**	16.7552	**3/16**	1.5159
1 3/4	1.7952	14.5143	4.5696	11/16	5.5585	16.9333	0.1855	1.5000
1.7952	1 3/4	14.1489	4.6182	0.6803	5.5000	**17**	0.1848	1.4941
1.8143	1.7316	**14**	**5**	0.6283	5.0800	**18**	0.1745	1.4111
1.9333	1 5/8	13.1382	5.0265	**5/8**	5.0532	**19**	0.1653	1.3368
1.9538	1.6079	**13**	5.0800	0.6184	**5**	**20**	0.1571	1.2700
2	1.5708	12.7000	5.5851	9/16	4.5478	**22**	0.1428	1.1545
2.0944	**1 1/2**	12.1276	5.6443	0.5566	4.5000	**24**	0.1309	1.0583
2.1167	1.4842	**12**	**6**	0.5236	4.2333	**25**	0.1257	1.0160
2 1/4	1.3963	11.2892	6.2832	**1/2**	4.0425	25.1327	**1/8**	1.0106
2.2848	1 3/8	11.1170	6.3500	0.4947	**4**	25.4000	0.1237	**1**
2.3091	1.3605	**11**	**7**	0.4488	3.6286	**26**	0.1208	0.9769
2 1/2	1.2566	10.1600	7.1808	**7/16**	3.5372	**28**	0.1122	0.9071
2.5133	**1 1/4**	10.1062	7.2571	0.4329	3.5000	**30**	0.1047	0.8467
2.5400	1.2368	**10**	**8**	0.3927	3.1750			
2 3/4	1.1424	9.2364	8.3776	**3/8**	3.0319			

GEARS

CALCULATION OF SPINDLE SPEEDS / NUMBER OF TEETH IN BELT / GEAR DRIVES

BELT DRIVE

Diameter x rpm of Driving Pulley = Diameter x rpm of Driven Pulley

$$D_1 \times N_1 = D_2 \times N_2$$

D_1 & D_3 = Diameters of Driving Pulleys

D_2 & D_4 = Diameters of Driven Pulleys

N_1 & N_3 = Speed(rpm) of Driving Pulleys

N_2 & N_4 = Speed(rpm) of Driven Pulleys

Single Drive

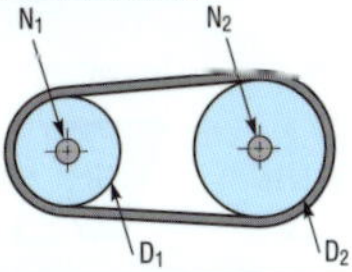

$$D_1 = \frac{D_2 \times N_2}{N_1} \qquad N_1 = \frac{D_2 \times N_2}{D_1}$$

$$D_2 = \frac{D_1 \times N_1}{N_2} \qquad N_2 = \frac{D_1 \times N_1}{D_2}$$

Compound Drive

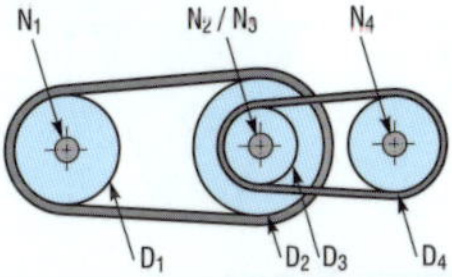

$$N_4 = N_1 \times \frac{D_1}{D_2} \times \frac{D_3}{D_4}$$

Note : Speed(rpm) N_2 always = Speed(rpm) N_3

GEAR DRIVE

No. of Teeth x rpm of Driving Gear = No. of Teeth x rpm of Driven Gear

$$Z_1 \times N_1 = Z_2 \times N_2$$

Z_1 & Z_3 = No of Teeth in Driving Gear

Z_2 & Z_4 = No of Teeth in Driven Gear

N_1 & N_3 = Speed(rpm) of Driving Gear

N_2 & N_4 = Speed(rpm) of Driven Gear

Single Gear Train

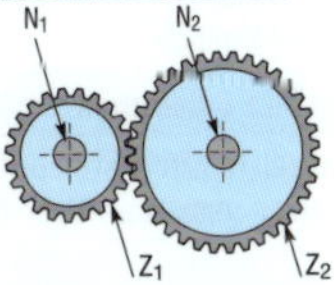

$$Z_1 = \frac{Z_2 \times N_2}{N_1} \qquad N_1 = \frac{Z_2 \times N_2}{Z_1}$$

$$Z_2 = \frac{Z_1 \times N_1}{N_2} \qquad N_2 = \frac{Z_1 \times N_1}{Z_2}$$

Double Gear Train

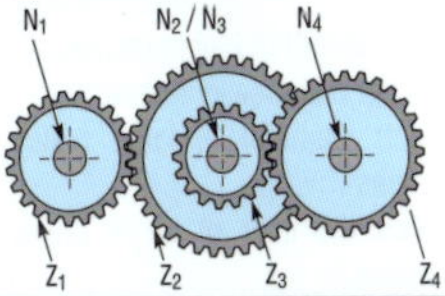

$$N_4 = N_1 \times \frac{Z_1}{Z_2} \times \frac{Z_3}{Z_4}$$

Note : Speed(rpm) N_2 always = Speed(rpm) N_3

V-BELTS CROSS SECTION DIMENSIONS

Conventional & Narrow Section V-Belts *(Actual Size)*

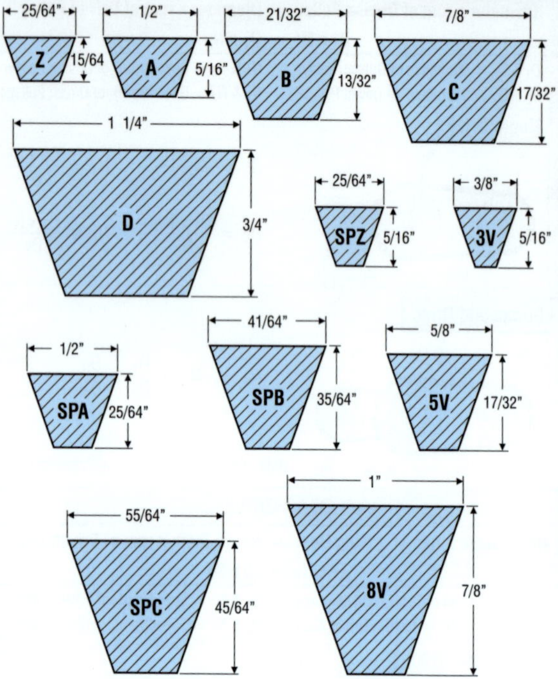

NOTE : See Comparison Chart on page 103 for other Equivalent Symbols

Calculation of Pitch Length

Section	Pitch Length exceeds inside length by inches.
Z	0.984"
A	1.299"
B	1.693"
C	2.441"
D	3.228"
SPZ/3V	1.653"
SPA	1.811"
SPB/5V	2.795"
SPC	3.425"

Center Distance Calculation

To determine belt length required given pulley diameters and center distance, use the following formula :

$$L = 2C + \frac{(D - d)^2}{4C} + 1.57\,(D + d) \text{ (inches)}$$

where

L = Pitch length of belt in inches.

C = Center distance in inches.

D = Pitch diameter of large pulley in inches.

d = Pitch diameter of small pulley in inches.

V-BELTS CROSS SECTION DIMENSIONS AND COMPARISON CHART

Conventional & Narrow Section V-Belts *(Dimensional Data)*

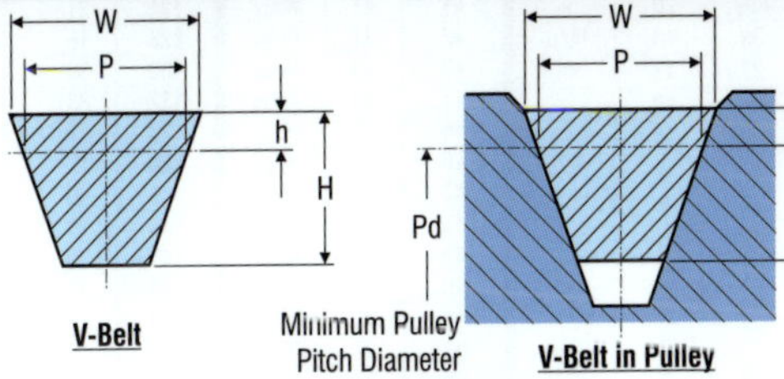

COMPARISON CHART

Dimensions in inches

BS / ISO	Z (ZX)	A (AX)	B (BX)	C (CX)	D (DX)	SPZ (XPZ)		SPA (XPA)	SPB (XPB)		SPC (XPC)	
DIN	10 (X10)	13 (X13)	17 (X17)	22 (X22)	32 (X32)	SPZ (XPZ)		SPA (XPA)	SPB (XPB)		SPC (XPC)	
RMA / MPTA		A (AX)	B (BX)	C (CX)	D (DX)		3V (3VX)			5V (6VX)		8V (8VX)
Top of Belt Width [W]	25/64	1/2	21/32	7/8	1 1/4	25/64	3/8	1/2	41/64	5/8	55/64	1
Pitch Width [P]	11/32	7/16	9/16	3/4	1 1/16	11/32	11/32	7/16	9/16	9/16	3/4	59/64
Height of Belt [H]	15/64	5/16	13/32	17/32	3/4	5/16	5/16	25/64	35/64	17/32	45/64	7/8
Pitch Height [h]	5/64	7/64	11/64	7/32	17/64	5/64	5/64	7/64	11/64	11/64	7/32	1/8
Min. Pulley Pitch Dia. [Pd]	2 (1 9/16)	3 (2 1/2)	4 1/2 (3 1/2)	7 (5 1/2)	14 (10)	2 1/2 (2 3/16)	2 1/2 (2 3/16)	3 1/2 (3)	5 1/2 (4 1/2)	5 1/2 (4 1/2)	8 3/8 (7)	12 1/2 (11 1/2)
Max. Belt Speed (ft/sec)	100	100	100	100	100	130	130	130	130	130	130	130

TENSIONING FORCES

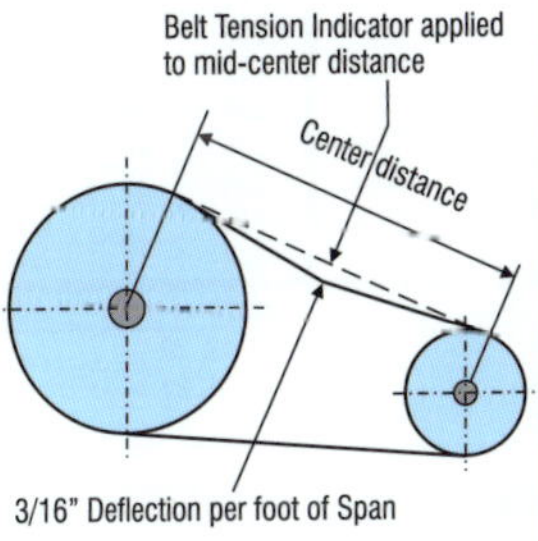

Belt Section	Force required to deflect belt 3/16" per foot of span: Small Pulley Dia. (inch)	Force required to deflect belt 3/16" per foot of span: Pound - Force (lbf)
Z	2 1/4 to 4	1.12 to 1.69
A	3 1/4 to 5 1/2	2.25 to 3.37
B	5 to 8	4.50 to 6.75
C	8 to 16	9.00 to 13.50
D	14 to 24	15.75 to 23.60
SPZ/3V	2 1/4 to 4	2.90 to 4.50
	4 to 5 1/2	4.50 to 5.60
SPA	3 1/4 to 5 1/4	5.60 to 7.87
	5 1/2 to 8	7.87 to 10.12
SPB/5V	4 1/2 to 9	10.12 to 14.61
	9 1/4 to 12 1/2	14.61 to 19.11
SPC	8 3/4 to 14	19.11 to 25.85
	14 1/4 to 22	25.85 to 33.72

DIVIDING HEAD - INDEXING CHART
WORM GEAR RATIO 40:1

No. of Divisions	Hole Circle	Complete Turns	Fraction of a Turn
2	Any	20	–
3	39	13	13/39
3	27	13	9/27
4	Any	10	–
5	Any	8	–
6	39	6	26/39
6	27	6	18/27
7	49	5	35/49
7	21	5	15/21
8	Any	5	–
9	27	4	12/27
9	18	4	8/18
10	Any	4	–
11	33	3	21/33
12	39	3	13/39
12	27	3	9/27
13	39	3	3/39
14	49	2	42/49
14	21	2	18/21
15	39	2	26/39
15	27	2	18/27
16	20	2	10/20
17	17	2	6/17
18	27	2	6/27
18	18	2	4/18
19	19	2	2/19
20	Any	2	–
21	21	1	19/21
22	33	1	27/33
23	23	1	17/23
24	39	1	26/39
24	27	1	18/27
25	20	1	12/20
26	39	1	21/39
27	27	1	13/27
28	49	1	21/49
28	21	1	9/21
29	29	1	11/29
30	39	1	13/39
30	27	1	9/27
31	31	1	9/31
32	20	1	5/20
33	33	1	7/33
34	17	1	3/17
35	49	1	7/49
35	21	1	3/21
36	27	1	3/27
37	37	1	3/37
38	19	1	1/19
39	39	1	1/39
40	Any	1	–

No. of Divisions	Hole Circle	Fraction of a Turn
41	41	40/41
42	21	20/21
43	43	40/43
44	33	30/33
45	27	24/27
46	23	20/23
47	47	40/47
48	18	15/18
49	49	40/49
50	20	16/20
52	39	30/39
54	27	20/27
55	33	24/33
56	49	35/49
58	29	20/29
60	39	26/39
60	27	18/27
62	31	20/31
64	16	10/16
65	39	24/39
66	33	20/33
68	17	10/17
70	49	28/49
72	27	15/27
74	37	20/37
75	15	8/15
76	19	10/19
78	39	20/39
80	20	10/20
82	41	20/41
84	21	10/21
85	17	8/17
86	43	20/43
88	33	15/33
90	27	12/27
92	23	10/23
94	47	20/47
95	19	8/19
98	49	20/49
100	20	8/20
104	39	15/39
105	21	8/21
115	23	8/23
116	29	10/29
120	39	13/39
120	27	9/27

No. of Divisions	Hole Circle	Fraction of a Turn
124	31	10/31
128	16	5/16
130	39	12/39
132	33	10/33
135	27	8/27
136	17	5/17
140	49	14/49
144	18	5/18
145	29	8/29
148	37	10/37
150	15	4/15
152	19	5/19
155	31	8/31
156	39	10/39
160	20	5/20
164	41	10/41
165	33	8/33
168	21	5/21
170	17	4/17
172	43	10/43
180	27	6/27
185	37	8/37
188	47	10/47
190	19	4/19
195	39	8/39
196	49	10/49
200	20	4/20
205	41	8/41
210	21	4/21
215	43	8/43
216	27	5/27
220	33	6/33
230	23	4/23
232	29	5/29
235	47	8/47
240	18	3/18
245	49	8/49
248	31	5/31
260	39	6/39
264	33	5/33
270	27	4/27
280	49	7/49
290	29	4/29
296	37	5/37
300	15	2/15
360	18	2/18

Index Plate	Number of Holes
A	15, 16, 17, 18, 19, 20
B	21, 23, 27, 29, 31, 33
C	37, 39, 41, 43, 47, 49

OR

Index Plate	Number of Holes
No. 1	15, 18, 20, 23, 27, 31, 37, 41, 47
No. 2	16, 17, 19, 21, 29, 33, 39, 43, 49

ROTARY TABLE - INDEXING CHART
WORM GEAR RATIO 90:1

No. of Divisions	Hole Circle	Complete Turns	Fraction of a Turn	No. of Divisions	Hole Circle	Complete Turns	Fraction of a Turn	No. of Divisions	Hole Circle	Complete Turns	Fraction of a Turn
2	Any	45	–	**32**	32	2	26/32	**65**	91	1	35/91
3	Any	30	–	**33**	99	2	72/99	**66**	44	1	16/44
4	26	22	13/26	**34**	34	2	22/34	**66**	99	1	36/99
4	28	22	14/28	**35**	28	2	16/28	**67**	67	1	23/67
5	Any	18	–	**35**	63	2	36/63	**68**	34	1	11/34
6	Any	15	–	**36**	26	2	13/26	**69**	46	1	14/46
7	28	12	24/28	**36**	28	2	14/28	**69**	69	1	21/69
7	77	12	66/77	**37**	37	2	16/37	**70**	28	1	8/28
8	28	11	7/28	**38**	38	2	14/38	**70**	63	1	18/63
8	44	11	11/44	**39**	26	2	8/26	**71**	71	1	19/71
9	Any	10	–	**39**	91	2	28/91	**72**	32	1	8/32
10	Any	9	–	**40**	28	2	7/28	**72**	44	1	11/44
11	44	8	8/44	**40**	44	2	11/44	**73**	73	1	17/73
11	77	8	14/77	**41**	41	2	8/41	**74**	37	1	8/37
12	26	7	13/26	**42**	28	2	4/28	**75**	30	1	6/30
12	28	7	14/28	**42**	63	2	9/63	**76**	38	1	7/38
13	26	6	24/26	**43**	43	2	4/43	**77**	77	1	13/77
13	91	6	84/91	**44**	44	2	2/44	**78**	39	1	6/39
14	28	6	12/28	**45**	Any	2	–	**78**	91	1	14/91
14	77	6	33/77	**46**	46	1	44/46	**79**	79	1	11/79
15	Any	6	–	**46**	69	1	66/69	**80**	32	1	4/32
16	32	5	20/32	**47**	47	1	43/47	**81**	63	1	7/63
17	34	5	10/34	**48**	32	1	28/32	**81**	81	1	9/81
18	Any	5	–	**49**	49	1	41/49	**82**	41	1	4/41
19	38	4	28/38	**50**	30	1	24/30	**83**	83	1	7/83
20	26	4	13/26	**51**	34	1	26/34	**84**	28	1	2/28
20	28	4	14/28	**52**	26	1	19/26	**85**	34	1	2/34
21	28	4	8/28	**53**	53	1	37/53	**86**	43	1	2/43
21	77	4	22/77	**54**	30	1	20/30	**87**	87	1	3/87
22	44	4	4/44	**54**	63	1	42/63	**88**	44	1	1/44
22	77	4	7/77	**55**	44	1	28/44	**89**	89	1	1/89
23	46	3	42/46	**55**	77	1	49/77	**90**	Any	1	–
23	69	3	63/69	**56**	28	1	17/28	**91**	91	–	90/91
24	28	3	21/28	**57**	38	1	22/38	**92**	46	–	45/46
24	44	3	33/44	**58**	87	1	48/87	**90**	93	–	90/93
25	30	3	18/30	**59**	59	1	31/59	**94**	47	–	45/47
26	26	3	12/26	**60**	34	1	17/34	**95**	38	–	36/38
26	91	3	42/91	**60**	32	1	16/32	**96**	32	–	30/32
27	30	3	10/30	**61**	61	1	29/61	**97**	97	–	90/97
27	63	3	21/63	**62**	93	1	42/93	**98**	49	–	45/49
28	28	3	6/28	**63**	49	1	21/49	**99**	44	–	40/44
29	87	3	9/87	**63**	77	1	33/77	**99**	99	–	90/99
30	Any	3	–	**64**	32	1	13/32	**100**	30	–	27/30
31	93	2	84/93	**65**	26	1	10/26				

Index Plate	Number of Holes
A	26, 28, 30, 32, 34, 37, 38, 39, 41, 43, 44, 46, 47, 49, 51, 53, 57, 59
B	61, 63, 67, 69, 71, 73, 77, 79, 81, 83, 87, 89, 91, 93, 97, 99

GEOMETRICAL CONSTRUCTION

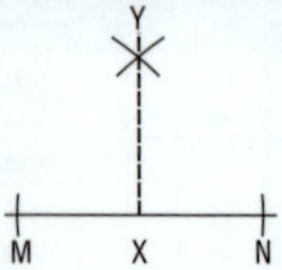

To construct a perpendicular to a line at point X

- Draw arcs on both sides of point **X** having the same radius.
- Draw intersecting arcs from **M** and **N**.
- Connect point of intersection **Y** with point **X**.
- **XY** is the perpendicular of **MN**.

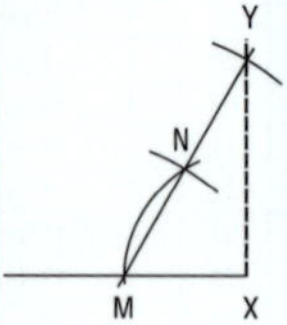

To construct a perpendicular to a straight line at its end point X

- With **X** as center, draw a sufficiently large arc cutting the base line at one end at **M** extending upwards.
- With **M** as center, draw an arc to cut the former at **N**.
- With **N** as center and with the same radius, draw an arc in the same direction.
- Draw a straight line through **M** and **N** to cut the last arc at **Y**.
- The line **XY** is perpendicular to **XM** at point **X**.

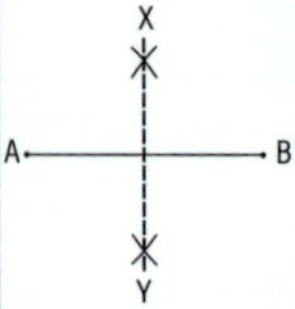

To bisect a line

- With any radius greater than one-half **AB**, draw two arcs from **A** on either side of the line **AB**. Similarly, using the same radius, draw two arcs from **B** intersecting the arcs drawn from point **A**. Draw a line joining points **X** and **Y**.
- The line **XY** is the perpendicular bisector of the line **AB**.

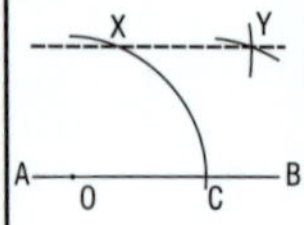

To construct a parallel line to AB through point X

- With **O** as center, at any point on **AB**, draw a large arc through point **X**, extending downwards to cut the line **AB** at **C**.
- With **X** and **C** as centers and with the same radius, draw arcs to intersect at **Y**.
- A straight line drawn through points **X** and **Y** is parallel to **AB**.

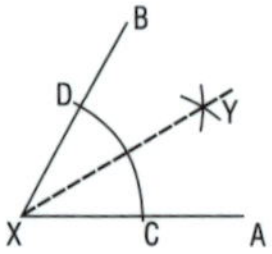

To bisect an angle

- With **X** as center, draw an arc intersecting **XA** and **XB** at **C** and **D** respectively.
- With **C** and **D** as centers, draw arcs of equal radii intersecting at **Y**.
- **XY** is the required bisector.

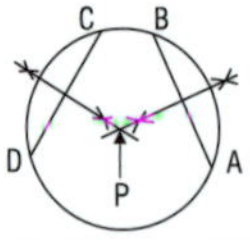

To locate the center of a circle/arc

- Draw any two chords **AB** and **CD** in the circle.
- The intersecting point **P** of the perpendicular bisectors of the two chords is the center point required.

GEOMETRICAL CONSTRUCTION

Square / Octagon

- Draw two lines **LN**, **MO** at right angles and a circle whose center is the point of intersection of these two lines.
- Join the points **LMNO** where the circle cuts the lines, to get the square.
- By drawing the two diagonals which bisect the lines **LM**, **MN**, **NO** and **OL**, the other four corners required to complete the regular octagon can be constructed.

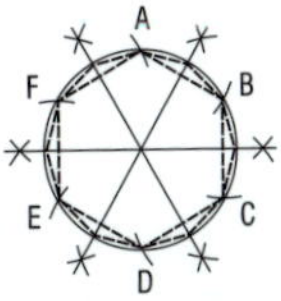

Hexagon / Dodecagon

- Starting at point **A**, mark 6 arcs around the circle equal to the radius of the circle.
- These points are the vertices of the required Hexagon.
- Draw lines bisecting **AB**, **BC** & **CD**, through the center of the circle. These points together with the vertices of the constructed hexagon, form the vertices of the dodecagon.

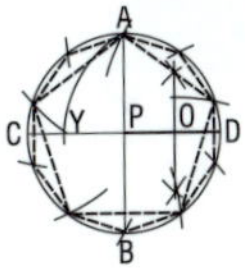

Pentagon / Decagon

- Draw two diameter lines **AB** & **CD** intersecting at right angles at the center of the circle.
- Bisect radius **PD** at **O**. With **O** as center and **OA** as radius, mark an arc cutting the horizontal line **CD** at **Y**.
- Length **AY** is the side length of the Pentagon which can be marked off along the circumference.
- To construct a Decagon, mark 5 additional arcs starting from point **B**.

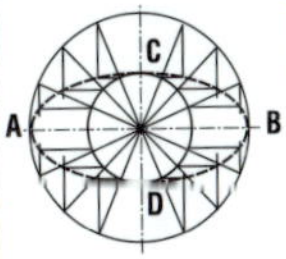

Ellipse by means of two circles

- Draw two concentric circles with diameters **AB** and **CD** representing the major and minor axis of the ellipse to be constructed. Draw additional diameters.
- Now draw vertical lines starting from the points of intersection of these diameters with the circumference of bigger circle.
- Then draw equal number of horizontal lines starting from the points of intersection of the diameters with the circumference of the smaller circle to meet the vertical lines.
- Joining these intersecting points will complete the ellipse.

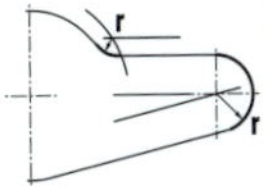

Curvature

- Draw lines parallel to the curvature at a distance = **r**.
- The point of intersection is the center of the required radii.

USEFUL FORMULAE FOR FINDING DIMENSIONS AND AREA OF PLANE FIGURES

Square

$c = D \times 0.14645$

$D = s \times 1.4142 = 1.414\sqrt{A}$

$s = D \times 0.7071 = \sqrt{A}$

Area A $= s^2 = \dfrac{D^2}{2}$

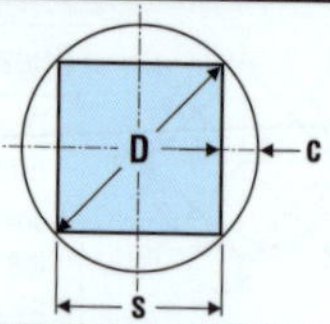

Rectangle

$a = \sqrt{d^2 - b^2} = A \div b$

$b = \sqrt{d^2 - a^2} = A \div a$

$d = \sqrt{a^2 + b^2}$

Area A $= ab = a\sqrt{d^2 - a^2} = b\sqrt{d^2 - b^2}$

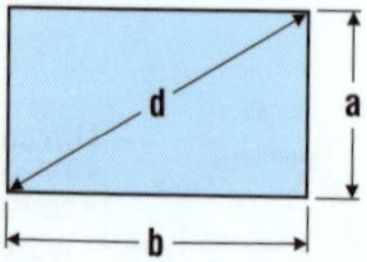

Equilateral Triangle

$c = d \times 0.5 = D \times 0.25$

$D = s \times 1.1547 = 2d$

$d = s \times 0.57735$

$h = s \times 0.8660255$

$s = \dfrac{D}{1.1547}$

Area A $= \left(\dfrac{s^2}{4}\right) \times \sqrt{3} = s^2 \times 0.43301275 = \dfrac{s \times h}{2}$

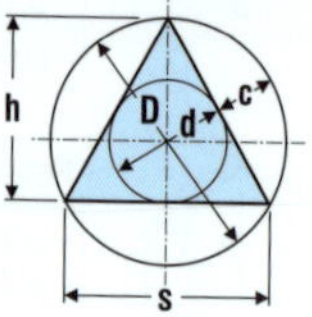

Right Angle Triangle

(also refer pages 70 & 71)

$a = \sqrt{b^2 + c^2}$

$b = \sqrt{a^2 - c^2}$

$c = \sqrt{a^2 - b^2}$

Area A $= \dfrac{b \times c}{2}$

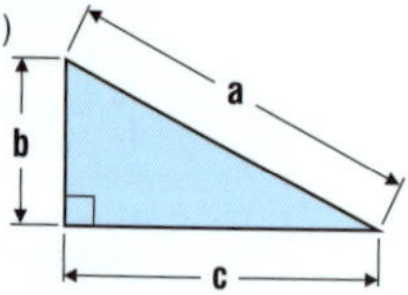

Acute Angle Triangle

Area A $= \dfrac{b \times h}{2}$

Also **Area A** $= \sqrt{S(S-a)(S-b)(S-c)}$

where $S = \dfrac{(a+b+c)}{2}$

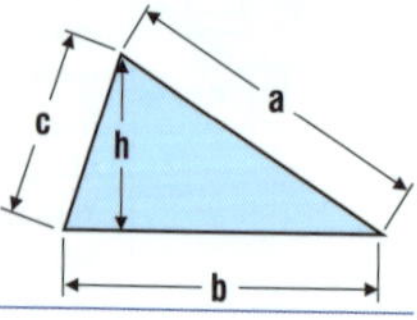

Obtuse Angle Triangle

Area A $= \dfrac{b \times h}{2}$

Also **Area A** $= \sqrt{S(S-a)(S-b)(S-c)}$

where $S = \dfrac{(a+b+c)}{2}$

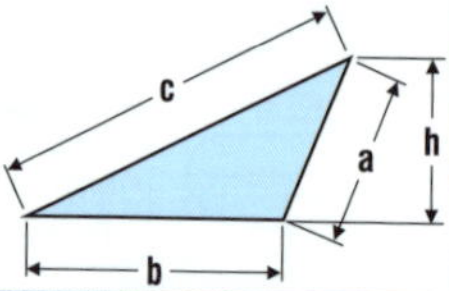

USEFUL FORMULAE FOR FINDING DIMENSIONS AND AREA OF PLANE FIGURES

Circle

Circumference C $= 2\pi r = 6.2832r = 3.1416d$

$D = C \div 3.1416 = A \div 0.7854 = 1.128\sqrt{A}$

$r = C \div 6.2832 = A \div 3.1416 = 0.564\sqrt{A}$

Area A $= \pi r^2 = 3.1416r^2$

Also **Area A** $= \dfrac{\pi D^2}{4} = 0.7854D^2$

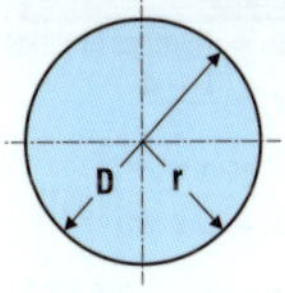

Hollow Circle

Area A $= 0.785(D^2 - d^2)$

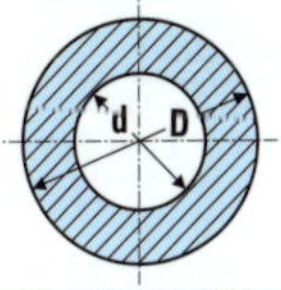

Sector of a Circle

Arc length $b = \dfrac{\pi \times D \times \theta}{360}$

Area A $= \dfrac{b \times r}{2}$

θ = decimal degrees

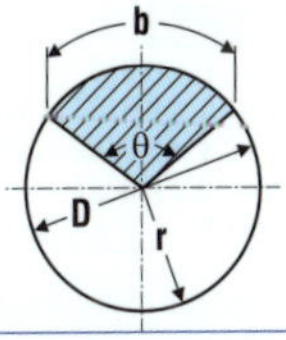

Segment of a Circle

Area A $= \dfrac{\pi \times r^2 \times \theta}{360} - \dfrac{s(r-h)}{2}$

where $s = 2r \times \sin\left(\dfrac{\theta}{2}\right)$

and $h = r\left(1 - \cos\left(\dfrac{\theta}{2}\right)\right)$

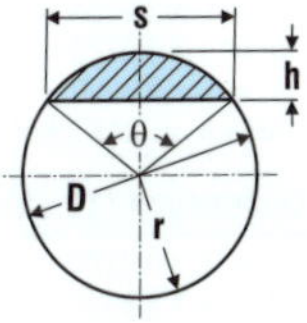

Trapezium / Trapezoid

Area A $= \dfrac{a+b}{2} \times h$

where $a = \dfrac{2A}{h} - b$, $b = \dfrac{2A}{h} - a$, $h = \dfrac{2A}{a+b}$

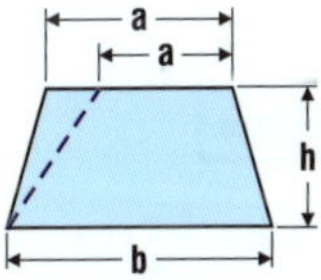

Parallelogram

Area A $= b \times h$

where $b = \dfrac{A}{h}$ $h = \dfrac{A}{b}$

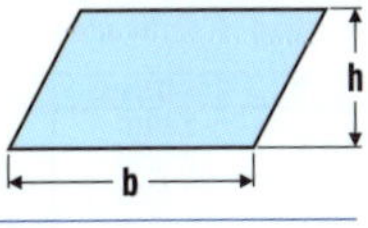

USEFUL FORMULAE FOR FINDING DIMENSIONS AND AREA OF PLANE FIGURES

Regular Polygon

n = Number of Sides

$c = \frac{D - d}{2}$

$D = \sqrt{d^2 + s^2}$ $\quad d = \sqrt{D^2 - s^2}$

$s = D \times \sin\left(\frac{180^\circ}{n}\right)$

Area A $= \frac{n \times s \times d}{4}$

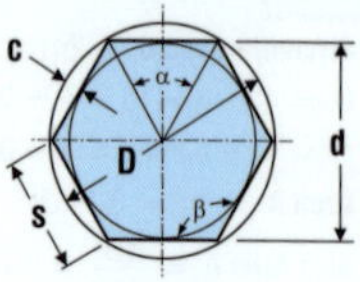

Hexagon

$c = D \times 0.067$
$D = s \times 2 = d \times 1.1547$
$d = s \times 1.7321 = D \times 0.866$
$s = D \times 0.5$
Area A $= s^2 \times 2.5981$

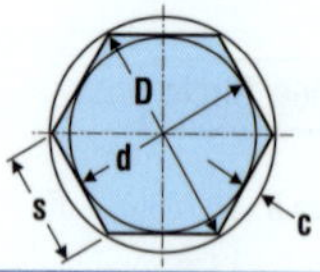

Octagon

$c = D \times 0.038$
$D = s \times 2.6131 = d \times 1.0824$
$d = s \times 2.4142 = D \times 0.9239$
$s = D \times 0.3827$
Area A $= s^2 \times 4.8284$

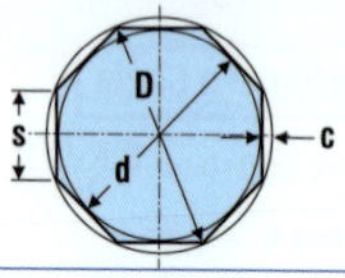

Decagon (10 sides)

$c = D \times 0.0245$
$D = s \times 3.2360 = d \times 1.0515$
$d = s \times 3.0776 = D \times 0.9511$
$s = D \times 0.3090$
Area A $= s^2 \times 7.6942$

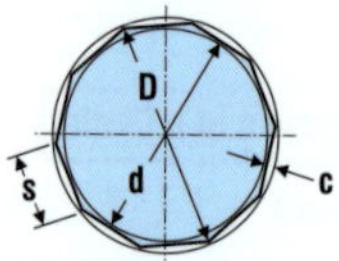

Dodecagon (12 sides)

$c = D \times 0.0170$
$D = s \times 3.8637 = d \times 1.0353$
$d = s \times 3.7320 = D \times 0.9659$
$s = D \times 0.2588$
Area A $= s^2 \times 11.1961$

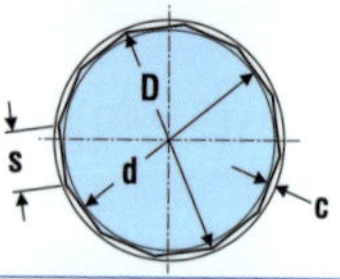

Ellipse

Area A $= \frac{\pi}{4} \times D \times d = 0.785 \times D \times d$

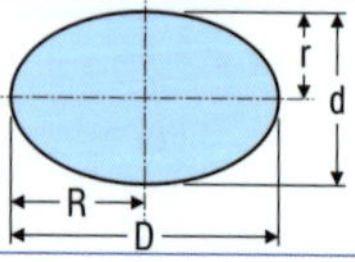

Circumference C can be calculated and depends on the ratio d : D (Refer table)

d : D	C = D times	d : D	C = D times
0.2	2.1010	0.6	2.5527
0.3	2.1930	0.7	2.6912
0.4	2.3013	0.8	2.9361
0.5	2.4221	0.9	2.9866

Example : To calculate C when d = 10 inches, D = 20 inches
ratio d : D = 10 ÷ 20 = 0.5

$\therefore C = D \times \text{ratio}$
$\therefore C = D \times 2.4221$
$= 20 \times 2.4221 = 48.44$ inches

DISTANCE ACROSS CORNERS OF SQUARES AND HEXAGONS

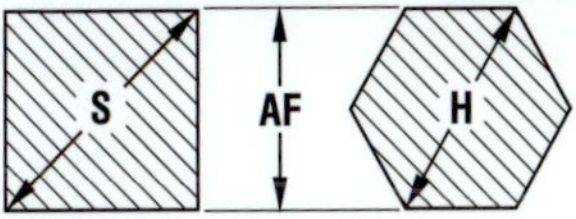

$S = 1.414214 \times AF$

$H = 1.154701 \times AF$

INCH			METRIC					
AF (inch)	S (inch)	H (inch)	AF (mm)	S (mm)	H (mm)	AF (mm)	S (mm)	H (mm)
1/32"	0.0442	0.0361	1	1.4142	1.1547	41	57.9828	47.3427
1/16"	0.0884	0.0722	2	2.8284	2.3094	42	59.3970	48.4974
3/32"	0.1326	0.1083	3	4.2426	3.4641	43	60.8112	49.6521
1/8"	0.1768	0.1443	4	5.6569	4.6188	44	62.2254	50.8068
5/32"	0.2210	0.1804	5	7.0711	5.7735	45	63.6396	51.9615
3/16"	0.2652	0.2165	6	8.4853	6.9282	46	65.0538	53.1162
7/32"	0.3094	0.2526	7	9.8995	8.0829	47	66.4681	54.2709
1/4"	0.3536	0.2887	8	11.3137	9.2376	48	67.8823	55.4256
9/32"	0.3977	0.3248	9	12.7279	10.3923	49	69.2965	56.5803
5/16"	0.4419	0.3608	10	14.1421	11.5470	50	70.7107	57.7351
11/32"	0.4861	0.3969	11	15.5564	12.7017	51	72.1249	58.8898
3/8"	0.5303	0.4330	12	16.9706	13.8564	52	73.5391	60.0445
13/32"	0.5745	0.4691	13	18.3848	15.0111	53	74.9533	61.1992
7/16"	0.6187	0.5052	14	19.7990	16.1658	54	76.3676	62.3539
15/32"	0.6629	0.5413	15	21.2132	17.3205	55	77.7818	63.5086
1/2"	0.7071	0.5774	16	22.6274	18.4752	56	79.1960	64.6633
17/32"	0.7513	0.6134	17	24.0416	19.6299	57	80.6102	65.8180
9/16"	0.7955	0.6495	18	25.4559	20.7846	58	82.0244	66.9727
19/32"	0.8397	0.6856	19	26.8701	21.9393	59	83.4386	68.1274
5/8"	0.8839	0.7217	20	28.2843	23.0940	60	84.8528	69.2821
21/32"	0.9281	0.7578	21	29.6985	24.2487	61	86.2671	70.4368
11/16"	0.9723	0.7939	22	31.1127	25.4034	62	87.6813	71.5915
23/32"	1.0165	0.8299	23	32.5269	26.5581	63	89.0955	72.7462
3/4"	1.0607	0.8660	24	33.9411	27.7128	64	90.5097	73.9009
25/32"	1.1049	0.9021	25	35.3554	28.8675	65	91.9239	75.0556
13/16"	1.1490	0.9382	26	36.7696	30.0222	66	93.3381	76.2103
27/32"	1.1932	0.9740	27	38.1838	31.1769	67	94.7523	77.3650
7/8"	1.2374	1.0104	28	39.5980	32.3316	68	96.1666	78.5197
29/32"	1.2816	1.0464	29	41.0122	33.4863	69	97.5808	79.6744
15/16"	1.3258	1.0825	30	42.4264	34.6410	70	98.9950	80.8291
31/32"	1.3700	1.1186	31	43.8406	35.7957	71	100.4092	81.9838
1"	1.4142	1.1547	32	45.2548	36.9504	72	101.8234	83.1385
1 1/8"	1.5910	1.2990	33	46.6691	38.1051	73	103.2376	84.2932
1 1/4"	1.7678	1.4434	34	48.0833	39.2598	74	104.6518	85.4479
1 3/8"	1.9445	1.5877	35	49.4975	40.4145	75	106.0661	86.6026
1 1/2"	2.1213	1.7321	36	50.9117	41.5692	76	107.4803	87.7573
1 5/8"	2.2981	1.8764	37	52.3259	42.7239	77	108.8945	88.9120
1 3/4"	2.4749	2.0207	38	53.7401	43.8786	78	110.3087	90.0667
1 7/8"	2.6517	2.1651	39	55.1543	45.0333	79	111.7229	91.2214
2"	2.8284	2.3094	40	56.5686	46.1880	80	113.1371	92.3761

USEFUL FORMULAE FOR FINDING AREA AND VOLUME OF SOLID FIGURES

Cube

Surface area $= 6 \times s \times s = 6s^2$

Volume $= s \times s \times s = s^3$

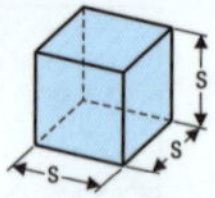

Prism

Surface area $= 2(a \times b) + 2(a \times h) + 2(b \times h)$

Volume $= a \times b \times h$

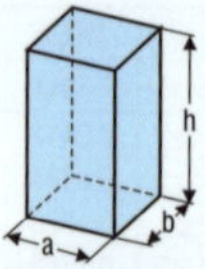

Prism

Surface area $= 2 \times$ area base A $+ (a \times h) + (b \times h) + (c \times h)$

Volume $=$ area base A $\times h$

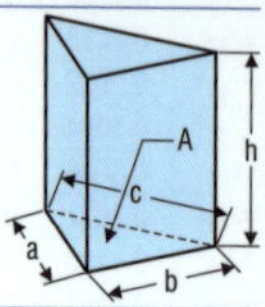

Sphere

Surface area $= \pi \times d^2 = 3.1416d^2$

Volume $= \dfrac{\pi \times d^3}{6} = 0.5236d^3$

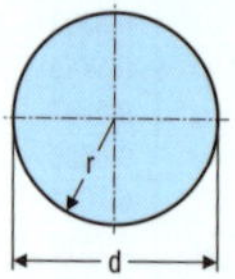

Spherical Segment

Surface area of segment $= 2\pi rh$

Volume $= \pi \times h^2 \left(r - \dfrac{h}{3}\right)$ or $= \pi \times h \left(\dfrac{s^2}{8} + \dfrac{h^2}{6}\right)$

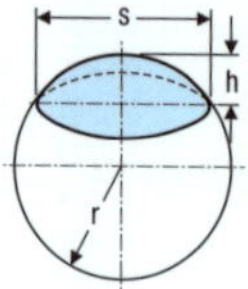

Cone

Surface area $= \dfrac{\pi}{4} \times D^2 (D + 2s)$

Volume $= \dfrac{\pi}{4} \times D^2 \times \dfrac{h}{3}$

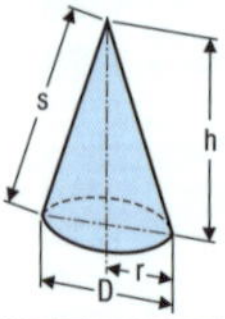

Frustum of a Cone

Surface area $= 0.7854D^2 + 0.7854d^2 + \pi \times \dfrac{D + d}{2} \times s$

Volume $= \dfrac{\pi}{12} \times h (D^2 + D \times d + d^2)$

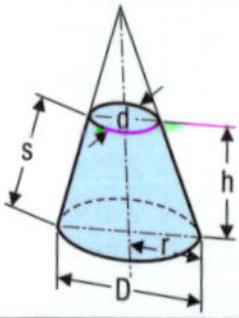

USEFUL FORMULAE FOR FINDING AREA AND VOLUME OF SOLID FIGURES

Cylinder

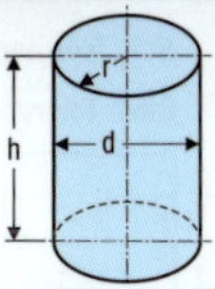

Surface area $= \pi \times d \times h + 2\pi r^2$

Volume $= \frac{\pi}{4} \times d^2 \times h$ *or* $= \pi r^2 h$

Frustum of a Cylinder

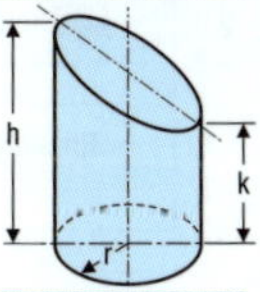

Area of cylindrical surface $= \pi \times r \times (h + k)$

Volume $= \frac{\pi}{2} \times r^2 \times (h + k)$

Pyramid (Square / Rectangular Base)

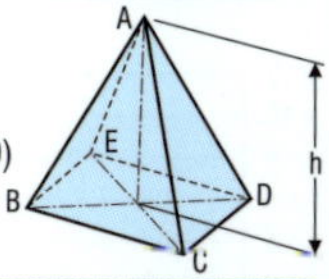

Surface area $= BC \times CD + 2(\text{area } \Delta ABC) + 2(\text{area } \Delta ACD)$

Volume $= \frac{BC \times CD \times h}{3}$

Pyramid (Triangular Base)

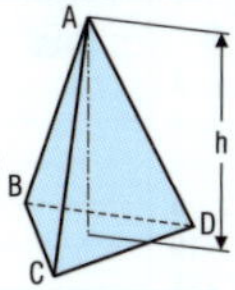

Surf. area $= \text{area } \Delta BCD + \text{area } \Delta ABC + \text{area } \Delta ACD + \text{area } \Delta ABD$

Volume $= \frac{\text{area } \Delta BCD \times h}{3}$

Frustum of a Pyramid (Square Base)

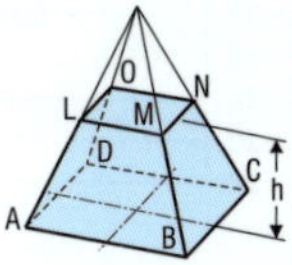

Surface area $= \text{area } ABCD + \text{area } LMNO + 4(\text{area } ABML)$

Volume $= \frac{h}{3} (LM)^2 + LM \times AB + (AB)^2$

Solids of Revolution

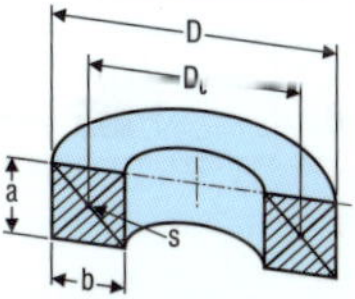

Surface area $= \pi^2 \times D \times D_C$

Volume $= a \times b \times \pi \times D_C$

NOTE: s = center of gravity of cross-section
D_0 = diameter of circle of center of gravity

Ring of Circular Cross-Section

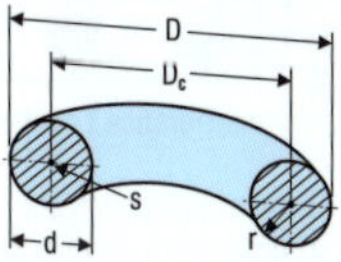

Surface area $= \pi^2 \times d \times D_C$

Volume $= \frac{\pi}{4} \times d^2 \times \pi \times D_C$

NOTE: s = center of gravity of cross-section
D_C = diameter of circle of center of gravity

GRAPHICAL REPRESENTATION OF ANSI STANDARD LIMITS AND FITS

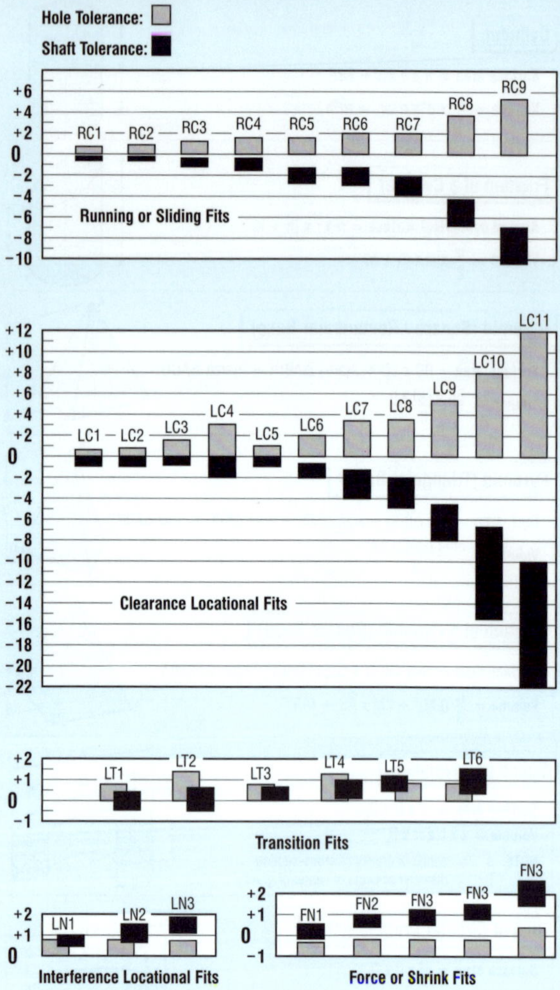

The above diagrams show the placement of hole and shaft tolerances (in thousands of an inch) with respect to basic size (0) for a diameter of 1 inch.

INCH PREFERRED LIMITS AND FITS

ACCORDING TO ANSI B4.1-1967 (R2004)

The following chart covers a selection of Preferred Fits (Hole Basis) for Holes & Shafts which will meet the needs of a large proportion of the requirements for Standard Engineering Components.

In selecting limits of size for any applications, first determine the type of fit required based on use or service required from the equipment being designed, resulting in the limits of size of the mating parts, to ensure that the desired fit will be produced.

Designation		Type of Fit / Application
Class	**Hole / Shaft Tolerance**	**CLEARANCE FITS**
RC8	**H10 / c9**	**LOOSE RUNNING FIT** - Wide commercial tolerances, used where accuracy is not essential.
RC7	**H9 / d8**	**FREE RUNNING FIT** - High running speeds, large temperature variations, heavy pressures and where accuracy is not essential.
RC4	**H8 / f7**	**CLOSE RUNNING FIT** - Accurate location at moderate speeds on accurate machines.
RC2	**H6 / g5**	**SLIDING FIT** - Accurate location where components are intended to move and turn freely but not run freely.
LC2	**H7 / h6**	**LOCATIONAL CLEARANCE FIT** - Snug fit for locating stationary components which need to be freely assembled and disassembled.
		TRANSITION FITS
LT3	**H7 / k6**	**LOCATIONAL TRANSITION FIT** (Tap Fit) - Accurate location where assembly requires gentle persuasion.
LT5	**H7 / n6**	**LOCATIONAL TRANSITION FIT** - More accurate location where assembly permits greater interference.
		INTERFERENCE FITS
LN2	**H7 / p6**	**LOCATIONAL INTERFERENCE** - Prime accuracy of location where assembly requires alignment and rigidity.
FN2	**H7 / s6**	**MEDIUM DRIVE FIT (Press Fit)** - For assembly of steel parts and shrink fits on light sections.
FN4	**H7 / u6**	**FORCE FIT (Press Fit)** - For assembly of components where high pressures and stresses are permitted.

NOTE : See pages 116 & 117 for **INCH** tolerances.

INCH PREFERRED LIMITS AND FITS
ACCORDING TO ANSI B4.1-1967 (R2004)

The following chart covers a simple selection of Preferred Fits (Hole Basis) for Holes & Shafts which will meet the needs of a large proportion of the requirements for Standard Engineering Components (Refer page 115 for Fit Designations).

To scale for diameter range 0.71" to 1.19"	Type of Fit – CLEARANCE			
	Free Running Class RC7	Close Running Class RC4	Sliding Class RC2	Locational Clearance Class LC2

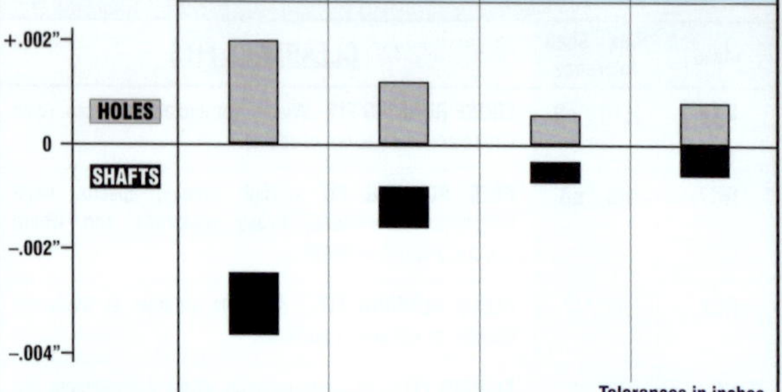

Tolerances in inches

Nominal Dia. (inch)		Hole	Shaft	Hole	Shaft	Hole	Shaft	Hole	Shaft
Above	Upto & incl.	H9	d8	H8	f7	H6	g5	H7	h6
0	0.12	+.0010 .0000	–.0010 –.0016	+.0006 .0000	–.0003 –.0007	+.00025 .00000	–.0001 –.0003	+.0004 .0000	.00000 –.00025
0.12	0.24	+.0012 .0000	–.0012 –.0019	+.0007 .0000	–.0004 –.0009	+.0003 .0000	–.00015 –.00035	+.0005 .0000	.0000 –.0003
0.24	0.40	+.0014 .0000	–.0016 –.0025	+.0009 .0000	–.0005 –.0011	+.0004 .0000	–.00020 –.00045	+.0006 .0000	.0000 –.0004
0.40	0.71	+.0016 .0000	–.0020 –.0030	+.0010 .0000	–.0006 –.0013	+.0004 .0000	–.00025 –.00055	+.0007 .0000	.0000 –.0004
0.71	1.19	+.002 .0000	–.0025 –.0037	+.0012 .0000	–.0008 –.0016	+.0005 .0000	–.0003 –.0007	+.0008 .0000	.0000 –.0005
1.19	1.97	+.0025 .0000	–.0030 –.0046	+.0016 .0000	–.0010 –.0020	+.0006 .0000	–.0004 –.0008	+.0010 .0000	.0000 –.0006
1.97	3.15	+.003 .0000	–.0040 –.0058	+.0018 .0000	–.0012 –.0024	+.0007 .0000	–.0004 –.0009	+.0012 .0000	.0000 –.0007
3.15	4.73	+.0035 .0000	–.0050 –.0072	+.0022 .0000	–.0014 –.0028	+.0009 .0000	–.0005 –.0011	+.0014 .0000	.0000 –.0009
4.73	7.09	+.004 .0000	–.0060 –.0085	+.0025 .0000	–.0016 –.0032	+.0010 .0000	–.0006 –.0013	+.0016 .0000	.0000 –.0010
7.09	9.85	+.0045 .0000	–.0070 –.0098	+.0028 .0000	–.0020 –.0038	+.0012 .0000	–.0006 –.0014	+.0018 .0000	.0000 –.0012
9.85	12.41	+.005 .0000	–.0080 –.0110	+.0030 .0000	–.0025 –.0045	+.0012 .0000	–.0008 –.0017	+.0020 .0000	.0000 –.0012
12.41	15.75	+.006 .0000	–.0100 –.0135	+.0035 .0000	–.0030 –.0052	+.0014 .0000	–.0010 –.0020	+.0022 .0000	.0000 –.0014
15.75	19.69	+.006 .0000	–.0120 –.0160	+.0040 .0000	–.0040 –.0065	+.0016 .0000	–.0012 –.0022	+.0025 .0000	.0000 –.0016

INCH PREFERRED LIMITS AND FITS
ACCORDING TO ANSI B4.1-1967 (R2004)

The following chart covers a simple selection of Preferred Fits (Hole Basis) for Holes & Shafts which will meet the needs of a large proportion of the requirements for Standard Engineering Components (Refer page 115 for Fit Designations).

To scale for diameter range 1.19" to 1.97"

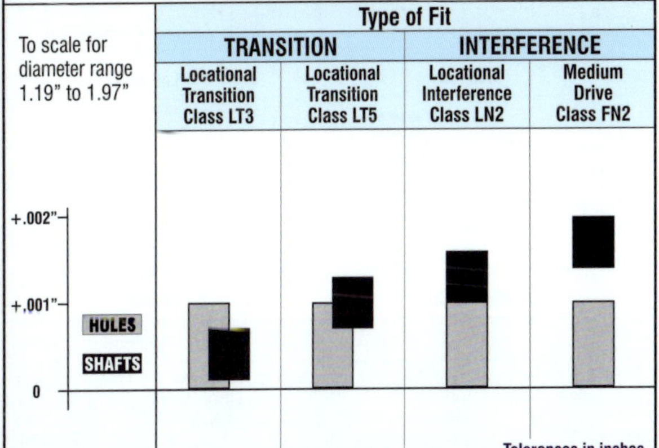

Tolerances in inches

Nominal Dia. (inch)		TRANSITION: Locational Transition Class LT3		TRANSITION: Locational Transition Class LT5		INTERFERENCE: Locational Interference Class LN2		INTERFERENCE: Medium Drive Class FN2	
Above	Upto & incl.	Hole H7	Shaft k6	Hole H7	Shaft n6	Hole H7	Shaft p6	Hole H7	Shaft s6
0	0.12	– –	– –	+.0004 .0000	+.00050 +.00025	+.0004 .0000	+.00065 +.00040	+.0004 .0000	+.00085 +.00060
0.12	0.24	– –	– –	+.0005 .0000	+.0006 +.0003	+.0005 .0000	+.0008 +.0005	+.0005 .0000	+.0010 +.0007
0.24	0.40	+.0006 .0000	+.0005 +.0001	+.0006 .0000	+.0008 +.0004	+.0006 .0000	+.0010 +.0006	+.0006 .0000	+.0014 +.0010
0.40	0.71	+.0007 .0000	+.0005 +.0001	+.0007 .0000	+.0009 +.0005	+.0007 .0000	+.0011 +.0007	+.0007 .0000	+.0016 +.0012
0.71	1.19	+.0008 .0000	+.0006 +.0001	+.0008 .0000	+.0010 +.0006	+.0008 .0000	+.0013 +.0008	+.0008 .0000	+.0016 +.0012
1.19	1.97	+.0010 .0000	+.0007 +.0001	+.0010 .0000	+.0013 +.0007	+.0010 .0000	+.0016 +.0010	+.0010 .0000	+.0019 +.0014
1.97	3.15	+.0012 .0000	+.0008 +.0001	+.0012 .0000	+.0015 +.0008	+.0012 .0000	+.0021 +.0014	+.0012 .0000	+.0019 +.0014
3.15	4.73	+.0014 .0000	+.0010 +.0001	+.0014 .0000	+.0019 +.0010	+.0014 .0000	+.0025 +.0016	+.0014 .0000	+.0024 +.0018
4.73	7.09	+.0016 .0000	+.0011 +.0001	+.0016 .0000	+.0022 +.0012	+.0016 .0000	+.0028 +.0010	+.0016 .0000	+.0024 +.0018
7.09	9.85	+.0018 .0000	+.0014 +.0002	+.0018 .0000	+.0026 +.0014	+.0018 .0000	+.0032 +.0020	+.0018 .0000	+.0027 +.0020
9.85	12.41	+.0020 .0000	+.0014 +.0002	+.0020 .0000	+.0026 +.0014	+.0020 .0000	+.0034 +.0022	+.0020 .0000	+.0029 +.0022
12.41	15.75	+.0022 .0000	+.0016 +.0002	+.0022 .0000	+.0030 +.0016	+.0022 .0000	+.0039 +.0025	+.0022 .0000	+.0037 +.0028
15.75	19.69	+.0025 .0000	+.0018 +.0002	+.0025 .0000	+.0034 +.0018	+.0025 .0000	+.0044 +.0028	+.0025 .0000	+.0039 +.0030

FLANGE DETAILS
ACCORDING TO ANSI B 16.5

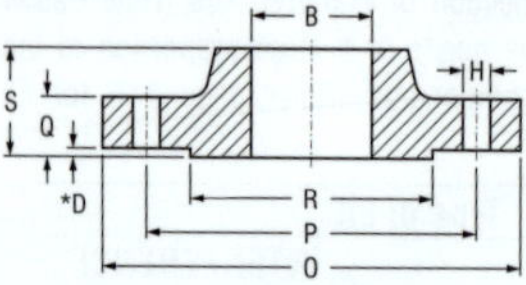

SLIP-ON

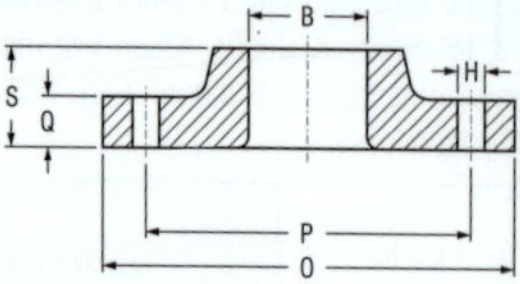

LAP JOINT

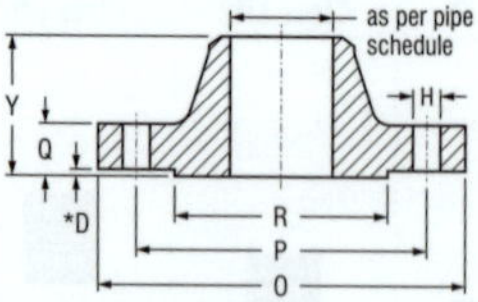

WELD NECK

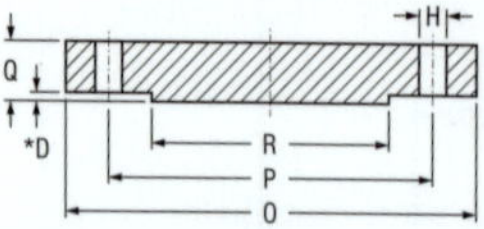

BLIND

ANSI 150

Dimensions in inches

Pipe Size (inch)	Slip on Bore B	Outside Dia. O	Raised Face Dia. R	Thickness Q	Weld Neck Y	Slip On S	Bolt Circle Dia. P	No. of Holes N	Hole Dia. H	Stud Bolt x Length RF (Steel Flanges)
1/2"	0.88	3.50	1.38	0.44	1.88	0.62	2.38	4	5/8	1/2 x 2 1/2
3/4"	1.09	3.88	1.69	0.50	2.06	0.62	2.75	4	5/8	1/2 x 2 1/2
1"	1.36	4.25	2.00	0.56	2.19	0.69	3.12	4	5/8	1/2 x 2 3/4
1 1/4"	1.70	4.62	2.50	0.62	2.25	0.81	3.50	4	5/8	1/2 x 3
1 1/2"	1.95	5.00	2.88	0.68	2.44	0.88	3.88	4	5/8	1/2 x 3
2"	2.44	6.00	3.62	0.75	2.50	1.00	4.75	4	3/4	5/8 x 3 1/4
2 1/2"	2.94	7.00	4.12	0.88	2.75	1.12	5.50	4	3/4	5/8 x 3 1/2
3"	3.57	7.50	5.00	0.94	2.75	1.19	6.00	4	3/4	5/8 x 3 3/4
3 1/2"	4.07	8.50	5.50	0.94	2.81	1.25	7.00	8	3/4	5/8 x 3 3/4
4"	4.57	9.00	6.19	0.94	3.00	1.31	7.50	8	3/4	5/8 x 3 3/4
5"	5.66	10.00	7.31	0.94	3.50	1.44	8.50	8	7/8	3/4 x 4
6"	6.72	11.00	8.50	1.00	3.50	1.56	9.50	8	7/8	3/4 x 4
8"	8.72	13.50	10.62	1.12	4.00	1.75	11.75	8	7/8	3/4 x 4 1/4
10"	10.88	16.00	12.75	1.19	4.00	1.94	14.25	12	1	7/8 x 4 3/4
12"	12.88	19.00	15.00	1.25	4.50	2.19	17.00	12	1	7/8 x 4 3/4
14"	14.14	21.00	16.25	1.38	5.00	2.25	18.75	12	1 1/8	1 x 5 1/4
16"	16.16	23.50	18.50	1.44	5.00	2.50	21.25	16	1 1/8	1 x 5 1/2

***D** = 1/16" for ANSI Class 150-300 and ***D** = 1/4" for ANSI Class 400-2500 for Steel Flanges only

FLANGE DETAILS
ACCORDING TO ANSI B 16.5

ANSI 300
Dimensions in inches

Pipe Size (inch)	Slip on Bore B	Outside Dia. O	Raised Face Dia. R	Thickness Q	Weld Neck Y	Slip On S	Bolt Circle Dia. P	No. of Holes N	Hole Dia. H	Stud Bolt x Length RF (Steel Flanges)
1/2"	0.88	3.75	1.38	0.56	2.06	0.88	2.62	4	5/8	1/2 x 2 3/4
3/4"	1.09	4.62	1.69	0.62	2.25	1.00	3.25	4	3/4	5/8 x 3
1"	1.36	4.88	2.00	0.69	2.44	1.06	3.50	4	3/4	5/8 x 3 1/4
1 1/4"	1.70	5.25	2.50	0.75	2.56	1.06	3.88	4	3/4	5/8 x 3 1/4
1 1/2"	1.95	6.12	2.88	0.81	2.69	1.19	4.50	4	7/8	3/4 x 3 3/4
2"	2.44	6.50	3.62	0.88	2.75	1.31	5.00	8	3/4	5/8 x 3 1/2
2 1/2"	2.94	7.50	4.12	1.00	3.00	1.50	5.88	8	7/8	3/4 x 4
3"	3.57	8.25	5.00	1.12	3.12	1.69	6.62	8	7/8	3/4 x 4 1/4
3 1/2"	4.07	9.00	5.50	1.19	3.19	1.75	7.25	8	7/8	3/4 x 4 1/2
4"	4.57	10.00	6.19	1.25	3.38	1.88	7.88	8	7/8	3/4 x 4 1/2
5"	5.66	11.00	7.31	1.38	3.88	2.00	9.25	8	7/8	3/4 x 4 3/4
6"	6.72	12.50	8.50	1.44	3.88	2.06	10.62	12	7/8	3/4 x 5
8"	8.72	15.00	10.62	1.62	4.38	2.44	13.00	12	1	7/8 x 5 1/2
10"	10.88	17.50	12.75	1.88	4.62	2.62	15.25	16	1 1/8	1 x 6 1/4
12"	12.88	20.50	15.00	2.00	5.12	2.88	17.75	16	1 1/4	1 1/8 x 6 3/4
14"	14.14	23.00	16.25	2.12	5.62	3.00	20.25	20	1 1/4	1 1/8 x 7
16"	16.16	25.50	18.50	2.25	5.75	3.25	22.50	20	1 3/8	1 1/4 x 7 1/2

ANSI 400
Dimensions in inches

Pipe Size (inch)	Slip on Bore B	Outside Dia. O	Raised Face Dia. R	Thickness Q	Weld Neck Y	Slip On S	Bolt Circle Dia. P	No. of Holes N	Hole Dia. H	Stud Bolt x Length RF (Steel Flanges)
1/2"	0.88	3.75	1.38	0.56	2.06	0.88	2.62	4	5/8	1/2 x 3 1/4
3/4"	1.09	4.62	1.69	0.62	2.25	1.00	3.25	4	3/4	5/8 x 3 1/2
1"	1.36	4.88	2.00	0.69	2.44	1.06	3.50	4	3/4	5/8 x 3 3/4
1 1/4"	1.70	5.25	2.50	0.81	2.62	1.12	3.88	4	3/4	5/8 x 4
1 1/2"	1.95	6.12	2.88	0.88	2.75	1.25	4.50	4	7/8	3/4 x 4 1/4
2"	2.44	6.50	3.62	1.00	2.88	1.44	5.00	8	3/4	5/8 x 4 1/4
2 1/2"	2.94	7.50	4.12	1.12	3.12	1.62	5.88	8	7/8	3/4 x 4 3/4
3"	3.57	8.25	5.00	1.25	3.25	1.81	6.62	8	7/8	3/4 x 5
3 1/2"	4.07	9.00	5.50	1.38	3.38	1.94	7.25	8	1	7/8 x 5 1/2
4"	4.57	10.00	6.19	1.30	3.50	2.00	7.88	8	1	7/8 x 5 1/2
5"	5.66	11.00	7.31	1.50	4.00	2.12	9.25	8	1	7/8 x 5 3/4
6"	6.72	12.50	8.50	1.62	4.06	2.25	10.62	12	1	7/8 x 6
8"	8.72	15.00	10.62	1.88	4.62	2.69	13.00	12	1 1/8	1 x 6 3/4
10"	10.88	17.50	12.75	2.12	4.88	2.88	15.25	16	1 1/4	1 1/8 x 7 1/2
12"	12.88	20.50	15.00	2.25	5.38	3.12	17.75	16	1 3/8	1 1/4 x 8
14"	14.14	23.00	16.25	2.38	5.88	3.31	20.25	20	1 3/8	1 1/4 x 8 1/4
16"	16.16	25.50	18.50	2.50	6.00	3.69	22.50	20	1 1/2	1 3/8 x 8 3/4

*D = 1/16" for ANSI Class 150-300 and *D = 1/4" for ANSI Class 400-2500 for Steel Flanges only

FLANGE DETAILS
ACCORDING TO ANSI B 16.5

ANSI 600
Dimensions in inches

Pipe Size (inch)	Slip on Bore B	Outside Dia. O	Raised Face Dia. R	Thickness Q	Weld Neck Y	Slip On S	Bolt Circle Dia. P	No. of Holes N	Hole Dia. H	Stud Bolt x Length RF (Steel Flanges)
1/2"	0.88	3.75	1.38	0.56	2.06	0.88	2.62	4	5/8	1/2 x 3 1/4
3/4"	1.09	4.62	1.69	0.62	2.25	1.00	3.25	4	3/4	5/8 x 3 1/2
1"	1.36	4.88	2.00	0.69	2.44	1.06	3.50	4	3/4	5/8 x 3 3/4
1 1/4"	1.70	5.25	2.50	0.81	2.62	1.12	3.88	4	3/4	5/8 x 4
1 1/2"	1.95	6.12	2.88	0.88	2.75	1.25	4.50	4	7/8	3/4 x 4 1/4
2"	2.44	6.50	3.62	1.00	2.88	1.44	5.00	8	3/4	5/8 x 4 1/4
2 1/2"	2.94	7.50	4.12	1.12	3.12	1.62	5.88	8	7/8	3/4 x 4 3/4
3"	3.57	8.25	5.00	1.25	3.25	1.81	6.62	8	7/8	3/4 x 5
3 1/2"	4.07	9.00	5.50	1.38	3.38	1.94	7.25	8	1	7/8 x 5 1/2
4"	4.57	10.75	6.19	1.50	4.00	2.12	8.50	8	1	7/8 x 5 3/4
5"	5.66	13.00	7.31	1.75	4.50	2.38	10.50	8	1 1/8	1 x 6 1/2
6"	6.72	14.00	8.50	1.88	4.62	2.62	11.50	12	1 1/8	1 x 6 3/4
8"	8.72	16.50	10.62	2.19	5.25	3.00	13.75	12	1 1/4	1 1/8 x 7 3/4
10"	10.88	20.00	12.75	2.50	6.00	3.38	17.00	16	1 3/8	1 1/4 x 8 1/2
12"	12.88	22.00	15.00	2.62	6.12	3.62	19.25	20	1 3/8	1 1/4 x 8 3/4
14"	14.14	23.75	16.25	2.75	6.50	3.69	20.75	20	1 1/2	1 3/8 x 9 1/4
16"	16.16	27.00	18.50	3.00	7.00	4.19	23.75	20	1 5/8	1 1/2 x 10

ANSI 900
Dimensions in inches

Pipe Size (inch)	Slip on Bore B	Outside Dia. O	Raised Face Dia. R	Thickness Q	Weld Neck Y	Slip On S	Bolt Circle Dia. P	No. of Holes N	Hole Dia. H	Stud Bolt x Length RF (Steel Flanges)
1/2"	0.88	4.75	1.38	0.88	2.38	1.25	3.25	4	7/8	3/4 x 4 1/4
3/4"	1.09	5.13	1.69	1.00	2.75	1.38	3.50	4	7/8	3/4 x 4 1/2
1"	1.36	5.88	2.00	1.13	2.88	1.63	4.00	4	1	7/8 x 5
1 1/4"	1.70	6.25	2.50	1.13	2.88	1.63	4.38	4	1	7/8 x 5
1 1/2"	1.95	7.00	2.88	1.25	3.25	1.75	4.88	4	1 1/8	1 x 5 1/2
2"	2.44	8.50	3.63	1.50	4.00	2.25	6.50	8	1	7/8 x 5 3/4
2 1/2"	2.94	9.63	4.13	1.63	4.13	2.50	7.50	8	1 1/8	1 x 6 1/4
3"	3.57	9.50	5.00	1.50	4.00	2.13	7.50	8	1	7/8 x 5 3/4
4"	4.57	11.50	6.19	1.75	4.50	2.75	9.25	8	1 1/4	1 1/8 x 6 3/4
5"	5.66	13.75	7.31	2.00	5.00	3.13	11.00	8	1 3/8	1 1/4 x 7 1/2
6"	6.72	15.00	8.50	2.19	5.50	3.38	12.50	12	1 1/4	1 1/8 x 7 3/4
8"	8.72	18.50	10.63	2.50	6.38	4.00	15.50	12	1 1/2	1 3/8 x 8 3/4
10"	10.88	21.50	12.75	2.75	7.25	4.25	18.50	16	1 1/2	1 3/8 x 9 1/4
12"	12.88	24.00	15.00	3.13	7.88	4.63	21.00	20	1 1/2	1 3/8 x 10
14"	14.14	25.25	16.25	3.38	8.38	5.13	22.00	20	1 5/8	1 1/2 x 10 3/4
16"	16.16	27.75	18.50	3.50	8.50	5.25	24.25	20	1 3/4	1 5/8 x 11 1/4

*D = 1/16" for ANSI Class 150-300 and *D = 1/4" for ANSI Class 400-2500 for Steel Flanges only

FLANGE DETAILS
ACCORDING TO ANSI B 16.5

ANSI 1500
Dimensions in inches

Pipe Size (inch)	Slip on Bore B	Outside Dia. O	Raised Face Dia. R	Thickness Q	Weld Neck Y	Slip On S	Bolt Circle Dia. P	No. of Holes N	Hole Dia. H	Stud Bolt x Length RF (Steel Flanges)
1/2"	0.88	4.75	1.38	0.88	2.38	1.25	3.25	4	7/8	3/4 x 4 1/4
3/4"	1.09	5.13	1.69	1.00	2.75	1.38	3.50	4	7/8	3/4 x 4 1/2
1"	1.36	5.88	2.00	1.13	2.88	1.63	4.00	4	1	7/8 x 5
1 1/4"	1.70	6.25	2.50	1.13	2.88	1.63	4.38	4	1	7/8 x 5
1 1/2"	1.95	7.00	2.88	1.25	3.25	1.75	4.88	4	1 1/8	1 x 5 1/2
2"	2.44	8.50	3.63	1.50	4.00	2.25	6.50	8	1	7/8 x 5 1/4
2 1/2"	2.94	9.63	4.13	1.63	4.13	2.50	7.50	8	1 1/8	1 x 6 1/4
3"	3.57	10.50	5.00	1.88	4.63	2.88	8.00	8	1 1/4	1 1/8 x 7
4"	4.57	12.25	6.19	2.13	4.88	3.56	9.50	8	1 3/8	1 1/4 x 7 3/4
5"	5.66	14.75	7.31	2.88	6.13	4.13	11.50	8	1 5/8	1 1/2 x 9 3/4
6"	6.72	15.50	8.50	3.25	6.75	4.69	12.50	12	1 1/2	1 3/8 x 10 1/4
8"	8.72	19.00	10.63	3.63	8.38	5.63	15.50	12	1 3/4	1 5/8 x 11 1/2
10"	10.88	23.00	12.75	4.25	10.00	6.25	19.00	12	2	1 7/8 x 13 1/2
12"	12.88	26.50	15.00	4.88	11.13	7.13	22.50	16	2 1/8	2 x 15
14"	–	29.50	16.25	5.25	11.75	–	25.00	16	2 3/8	2 1/4 x 16 1/4
16"	–	32.50	18.50	5.75	12.25	–	27.75	16	2 5/8	2 1/2 x 17 3/4

ANSI 2500
Dimensions in inches

Pipe Size (inch)	Slip on Bore B	Outside Dia. O	Raised Face Dia. R	Thickness Q	Weld Neck Y	Slip On S	Bolt Circle Dia. P	No. of Holes N	Hole Dia. H	Stud Bolt x Length RF (Steel Flanges)
1/2"	0.88	5.25	1.38	1.19	2.88	1.56	3.50	4	7/8	3/4 x 5
3/4"	1.09	5.50	1.69	1.25	3.13	1.69	3.75	4	7/8	3/4 x 5
1"	1.36	6.25	2.00	1.38	3.50	1.88	4.25	4	1	7/8 x 5 1/2
1 1/4"	1.70	7.25	2.50	1.50	3.75	2.06	5.13	4	1 1/8	1 x 6
1 1/2"	1.95	8.00	2.88	1.75	4.38	2.38	5.75	4	1 1/4	1 1/8 x 6 3/4
2"	2.44	9.25	3.63	2.00	5.00	2.75	6.75	8	1 1/8	1 x 7
2 1/2"	2.94	10.50	4.13	2.25	5.63	3.13	7.75	8	1 1/4	1 1/8 x 7 3/4
3"	3.57	12.00	5.00	2.63	6.63	3.63	9.00	8	1 3/8	1 1/4 x 8 3/4
4"	4.57	14.00	6.19	3.00	7.50	4.25	10.75	8	1 5/8	1 1/2 x 10
5"	5.66	16.50	7.31	3.63	9.00	5.13	12.75	8	1 7/8	1 3/4 x 11 3/4
6"	6.72	19.00	8.50	4.25	10.75	6.00	14.50	8	2 1/8	2 x 13 3/4
8"	8.72	21.75	10.63	5.00	12.50	7.00	17.25	12	2 1/8	2 x 15 1/4
10"	10.88	26.50	12.75	6.50	16.50	9.00	21.25	12	2 5/8	2 1/2 x 19 1/4
12"	12.88	30.00	15.00	7.25	18.25	10.00	24.38	12	2 7/8	2 3/4 x 21 1/4

*D = 1/16" for ANSI Class 150-300 and *D = 1/4" for ANSI Class 400-2500 for Steel Flanges only

COMMON SCREW AND BOLT HEADS

COMMON MACHINE SCREW HEADS

COMMON SELF-TAPPING SCREW HEADS

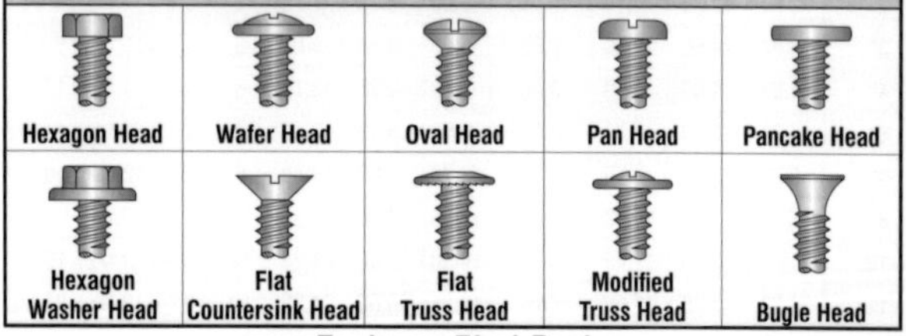

COMMON WORKSHOP NUTS

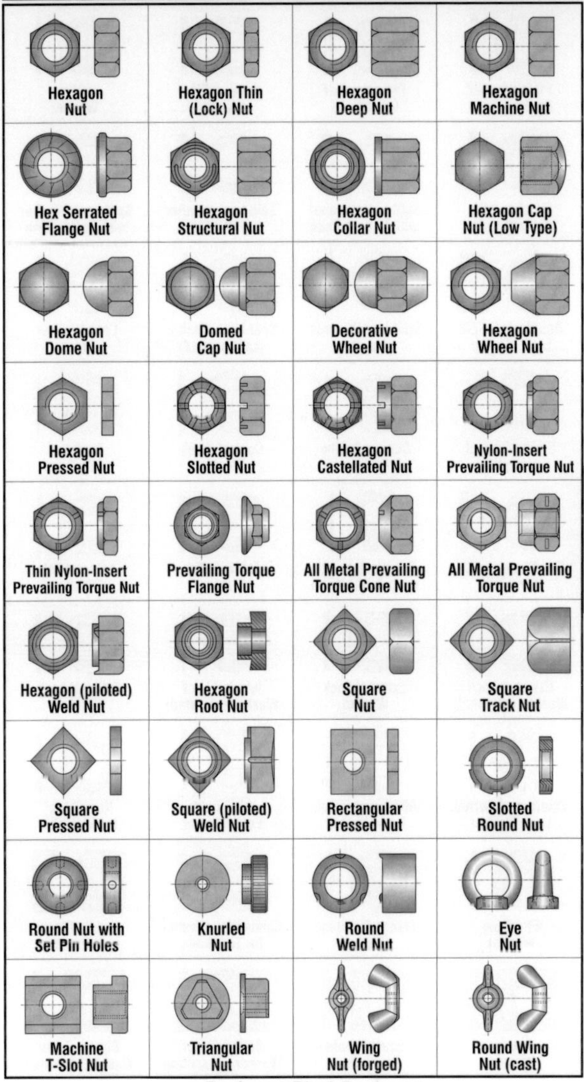

COMMON WORKSHOP WASHERS

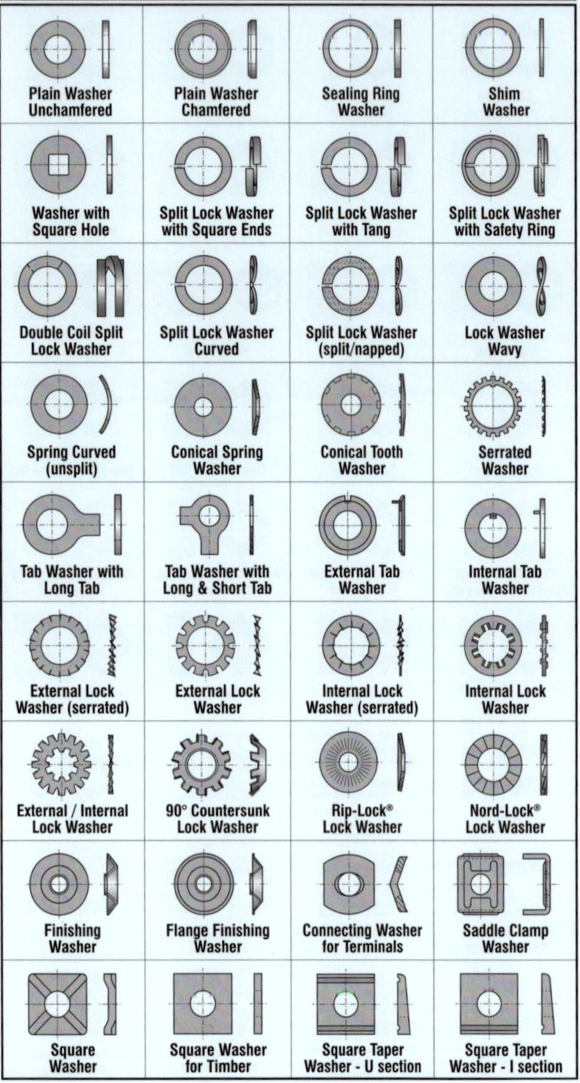

COMMON SCREW THREAD POINTS

Type A 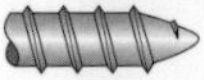 **Use :** Thin sheet metal applications.	**Type AB**  **Use :** Heavier sheet metal applications.	**Type B / Type Z**  **Use :** Shallow holes in composite materials.
Type BF  **Use :** Plastics, fiber cement and similar composites	**Type BP**  **Use :** Where holes are slightly misaligned	**Type BT / Type 25**  **Use :** Plastics & composites
Type C  **Use :** Heavy sheet metal & die castings	**Type CA** 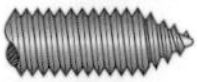 **Use :** Thread forming screws.	**Drill Point**  **Use :** Self drilling applications.
Type D / Type 1  **Use :** General machine screw thread fastener.	**Type F**  **Use :** Heavy gauge sheet metal, diecastings, etc.	**Type G**  **Use :** Low strength metals & plastics.
Type T / Type 23 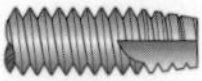 **Use :** Soft metal castings.	**Type U**  **Use :** Quick permanent assembly in castings, etc.	**Type 17**  **Use :** Self-drilling in thin metals.
Needle Point  **Use :** Self piercing point used in light metals.	**Spoon Point**  **Use :** Aluminum window frames.	**Nail Point**  **Use :** For locking against soft material.
Reamer (Wings)  **Use :** Wings ream clearance hole in wood & snap off on entering steel.	**Rolled Point**  **Use :** Mass produced fasteners.	**Round Point**  **Use :** Where end friction without cutting action is desireable.
MAT Point ®  **Use :** Designed to eliminate cross threading.	**Taptite ®**  **Use :** Thread forming screws.	**CTP Point ®**  **Use :** Cross thread prevention.

COMMON SCREW THREAD POINTS

Socket Set Screw

Cup Point

Use : Against hardened shafts.

Oval Point

Use : For Frequent adjustment.

Half Dog Point

Use : For permanent location.

Cone Point

Use : For permanent location.

Flat Point

Use : Where parts must be often reset.

Dog Point

Use : For permanent location.

knurled Cup Point

Use : For permanent location.

Radius (P) Point

Use : For permanent location.

POINTS FOR SELF DRILLING SCREWS

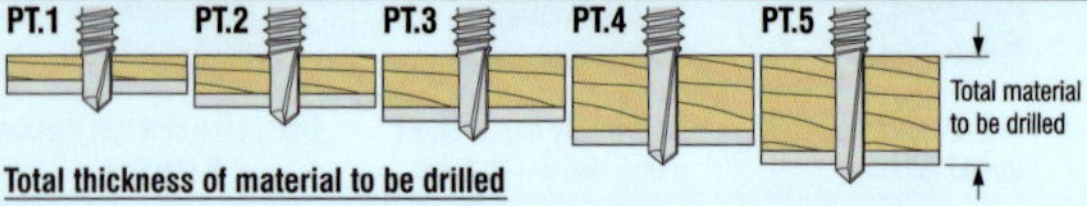

Total thickness of material to be drilled

Note: Thread should engage only after total material thickness has been drilled

Both top and bottom material to be drilled.

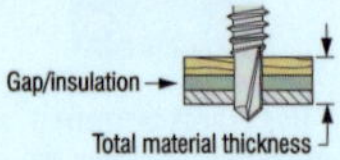

Both top and bottom material to be drilled with gap in between.

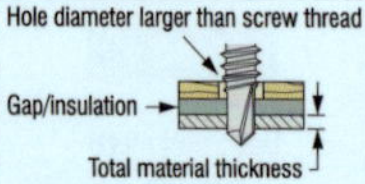

Bottom material to be drilled only. Top material predrilled to clear thread.

SELF-DRILLING SCREW POINT SELECTION GUIDE

Screw Size	PT 1		PT 2		PT 3		PT 4		PT 5	
	Total material to be drilled									
	max	min	max	min	max	min	max	min	max	min
No.6 (inch)	0.100	0.035								
3.5 (mm)	2.5	0.9								
No. 8 (inch)			0.100	0.035	0.140	0.100				
4.2 (mm)			2.5	0.9	3.5	2.5				
No.10 (inch)			0.110	0.035	0.175	0.110				
4.8 (mm)			2.8	0.9	4.4	2.8				
No.12 (inch)					0.210	0.110	0.220	0.175	0.500	0.250
5.5 (mm)					5.3	2.8	5.6	4.4	12.7	6.3
1/4" (inch)					0.220	0.110	0.250	0.175		
6.3 (mm)					5.6	2.8	6.3	4.4		

COMMON SCREW HEAD DRIVES

STANDARD SCREW HEAD DRIVES

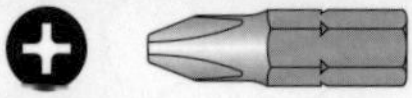
Phillips®

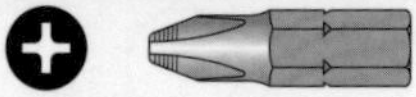
ACR Phillips ACR®

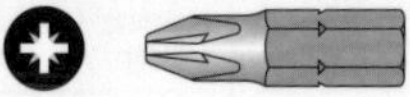
Pozidriv®

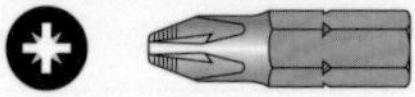
ACR Pozidriv ACR®

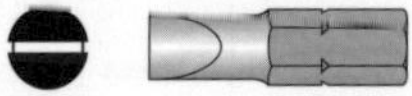
Slotted

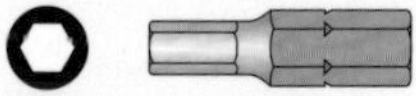
Hexagon (Internal Drive)

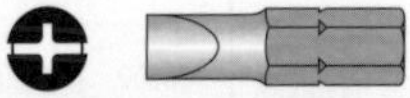
Combination Phillips / Slotted

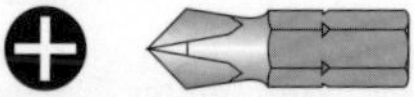
Frearson

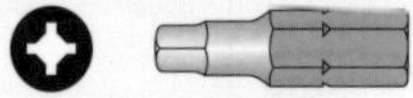
Quadrex®

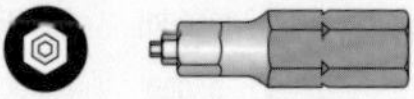
Uni-Screw ™

Hexagon (External Drive)

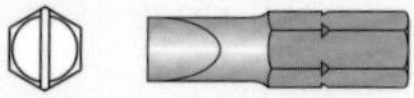
Combination Slotted / Hexagon

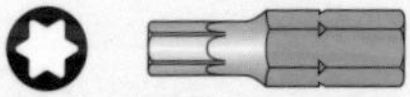
Torx®

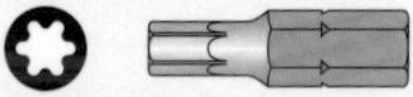
Torxalign®

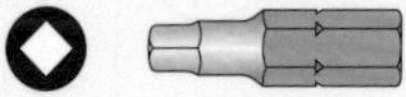
Square Recess (Robertson®) (Scrulox®)

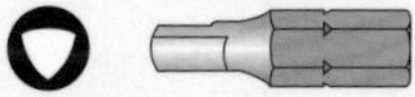
Tri-Lobular

TAMPER RESISTANT SCREW HEAD DRIVES

Spanner (Snake Eye)

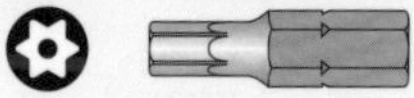
Torx® Tamper Resistant

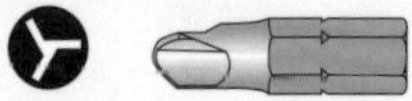
Tri-Wing®

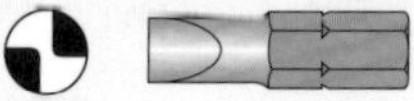
One Way

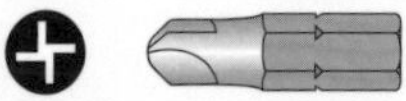
Torq-Set®

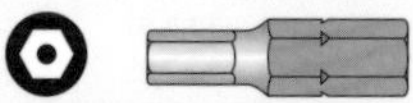
Internal Hex (Tamper Resistant)

COMMON WORKSHOP BOLTS IDENTIFICATION AND MECHANICAL PROPERTIES

	Common INCH Steel Bolts				Stainless Steel	
Identification Mark						
					M denotes Metric Thread No mark denotes Inch Thread	
Product Specifications	Grade 2	Grade 5	Grade 8	Soc. Hd. Cap Scrs.	Stainless Steel	Stainless Steel
Nominal Size Range	1/4 - 3/4 (>3/4 - 1 1/2)	1/4 - 1 (>1-1 1/2)	1/4-1 1/2	<1/2 (>1/2 - 2)	–	–
Material	Low/Medium Carbon Steel	Medium Carbon Steel Heat Treated	Medium Carbon Steel Heat Treated	Alloy Steel Heat Treated	Stainless Steel 304	Stainless Steel 316
Tensile Strength Minimum [psi]	74000 (60000)	120000 (105000)	150000	180000 (170000)	85000	85000
Yield Strength Minimum [psi]	57000 (36000)	91000 (81000)	130000	162000 (153000)	35000	35000
Proof Load [psi]	55000 (33000)	85000 (74000)	120000	140000 (135000)	–	–
Hardness Min / Max	HRB-70 / HRB-95	HRC-19 / HRC-34	HRC-33 / HRC-39	HRC-38 / HRC-45	HRB-80	HRB-80
Compatible Nut	Mild Steel	Grade 5	Grade 8	–	304	316

	Common METRIC Steel Bolts				
Identification Mark					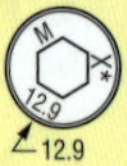
Product Specification	Class 4.6	Class 5.8	Class 8.8	Class 10.9	Class 12.9 Socket Head Cap Screws
Material	Low/Medium Carbon Steel	Low/Medium Carbon Steel Heat Treated	Medium Carbon Steel Heat Treated	Alloy Steel Heat Treated	Alloy Steel Heat Treated
Tensile Strength Minimum	400 MPa (58000 psi)	520 MPa (75000 psi)	830 MPa (120000 psi)	1040 MPa (150000 psi)	1220 MPa (176000 psi)
Yield Strength Minimum	240 MPa	420 MPa	640 MPa	940 MPa	1100 MPa
Proof Load	220 MPa	380 MPa	600 MPa	830 MPa	970 MPa
Hardness Min / Max	HRB-67 / HRB-95	HRB-82 / HRB-95	HRC-22 / HRC-34	HRC-32 / HRC-39	HRC-39 / HRC-44
Compatible Nut	Class 5	Class 5	Class 8	Class 10	–

X* = Manufacturer's identification

INCH - METRIC FASTENER SIZES / GRADE EQUIVALENTS

Diameter / Decimal / Metric Equivalents for Small Diameters

Gauge or Diameter	Decimal inches	Decimal millimeters
No. 0000	0.021	0.530
No. 000	0.034	0.860
No. 00	0.047	1.190
No. 0	0.060	1.524
No. 1	0.073	1.854
No. 2	0.086	2.184
No. 3	0.099	2.515
No. 4	0.112	2.845
No. 5	0.125	3.175
No. 6	0.138	3.505
No. 8	0.164	4.166
No. 10	0.190	4.826
No. 12	0.216	5.484

INCH to METRIC Fastener Size Crossover Chart

INCH UNC	INCH UNF	ISO METRIC Equivalent
1 - 64	1 - 72	M2 x 0.4
3 - 48	3 - 56	M2.5 x 0.45
5 - 40	5 - 44	M3 x 0.5
6 - 32	6 - 40	M3.5 x 0.6
8 - 32	8 - 36	M4 x 0.7
10 - 24	10 - 32	M5 x 0.8
1/4 - 20	1/4 - 28	M6 x 1
5/16 - 18	5/16 - 24	M8 x 1.25
3/8 - 16	3/8 - 24	M10 x 1.5
7/16 - 14	7/16 - 20	–
1/2 - 13	1/2 - 20	M12 x 1.75
9/16 - 12	9/16 - 18	M14 x 2
5/8 - 11	5/8 - 18	M16 x 2
3/4 - 10	3/4 - 16	M20 x 2.5
1 - 8	1 - 12	M24 x 3
1 - 1/4 - 7	1 - 1/8 - 12	M30 x 3.5
1 - 1/2 - 6	1 - 1/4 - 12	M36 x 4

INCH / METRIC *Fastener Equivalents

INCH (SAE J429)		METRIC (ISO 898)
	=	
Grade: 2 Over 3/4" (60,000 psi)		**Class: 4.8** 420 MPa (60,900 psi)
		Class: 5.6 500 MPa (72,519 psi)
Upto 3/4" (74,000 psi)		**Class: 5.8** 520 MPa (75,420 psi)
	=	
Grade: 5 (120,000 psi)		**Class: 8.8** 830 MPa (120,381 psi)
	=	
Grade: 8 (150,000 psi)		**Class: 10.9** 1040 MPa (150,839 psi)
 **ASTM A574** ≤ 1/2" (180,000 psi) > 1/2" (170,000 psi)	=	 **Class: 12.9** 1220 MPa (176,946 psi)
 **L'9® (Grade 9)** (180,000 psi) ®(Proprietory product)	=	1240 MPa

* Equivalents based on tensile strength

X* = Manufacturer's identification

® Use strictly as per manufacturer's recommendations

SCREWS & BOLTS

INCH - METRIC BREAKING-LOAD COMPARISON CHART FOR ALLOY FASTENERS

INCH - METRIC Steel/Alloy Fastener (Bolts) Breaking (Tensile) Load Comparison Chart

INCH						METRIC								
US Std. Thread Size UNC	Major Dia. in mm	Stress Area in mm^2 As	SAE Grade 2 kN	SAE Grade 5 kN	SAE Grade 8 kN	Metric Threads Coarse	Stress Area in mm^2 As	Metric Class 4.6 kN	Metric Class 4.8 kN	Metric Class 5.8 kN	Metric Class 8.8 kN	Metric Class 9.8 kN	Metric Class 10.9 kN	Metric Class 12.9 kN
						M1.6	1.27	0.51	0.53	0.66	1.05	1.14	1.32	1.55
No. 1 - 64	1.854	1.697	0.87	1.40	1.75	M2.0	2.07	0.83	0.87	1.08	1.72	1.86	2.15	2.53
No. 2 - 56	2.184	2.387	1.22	1.98	2.47									
No. 3 - 48	2.515	3.142	1.60	2.60	3.25	M2.5	3.39	1.36	1.42	1.76	2.81	3.05	3.53	4.14
No. 4 - 40	2.845	3.897	1.99	3.22	4.03									
No. 5 - 40	3.175	5.135	2.62	4.25	5.31	M3.0	5.03	2.01	2.11	2.62	4.17	4.53	5.23	6.14
No. 6 - 32	3.505	5.865	2.99	4.85	6.07	M3.5	6.78	2.71	2.85	3.53	5.63	6.10	7.05	8.27
No. 8 - 32	4.166	9.032	4.61	7.47	9.34	M4.0	8.78	3.51	3.69	4.57	7.29	7.90	9.13	10.71
No. 10 - 24	4.826	11.290	5.76	9.34	11.68	M5.0	14.20	5.68	5.96	7.38	11.79	12.78	14.77	17.32
No. 12 - 24	5.486	15.613	7.97	12.92	16.15									
1/4 - 20	6.350	20.516	10.47	16.98	21.22	M6.0	20.10	8.04	8.44	10.45	16.68	18.09	20.90	24.52
5/16 - 18	7.938	33.806	17.25	27.97	34.96	M8.0	36.60	14.64	15.37	19.03	30.38	32.94	38.06	44.65
3/8 - 16	9.525	50.000	25.51	41.37	51.71	M10.0	58.00	23.20	24.36	30.16	48.14	52.20	60.32	70.76
7/16 - 14	11.112	68.581	34.99	56.74	70.93									
1/2 - 13	12.700	91.548	46.71	75.75	94.68	M12.0	84.30	33.72	35.41	43.84	69.97	75.87	87.67	102.85
9/16 - 12	14.288	117.419	59.91	97.15	121.44	M14.0	115.00	46.00	48.30	59.80	95.45	103.50	119.60	140.30
5/8 - 11	15.875	145.806	74.40	120.64	150.80	M16.0	157.00	62.80	65.94	81.64	130.31	141.30	163.28	191.54

INCH - METRIC BREAKING-LOAD COMPARISON CHART FOR ALLOY FASTENERS

						M18.0	192.00	76.80	80.64	99.84	159.36	172.80	199.68	234.24
3/4 - 10	19.050	215.483	109.95	178.29	222.86	M20.0	245.00	98.00	102.90	127.40	203.35	220.50	254.80	298.90
7/8 - 9	22.225	298.064	123.31	246.62	308.27	M22.0	303.00	121.20	127.26	157.56	251.49	272.70	315.12	369.66
1 - 8	25.400	390.967	161.74	323.49	404.36	M24.0	353.00	141.20	148.26	183.56	292.99	317.70	367.12	430.66
						M27.0	459.00	183.60	192.78	238.68	380.97	413.10	477.36	559.98
1 1/8 - 7	28.575	492.257	203.65	356.38	509.12									
						M30.0	561.00	224.40	235.62	291.72	465.63	504.90	583.44	684.42
1 1/4 - 7	31.750	625.160	258.33	452.60	646.57									
1 3/8 - 6	34.925	745.160	308.27	539.48	770.69									
						M36.0	817.00	326.80	343.14	424.84	678.11	735.30	849.68	996.74
1 1/2 - 6	38.100	906.450	375.00	656.25	937.50									
						M42.0	1120.00	448.00	470.40	582.40	929.60	1008.00	1164.80	1366.40
1 3/4 - 5	44.450	1225.804	507.12	887.46	1267.79									
						M48.0	1470.00	588.00	617.40	764.40	1220.10	1323.00	1528.80	1793.40
2 - 4 1/2	50.800	1612.900	667.26	1167.70	1668.15									
2 1/4 - 4 1/2	57.150	2096.770	867.44	1518.02	2168.59	M56.0	2030.00	812.00	852.60	1055.60	1684.90	1827.00	2111.20	2476.60
2 1/2 - 4	63.500	2580.640	1067.62	1863.33	2669.04	M64.0	2680.00	1072.00	1125.60	1393.60	2224.40	2412.00	2787.20	3269.60

Note: For comparison sake, INCH units have been converted to METRIC.

Breaking (Tensile) Load = Stress area (As) x Tensile strength

Calculation based on the tensile strength of:

- 1 kN = 224.81 lbf
- 1 MPa = 145.04 psi
- 1 psi = 0.006894 MPa
- 1 lbf = 0.004448 kN

Grade 2 - 74000 psi upto 3/4" - 60000 psi over 3/4" | **Grade 5** - 120000 psi upto 1" - 105000 psi over 1" | **Grade 8** - 150000 psi.

Class 4.6 - 400 MPa | **Class 4.8** - 420 MPa | **Class 5.8** - 520 MPa | **Class 8.8** - 830 MPa | **Class 9.8** - 900 MPa | **Class 10.9** - 1040 MPa | **Class 12.9** - 1220 MPa

SCREWS & BOLTS

132 DRILL SIZES FOR SCREW EXTRACTORS, WOOD SCRS, SELF TAPPERS, BLIND RIVETS AND LAG SCREWS

Recommended Drill Sizes for Standard Screw Extractors

Screw Extractor No.	Outside Diameter of Screw		Recommended Drill Size	
	(inch)	(mm)	(inch)	(mm)
1	5/32 - 3/16	M4 - M5	# 47	2.0
2	7/32 - 1/4	M6 - M7	# 35	2.8
3	5/16 - 3/8	M8 - M10	5/32"	4.0
4	1/2 - 9/16	M12 - M14	7/32"	5.5
5	5/8 - 3/4	M16 - M20	19/64"	7.5
6	7/8 - 1	M22 - M24	7/16"	11.0
7	1 1/4 - 1 1/2	M28 - M36	17/32"	13.5
8	1 5/8 - 1 3/4	M40 - M45	13/16"	20.0
9	1 7/8 - 2 1/8	M48 - M52	1 1/16"	27.0
10	2 1/4 - 2 1/2	M56 - M60	1 5/16"	33.0

Recommended Drill Sizes for Self Tapping Screws Type 'A'

Screw Size No.	Sheet Thickness		Drill Size
	Gauge No.	(inch)	(inch)
No. 4 (0.112")	28	0.0156	# 44
	26	0.0188	# 44
	24	0.0250	# 42
	22	0.0312	# 42
	20	0.0375	# 40
No. 6 (0.138")	28	0.0156	# 37
	26	0.0188	# 37
	24	0.0250	# 37
	22	0.0312	# 37
	20	0.0375	# 36
No. 8 (0.164")	26	0.0188	1/8"
	24	0.0250	1/8"
	22	0.0312	1/8"
	20	0.0375	1/8"
	18	0.0500	# 30
No. 10 (0.190")	26	0.0188	# 29
	24	0.0250	# 29
	22	0.0312	# 29
	20	0.0375	# 29
	18	0.0500	# 25

Recommended Drill Sizes for Wood Screws
(For use on Hardwoods)

Wood Screw Gauge Size	Recommended Drill Size	
	For Screw Shank (inch)	For Screw Thread (inch)
2	# 46	3/64"
3	# 41	1/16"
4	# 34	1/16"
5	# 30	5/64"
6	# 27	5/64"
7	# 21	3/32"
8	# 16	3/32"
9	# 12	7/64"
12	# 2	9/64"
14	F	5/32"
16	L	11/64"
18	O	3/16"
20	S	7/32"
22	W	15/64"
24	Z	1/4"

Recommended Drill Sizes for Lag Screws

Size	For Screw Shank (inch)	For Screw Thread (inch)
1/4"	19/64"	11/64"
5/16"	23/64"	15/64"
3/8"	27/64"	17/64"
7/16"	15/32"	21/64"
1/2"	17/32"	3/8"
5/8"	11/16"	31/64"
3/4"	27/32"	19/32"
7/8"	15/16"	11/16"
1"	1 1/16"	25/32"

Recommended Drill Sizes for Blind Rivets

Rivet Diameter	Drill Diameter	
	(inch)	(mm)
3/32"	# 40	2.45
7/64"	# 33	2.85
1/8"	# 30	3.30
5/32"	# 20	4.10
3/16"	# 10	4.90
7/32"	# 2	5.60
1/4"	F	6.50

WEIGHTS OF ROUND, SQUARE AND HEXAGON STEEL AND BRASS BARS

Size - X		Inch (Pounds per foot) / Metric (Kilograms per meter)					
		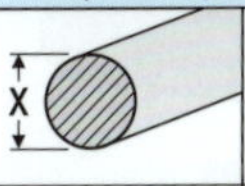		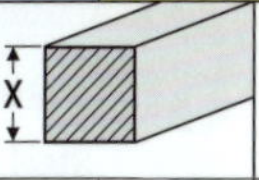		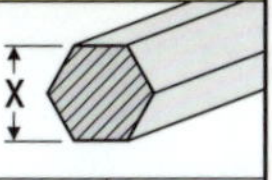	
(inch)	(mm)	Steel	Brass	Steel	Brass	Steel	Brass
	6	0.222	0.240	0.283	0.306	0.244	0.265
1/4"		0.167	0.182	0.213	0.232	0.184	0.200
5/16"		0.261	0.284	0.332	0.362	0.287	0.313
	8	0.394	0.427	0.502	0.544	0.435	0.471
3/8"		0.375	0.409	0.478	0.521	0.414	0.451
	10	0.616	0.667	0.785	0.850	0.679	0.735
7/16"		0.511	0.557	0.651	0.710	0.563	0.614
	12	0.887	0.961	1.130	1.224	0.978	1.059
1/2"		0.667	0.728	0.850	0.927	0.735	0.802
	14	1.208	1.308	1.539	1.666	1.331	1.441
9/16"		0.844	0.921	1.076	1.173	0.931	1.015
	15	1.387	1.501	1.766	1.913	1.528	1.654
5/8"		1.043	1.137	1.328	1.449	1.149	1.253
	16	1.578	1.708	2.010	2.176	1.738	1.002
	17	1.781	1.928	2.269	2.457	1.962	2.125
	18	1.997	2.162	2.543	2.754	2.200	2.382
	19	2.225	2.409	2.834	3.069	2.451	2.654
3/4"		1.501	1.637	1.913	2.086	1.654	1.804
	20	2.465	2.669	3.140	3.400	2.716	2.941
	22	2.983	3.229	3.799	4.114	3.286	3.559
7/8"		2.043	2.229	2.603	2.839	2.252	2.456
	24	3.549	3.843	4.522	4.896	3.911	4.235
	25	3.851	4.170	4.906	5.313	4.244	4.595
1"		2.669	2.911	3.400	3.708	2.941	3.208
	27	4.492	4.864	5.723	6.197	4.950	5.360
	28	4.831	5.231	6.154	6.664	5.324	5.764
1 1/8"		3.378	3.684	4.303	4.693	3.722	4.060
	30	5.546	6.005	7.065	7.650	6.111	6.617
1 1/4"		4.170	4.549	5.313	5.794	4.595	5.012
	32	6.310	6.833	8.038	8.704	6.953	7.529
	33	6.711	7.266	8.549	9.257	7.395	8.007
1 3/8"		5.046	5.504	6.428	7.011	5.560	6.065
	35	7.549	8.174	9.616	10.413	8.318	9.007
	36	7.986	8.648	10.174	11.016	8.800	9.529
1 1/2"		6.005	6.550	7.650	8.344	6.617	7.217
	39	9.373	10.149	11.940	12.929	10.328	11.183
	40	9.860	10.676	12.560	13.600	10.864	11.764
1 5/8"		7.048	7.687	8.978	9.792	7.766	8.470
	42	10.870	11.770	13.847	14.994	11.978	12.970
1 3/4"		8.174	8.915	10.413	11.357	9.007	9.824
	45	12.479	13.512	15.896	17.213	13.750	14.889
	50	15.406	16.681	19.625	21.250	16.976	18.381
2"		10.676	11.644	13.600	14.833	11.764	12.831

AMERICAN NATIONAL STD. WEIGHTS AND DIMENSIONS OF WELDED AND SEAMLESS STEEL PIPES

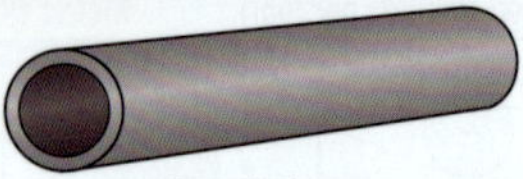

Steel Pipes

Nominal Size (inch)	Outside Diameter (inch)	Schedule	Wall Thickness (inch)	Inside Diameter (inch)	Weight lb/ft
1/8"	0.405	40	0.068	0.269	0.24
		80	0.095	0.215	0.31
1/4"	0.540	40	0.088	0.364	0.42
		80	0.119	0.302	0.54
3/8"	0.675	40	0.091	0.493	0.57
		80	0.126	0.423	0.74
1/2"	0.840	40	0.109	0.622	0.85
		80	0.147	0.546	1.09
		160	0.188	0.466	1.31
3/4"	1.050	40	0.113	0.824	1.13
		80	0.154	0.742	1.47
		160	0.219	0.614	1.94
1"	1.315	40	0.133	1.049	1.68
		80	0.179	0.957	2.17
		160	0.250	0.815	2.84
1 1/4"	1.660	40	0.140	1.380	2.27
		80	0.191	1.278	3.00
		160	0.250	1.660	3.76
1 1/2"	1.900	40	0.145	1.610	2.72
		80	0.200	1.500	3.63
		160	0.281	1.338	4.86
2"	2.375	40	0.154	2.067	3.65
		80	0.218	1.939	5.02
		160	0.344	1.689	7.46
2 1/2"	2.875	40	0.203	2.469	5.79
		80	0.276	2.323	7.66
		160	0.375	2.125	10.01
3"	3.500	40	0.216	3.068	7.58
		80	0.300	2.900	10.25
		160	0.438	2.624	14.32
3 1/2"	4.000	40	0.226	3.548	9.11
		80	0.318	3.364	12.50
4"	4.500	40	0.237	4.026	10.79
		80	0.337	3.826	14.98
		120	0.438	3.624	19.00
		160	0.531	3.438	22.51
5"	5.563	40	0.258	5.047	14.62
		80	0.375	4.813	20.78
		120	0.500	4.563	27.04
		160	0.625	4.313	32.96
6"	6.625	40	0.280	6.065	18.97
		80	0.432	5.761	28.57
		120	0.562	5.501	36.39
		160	0.719	5.187	45.35

WEIGHTS OF SHEET METAL
STEEL, ALUMINUM, BRASS, COPPER AND TIN PLATE

Sheet Steel (inch)

U.S. Gauge No.	(inch)	(mm)	Weight lb/ft²
0	0.3125	7.938	12.750
1	0.2812	7.144	11.473
2	0.2656	6.747	10.836
3	0.2500	6.350	10.200
4	0.2344	5.953	9.563
5	0.2187	5.556	8.923
6	0.2031	5.159	8.286
7	0.1875	4.762	7.650
8	0.1719	4.366	7.013
9	0.1562	3.969	6.373
10	0.1406	3.572	5.736
11	0.1250	3.175	5.100
12	0.1094	2.778	4.463
13	0.0937	2.381	3.823
14	0.0781	1.984	3.186
15	0.0703	1.786	2.868
16	0.0625	1.588	2.550
17	0.0562	1.429	2.293
18	0.0500	1.270	2.040
19	0.0437	1.111	1.783
20	0.0375	0.952	1.530
21	0.0344	0.873	1.403
22	0.0312	0.794	1.273
23	0.0281	0.714	1.146
24	0.0250	0.635	1.020
25	0.0219	0.556	0.893
26	0.0188	0.476	0.767
27	0.0172	0.437	0.702
28	0.0156	0.397	0.636
29	0.0141	0.357	0.575
30	0.0125	0.318	0.510
31	0.0109	0.278	0.443
32	0.0102	0.258	0.416
33	0.0094	0.238	0.383
34	0.0086	0.218	0.351
35	0.0078	0.198	0.318
36	0.0070	0.179	0.286

Sheet Steel (mm)

Sheet Thickness		Weight lb/ft²
(mm)	(inch)	
0.18	0.0070	0.286
0.20	0.0079	0.322
0.22	0.0087	0.355
0.24	0.0094	0.383
0.28	0.0110	0.449
0.32	0.0126	0.514
0.38	0.0150	0.612
0.44	0.0173	0.706
0.50	0.0197	0.804
0.56	0.0220	0.899
0.63	0.0248	1.012
0.75	0.0295	1.204
0.88	0.0346	1.412
1.00	0.0394	1.607
1.13	0.0445	1.816
1.25	0.0492	2.007
1.38	0.0543	2.215
1.50	0.0590	2.407
1.75	0.0689	2.811
2.00	0.0787	3.211
2.25	0.0886	3.615
2.50	0.0984	4.015
2.75	0.1083	4.419
3.00	0.1181	4.818
3.50	0.1378	5.622
4.00	0.1575	6.426
4.50	0.1772	7.230
4.75	0.1870	7.630
5.00	0.1968	8.029
6.00	0.2362	9.637
7.00	0.2756	11.244
8.00	0.3150	12.852

Al, Brass, Copper Sheet (mm)

Sheet Thickness		Alu. lb/ft²	Brass lb/ft²	Copper lb/ft²
(mm)	(inch)			
0.10	0.0039	0.054	0.169	0.181
0.15	0.0059	0.081	0.256	0.274
0.20	0.0079	0.109	0.342	0.367
0.25	0.0098	0.135	0.425	0.455
0.30	0.0118	0.162	0.511	0.548
0.35	0.0138	0.190	0.598	0.640
0.40	0.0157	0.216	0.680	0.729
0.45	0.0177	0.243	0.767	0.821
0.50	0.0197	0.271	0.854	0.914
0.55	0.0216	0.297	0.936	1.002
0.60	0.0236	0.324	1.023	1.095
0.65	0.0256	0.352	1.109	1.188
0.70	0.0275	0.378	1.192	1.276
0.75	0.0295	0.406	1.278	1.369
0.80	0.0315	0.433	1.365	1.462
0.85	0.0335	0.461	1.452	1.555
0.90	0.0354	0.487	1.534	1.643
1.00	0.0394	0.542	1.707	1.829
1.10	0.0433	0.595	1.876	2.010
1.20	0.0472	0.649	2.045	2.191
1.30	0.0512	0.704	2.219	2.376
1.40	0.0551	0.758	2.388	2.557
1.50	0.0590	0.811	2.557	2.738
1.60	0.0630	0.866	2.730	2.924
1.70	0.0669	0.920	2.899	3.105
1.80	0.0709	0.975	3.072	3.291
2.00	0.0787	1.082	3.410	3.653
2.50	0.0984	1.353	4.264	4.567
2.80	0.1102	1.515	4.775	5.115
3.00	0.1181	1.624	5.118	5.482
3.20	0.1260	1.732	5.460	5.848
3.50	0.1378	1.895	5.971	6.396

Tin Plate (mm)

Plate		Wt. lb/ft²	Plate		Wt. lb/ft²
(mm)	(inch)		(mm)	(inch)	
0.15	0.0059	0.225	0.46	0.0181	0.691
0.19	0.0075	0.286	0.50	0.0197	0.752
0.22	0.0087	0.332	0.52	0.0205	0.782
0.24	0.0094	0.359	0.57	0.0224	0.855
0.27	0.0106	0.404	0.58	0.0228	0.870
0.28	0.0110	0.420	0.64	0.0252	0.962
0.31	0.0122	0.466	0.66	0.0260	0.992
0.32	0.0126	0.481	0.70	0.0275	1.049
0.36	0.0142	0.542	0.75	0.0295	1.126
0.37	0.0146	0.557	0.80	0.0315	1.202
0.41	0.0161	0.614	0.90	0.0354	1.351
0.43	0.0169	0.645	1.00	0.0394	1.504

SHEET METAL AND WIRE GAUGES

Gauge No.	Sheet-Gauge					Wire-Gauge					
	US Std. Sheet Gauge	Galvanized Sheet Gauge		Zinc Gauge		Steel Wire Gauge (US)*		British Std. Wire Gauge (SWG)		Music or Piano Wire Gauge	
	(inch)	(inch)	(mm)	(inch)	(mm)	(inch)	(mm)	(inch)	(mm)	(inch)	(mm)
1	0.281	–	–	–	–	0.283	7.19	0.300	7.62	0.010	0.25
2	0.266	–	–	–	–	0.263	6.67	0.276	7.01	0.011	0.28
3	0.250	–	–	0.006	0.15	0.244	6.19	0.252	6.40	0.012	0.30
4	0.234	–	–	0.008	0.20	0.225	5.72	0.232	5.89	0.013	0.33
5	0.219	–	–	0.010	0.25	0.207	5.26	0.212	5.38	0.014	0.36
6	0.203	–	–	0.012	0.30	0.192	4.88	0.192	4.88	0.016	0.41
7	0.187	–	–	0.014	0.36	0.177	4.50	0.176	4.47	0.018	0.46
8	0.172	0.168	4.27	0.016	0.41	0.162	4.11	0.160	4.06	0.020	0.51
9	0.156	0.153	3.89	0.018	0.46	0.148	3.77	0.144	3.66	0.022	0.56
10	0.141	0.138	3.51	0.020	0.51	0.135	3.43	0.128	3.25	0.024	0.61
11	0.125	0.123	3.13	0.024	0.61	0.121	3.06	0.116	2.95	0.026	0.66
12	0.109	0.108	2.75	0.028	0.71	0.106	2.68	0.104	2.64	0.029	0.74
13	0.094	0.093	2.37	0.032	0.81	0.092	2.32	0.092	2.34	0.031	0.79
14	0.078	0.079	1.99	0.036	0.91	0.080	2.03	0.080	2.03	0.033	0.84
15	0.070	0.071	1.80	0.040	1.02	0.072	1.83	0.072	1.83	0.035	0.89
16	0.062	0.064	1.61	0.045	1.14	0.063	1.59	0.064	1.63	0.037	0.94
17	0.056	0.058	1.46	0.050	1.27	0.054	1.37	0.056	1.42	0.039	0.99
18	0.050	0.052	1.31	0.055	1.40	0.048	1.21	0.048	1.22	0.041	1.04
19	0.044	0.046	1.16	0.060	1.52	0.041	1.04	0.040	1.02	0.043	1.09
20	0.037	0.040	1.01	0.070	1.78	0.035	0.88	0.036	0.91	0.045	1.14
21	0.034	0.037	0.93	0.080	2.03	0.032	0.81	0.032	0.81	0.047	1.19
22	0.031	0.034	0.85	0.090	2.29	0.029	0.73	0.028	0.71	0.049	1.24
23	0.028	0.031	0.78	0.100	2.54	0.026	0.66	0.024	0.61	0.051	1.30
24	0.025	0.028	0.70	0.125	3.18	0.023	0.58	0.022	0.56	0.055	1.40
25	0.022	0.025	0.63	–	–	0.020	0.52	0.020	0.51	0.059	1.50
26	0.019	0.022	0.55	–	–	0.018	0.46	0.018	0.46	0.063	1.60
27	0.017	0.020	0.51	–	–	0.017	0.44	0.016	0.42	0.067	1.70
28	0.016	0.019	0.47	–	–	0.016	0.41	0.015	0.38	0.071	1.80
29	0.014	0.017	0.44	–	–	0.015	0.38	0.014	0.35	0.075	1.91
30	0.012	0.016	0.40	–	–	0.014	0.36	0.012	0.31	0.080	2.03
31	0.011	0.014	0.36	–	–	0.013	0.34	0.012	0.29	0.085	2.16
32	0.010	0.013	0.34	–	–	0.013	0.33	0.011	0.27	0.090	2.29

NOTE : The above Gauge Nos. are provided for reference only.
Much confusion exists in the use of various gauge numbers and therefore it is best to specify the exact decimal size in either inches or mm.

* Also known as Washburn & Moen, American Steel & Wire Co. & Roebling Wire Gauges.

SHEET METAL BENDING ALLOWANCE

To Calculate the Flat Blank Length (L) - 90° Bends

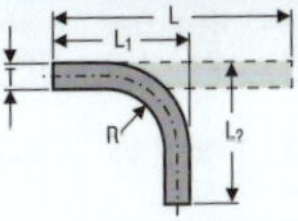

$L = L_1 + L_2 - Q$ (1 Bend)

Also $L = L_1 + L_2 + L_3 - 2Q$ (2 Bends)

Also $L = L_1 + L_2 + L_3 + L_4 - 3Q$ (3 Bends) ..., etc.

Where $Q = 0.43 \times R + 1.482 \times T$

Table for Correction Factor 'Q' (per bend)

Radius 'R'	Material Thickness 'T' (inch)											
	1/64	1/32	3/64	1/16	5/64	3/32	7/64	1/8	9/64	5/32	3/16	1/4
1/32	0.037	0.060	0.083	0.106	0.129	0.152	0.176	0.199	0.222	0.245	0.291	0.384
1/16	0.050	0.073	0.096	0.119	0.143	0.166	0.189	0.212	0.235	0.258	0.305	0.397
1/8	0.077	0.100	0.123	0.146	0.170	0.193	0.216	0.239	0.262	0.285	0.332	0.424
3/16	0.104	0.127	0.150	0.173	0.196	0.220	0.243	0.266	0.289	0.312	0.359	0.451
1/4	0.131	0.154	0.177	0.200	0.223	0.246	0.270	0.293	0.316	0.339	0.385	0.478
3/8	0.185	0.208	0.231	0.254	0.277	0.300	0.323	0.347	0.370	0.393	0.439	0.532
1/2	0.238	0.261	0.285	0.308	0.331	0.354	0.377	0.400	0.423	0.447	0.493	0.586
5/8	0.292	0.315	0.338	0.361	0.385	0.408	0.431	0.454	0.477	0.500	0.547	0.639
3/4	0.346	0.369	0.392	0.415	0.438	0.461	0.485	0.508	0.531	0.554	0.600	0.693
	Correction Factor 'Q' (inches) - per bend											

Examples

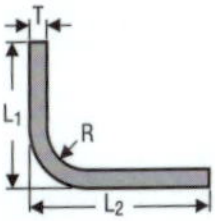

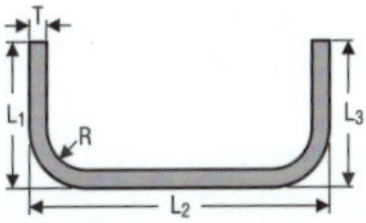

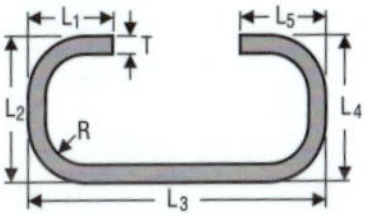

$L_1 = 1"$
$L_2 = 2"$
$R = 1/4"$ $T = 1/8"$
From Table, $Q = 0.293"$
$L = L_1 + L_2 - Q$
$L = 1 + 2 - 0.293$
$L = 2.707"$ Flat Blank Length

$L_1 = 1"$ $L_2 = 3"$
$L_3 = 1"$
$R = 1/4"$ $T = 3/16"$
From Table, $Q = 0.385"$
$L = L_1 + L_2 + L_3 - 2 \times Q$
$L = 1 + 3 + 1 - 2 \times 0.385$
$L = 4.23"$ Flat Blank Length

$L_1 = 1/2"$ $L_2 = 1"$ $L_3 = 3"$
$L_4 = 1"$ $L_5 = 1/2"$
$R = 3/8"$ $T = 1/8"$
From Table, $Q = 0.347"$
$L = L_1 + L_2 + L_3 + L_4 + L_5 - 4 \times Q$
$L = .5 + 1 + 3 + 1 + .5 - 4 \times 0.347$
$L = 4.612"$ Flat Blank Length

Minimum Bend Radius 'R'

Material	Material Thickness 'T' (inch)											
	1/64	1/32	3/64	1/16	5/64	3/32	7/64	1/8	9/64	5/32	3/16	1/4
Soft Brass Soft Copper	1/32	1/16	1/16	3/32	3/32	1/8	3/16	3/16	3/16	1/4	1/4	3/8
Aluminum Soft Steel	1/16	3/32	1/8	5/32	3/16	1/4	1/4	3/8	3/8	3/8	5/8	5/8
Cold Roll St. Stainless St.	1/16	1/8	5/32	3/16	1/4	3/8	3/8	3/8	5/8	5/8	3/4	3/4
	Minimum Bend Radius 'R' (inches)											

PHYSICAL PROPERTIES OF COMMON METALS / METALLOIDS

METAL / METALLOID	Weight lb. per cu ft	Weight gm. per cm³	Melting Point F°	Melting Point C°	Tensile Strength (approx.) tons/in.²	Brinell Hardness (approx.) HB	Conductivity Heat Cu = 100	Conductivity Elect. Cu = 100	Expansion per cent per 100°C
ALUMINUM (Al)									
Pure Aluminum (Al 99%)	165	2.6	1220	660	–	23	55	59	0.26
Silicon Aluminum (Si 5%)	160	2.5	1165	630	–	–	47	–	–
Silicon Aluminum (Si 10%)	155	2.4	1075	580	–	–	–	–	–
BRONZE (Cu 90%, Sn 10%)	481	7.7	1905	1040	35	125	20	15	0.30
Manganese Bronze (Cu 60%, Zn 39%, Mn 1%)	522	8.3	1600	871	28	100	31	25	0.21
Phosphor Bronze (Cu 95%, Sn 4%)	550	8.8	1942	1047	25	–	21	–	–
Tobin Bronze (Cu 60%, Zn 39%, Sn 1%)	525	8.4	1635	890	26	100	29	25	0.21
ANTIMONY (Sb)	417	6.7	1166	636	–	30	4.8	4.0	0.12
BRASS									
Yellow Brass (Cu 65%, Zn 35%)	517	8.3	1706	930	45	60	29	26	
Cartridge Brass (Cu 70%, Zn.30%)	525	8.4	1760	960	43	–	31	–	–
CADMIUM (Cd)	540	8.6	610	321	4	20	23	34	0.30
CALCIUM (Ca)	101	1.6	1562	850	–	–	–	–	–
CARBON (C)	134	2.2	6455	3550	–	–	–	–	–
CHROMIUM (Cr)	419	6.7	2740	1505	–	350	–	55	0.08
COBALT (Co)	550	8.8	2700	1483	16	125	18	16	0.22
COPPER Electrolytic (Cu)	557	8.9	1981	1082	30	30	100	100	0.18
DURALUMIN (Cu 4.5%, Mn 2%, Mg, Si, Fe, bal. Al.)	174	2.8	1112	600	25	–	34	–	–
GOLD (Au)	1204	19.2	1945	1060	8	25	77	71	0.13
GUN METAL (Cu 88%, Sn 10%, Zn 2%)	534	8.5	1900	1036	9	–	13.4	–	–
IRON (Fe)									
Pure Iron	490	7.8	2795	1535	12	67	20.6	17.5	0.12
Wrought Iron	480	7.7	2750	1510	18	100	15.5	15	0.11
Cast Iron	450	7.2	2300	1259	15	180	12.2	2.9	0.11
Malleable Iron	462	7.4	2300	1259	24	110	13.3	–	0.12
LEAD (Pb)	707	11.3	620	326	1	5	8.9	8	0.30
MAGNESIUM (Mg)	108	1.7	1240	670	11	40	40	37	0.25
MANGANESE (Mn)	475	7.6	2246	1229	58	200	40	35	0.17
MOLYBDENUM (Mo)	651	10.4	4532	2500	63	147	38	35	0.06
MONEL (Ni 67%, Cu 30%, C 3%, bal. Fe, Mn, Si)	551	8.8	2400	1315	45	125	6.7	3.6	0.13
MUNTZ METAL (Cu 60%, Zn 40%)	525	8.4	1652	900	22	60	31		–
NICKEL (Ni)	556	8.9	2650	1454	30	85	15	18.8	0.13
PHOSPHOROUS (P)	114	1.82	111	44	–	–	–	–	–
PLATINUM (Pt)	1330	21.3	3212	1770	11	25	18	15	0.09

WORKSHOP METALS

PHYSICAL PROPERTIES OF COMMON METALS / METALLOIDS

METAL / METALLOID	Weight		Melting Point		Tensile Strength (approx.)	Brinell Hardness (approx.)	Conductivity		Expansion per cent per 100°C
	lb. per cu ft	gm. per cm³	F°	C°	tons/in.²	HB	Heat (Cu = 100)	Elect. (Cu = 100)	
SILICON (Si)	147	2.4	2588	1420	–	–	–	1.6	0.06
SILVER (Ag)	655	10.5	1762	960	15	30	111	106	0.19
STEEL									
Low Carbon Steel	490	7.8	2700	1483	25	130	13.3	14.5	0.12
Medium Carbon Steel	490	7.8	2600	1427	34	170	13.3	14.5	0.12
High Carbon Steel	490	7.8	2500	1371	42	201	13.3	14.5	0.12
Manganese Steel (Mn 15%, C 1.5%, Si 0.2%)	495	7.9	2450	1342	45	300	–	–	0.12
Nickel Steel (Ni 3.5%, C 0.25%, Mn 0.8%)	490	7.8	2600	1427	40	217	–	–	0.12
Cast Steel	490	7.8	2600	1427	30	140	15	16	0.12
Stainless Steel	495	7.9	2550	1398	50	160	6	0.2	0.17
TIN (Sn)	458	7.3	450	232	4	6	17	13.5	0.21
TITANIUM (Ti)	281	4.5	3270	1800	32	230	–	–	0.08
TUNGSTEN (W)	1191	19	6152	3400	142	294	42	31	0.04
VANADIUM (V)	300	6.1	3182	1750	–	–	–	–	0.10
ZINC (Zn)	447	7.2	786	417	13	35	29	30	0.33

COMMON CHEMICAL COMPOUNDS

Common Name	Chemical Name	Formula	Description
Baking soda	Sodium Bicarbonate	$NaHCO_3$	White powder
Black lead	Carbon or Graphite	C	Black powder
Bleaching powder	Calcium Chloro-hypo Chlorite	$CaOCl_2$	White chlorine-smelling powder
Blue stone / vitriol	Copper Sulphate	$CuSO45H_2O$	Large blue crystals
Caustic Potash	Potassium Hydroxide	KOH	White deliquescent powder
Caustic Soda	Sodium Hydroxide	$NaOH$	White deliquescent powder
Chalk, Limestone	Calcium Carbonate	$CaCO_3$	White powder
Chile Saltpeter	Sodium Nitrate	$NaNO_3$	White crystalline substance
Condy's Crystals	Potassium Permanganate	$KMnO_4$	Small purple crystals
Epsom Salts	Magnesium Sulphate	$MgSO_47H_2O$	Small colorless crystals
Green Vitriol	Iron Sulphate	$FeSO_47H_2O$	Green crystals
Laughing Gas	Nitrous Oxide	N_2O	Colorless gas
Lime (quicklime)	Calcium Oxide	CaO	White powder
Limewater	Soln. of Calcium Hydroxide	$Ca(OH)_2$	Clear, bitter liquid
Liquid Ammonia	Ammonium Hydroxide	NH_4OH	Strongly fuming liquid
Litharge	Lead Monoxide	PbO	Orange powder
Muriatic Acid (Spirits of Salts)	Hydrochloric Acid	HCl	Strongly fuming colorless liquid
Oil of Vitriol	Sulphuric Acid	H_2SO_4	Heavy, colorless, viscous liquid
Red Lead	Triplumbic Tetroxide	Pb_3O_4	Fine, heavy red powder
Sal Ammoniac	Ammonium Chloride	NH_4Cl	White crystalline solid
Saltpeter (Nitre)	Potassium Nitrate	KNO_3	Colorless crystals
Slaked Lime	Calcium Hydroxide	$Ca(OH)_2$	White powder
Vinegar	Acetic Acid (weak)	CH_3COOH	Brown liquid
Washing Soda	Sodium Carbonate	$Na_2CO_310H_2O$	Large, white crystals

WORKSHOP METALS

SYMBOLS FOR ELEMENTS

Symbol	Name	Symbol	Name	Symbol	Name	Symbol	Name
Ac	Actinium	Er	Erbium	Mn	Manganese	Ru	Ruthenium
Ag	Silver	Es	Einsteinium	Mo	Molybdenum	S	Sulphur
Al	Aluminum	Eu	Europium	N	Nitrogen	Sb	Antimony
Am	Americium	F	Fluorine	Na	Sodium	Sc	Scandium
Ar	Argon	Fe	Iron	Nb	Niobium	Se	Selenium
As	Arsenic	Fm	Fermium	Nd	Neodymium	Si	Silicon
At	Astatine	Fr	Francium	Ne	Neon	Sm	Samarium
Au	Gold	Ga	Gallium	Ni	Nickel	Sn	Tin
B	Boron	Gd	Gadolinium	No	Nobelium	Sr	Strontium
Ba	Barium	Ge	Germanium	Np	Neptunium	Ta	Tantalum
Be	Beryllium	H	Hydrogen	O	Oxygen	Tb	Terbium
Bi	Bismuth	He	Helium	Os	Osmium	Tc	Technetium
Bk	Berkelium	Hf	Hafnium	P	Phosphorus	Te	Tellurium
Br	Bromine	Hg	Mercury	Pa	Protactinium	Th	Thorium
C	Carbon	Ho	Holmium	Pb	Lead	Ti	Titanium
Ca	Calcium	I	Iodine	Pd	Palladium	Tl	Thallium
Cd	Cadmium	In	Indium	Pm	Promethium	Tm	Thulium
Ce	Cerium	Ir	Iridium	Po	Polonium	U	Uranium
Cf	Californium	K	Potassium	Pr	Praseodymium	V	Vanadium
Cl	Chlorine	Kr	Krypton	Pt	Platinum	W	Tungsten
Cm	Curium	La	Lanthanum	Pu	Plutonium	Xe	Xenon
Co	Cobalt	Li	Lithium	Ra	Radium	Y	Yttrium
Cr	Chromium	Lr	Lawrencium	Rb	Rubidium	Yb	Ytterbium
Cs	Caesium	Lu	Lutetium	Re	Rhenium	Zn	Zinc
Cu	Copper	Md	Mendelevium	Rh	Rhodium	Zr	Zirconium
Dy	Dysprosium	Mg	Magnesium	Rn	Radon		

Viscosity Chart for Common Fluids

Viscosity : A measure of the resistance of a fluid to flow (usually through a specific orifice).

Thixotropic : Describes materials that are gel-like (e.g., Toothpaste) at rest but fluid when agitated.

Centipoise (cps) : It is a relative comparison between the viscosity of water and a fluid. Water is the standard by which viscosity of all fluids are measured. Water is 1 cps at 70°F (21°C).

Fluid	Approximate Viscosity (cps)
Water @ 70°F (21°C)	1
Milk	3
Kerosene / Blood	10
Ethylene Glycol / Antifreeze	15
Corn Oil	65
Motor Oil - SAE 30 / Maple Syrup	150
Motor Oil - SAE 40 / Castor Oil	250
Glycerine	1000
Corn Syrup / Honey	2000
Molasses	5000
Chocolate Syrup	10000
Tomato Sauce	50000
Sour Cream	100000
Peanut Butter	150000
Lard	1000000
Caulking Compound	5000000
Window Putty	100000000

NOTE: The above table is intended as a general guide only. Viscosity values vary with change in temperature.

WORKSHOP METALS

COEFFICIENT OF FRICTION BETWEEN VARIOUS MATERIALS

Static friction (μ_s) is friction between two or more solid objects that are not moving relative to each other.

Kinetic/Sliding friction (μ_k) is is friction between two or more solid objects that are moving relative to each other.

Material 1	Material 2	Coefficient of Friction			
		Dry		Greasy / Lubricated	
		Static (μ_s)	Sliding (μ_k)	Static (μ_s)	Sliding (μ_k)
Aluminum	Aluminum	1.05 - 1.35	1.40	0.30	
Aluminum	Mild Steel	0.61	0.47		
Brass	Cast Iron		0.30		
Bronze	Steel			0.16	
Cast Iron	Cast Iron	1.10	0.15		0.07
Copper	Copper	1.00		0.08	
Copper	Mild Steel	0.53	0.36		0.18
Copper	Steel		0.80		
Copper	Stainless Steel	0.23	0.21		
Diamond	Diamond	0.10		0.05 - 0.10	
Diamond	Metal	0.10 - 0.15		0.10	
Graphite	Graphite	0.10		0.10	
Graphite	Steel	0.10		0.10	
Lead	Steel		1.40		
Magnesium	Magnesium	0.60		0.08	
Nickel	Mild Steel		0.64		0.18
Nylon	Nylon	0.15 - 0.25			
Platinum	Platinum	1.20		0.25	
Polystyrene	Steel	0.30 - 0.35		0.30 - 0.35	
Polythene	Steel	0.20		0.20	
Rubber	Asphalt (Dry)		0.50 - 0.80		
Rubber	Concrete (Dry)		0.60 - 0.85		
Silver	Silver	1.40	0.55		
Sintered Bronze	Steel			0.13	
Steel	Brass	0.51	0.44		
Steel	Cast Iron		0.23	0.18	0.133
Steel	Copper Lead Alloy	0.22		0.16	0.145
Steel	Graphite	0.10		0.10	
Steel	Lead	0.95	0.95	0.50	0.30
Steel	Phosphor Bronze	0.35			
Steel	Steel (Mild)	0.74	0.57		0.09 - 0.19
Steel	Steel (Hard)	0.78	0.42	0.05 - 0.11	0.03 - 0.12
Steel	Zinc (Plated on steel)	0.50	0.45		
Teflon	Steel	0.04		0.04	0.04
Tin	Cast Iron		0.32		
Tungsten Carbide	Tungsten Carbide	0.20 - 0.25		0.12	
Tungsten Carbide	Steel	0.40 - 0.60		0.08 - 0.02	
Tungsten Carbide	Copper	0.35			
Tungsten Carbide	Iron	0.80			
Zinc	Zinc	0.60		0.04	
Zinc	Cast Iron	0.85	0.21		

COMMON WORKSHOP METALS

The following is a compilation of metals used in day-to-day applications. Its aim is to provide a basic knowledge of the metal's properties and general uses. It is always advisable to seek detailed information from the metal supplier.

Aluminum

It is a silver color low density light weight industrial metal which is produced in many different shapes and grades and used extensively in the engineering industry. In its pure form it is ductile, exhibits moderate strength, is very resistant to corrosion under most circumstances and can be dramatically strengthened by the addition of alloying elements such as Cu, Mg, Mn and Si. It is used extensively in the transport, aerospace, food, chemical and architectural industries.

Designation	Description and Uses
Alloy 1100	A commercially pure aluminum grade, soft, ductile and has excellent workability. It can be easily welded and is used extensively in the general fabrication, sheet-metal, drawn and spun parts, chemical and food processing industries.
Alloy 2011	This is a free machining alloy with high mechanical strength and formability, suitable for high speed/high feed machining on automatic lathes for complex parts. It has high strength but poor weldability and corrosion resistance and can be heat treated.
Alloy 2024	One of the best known of high strength aluminum alloys. It has good workability but poor weldability, excellent strength and fatigue resistance and used for structures and parts where a good strength-to-weight ratio is desired.
Alloy 3003	The most widely used of all aluminum alloys. It is essentially commercially pure aluminum with the addition of manganese which increases its strength. Has excellent corrosion resistance and workability, can be deep drawn or spun, welded and used for the manufacture of cooking utensils, decorative trim, tanks and chemical equipment.
Alloy 5052	This is the highest strength alloy of the more common non-heat-treatable aluminum grades, has good weldability, excellent workability and higher fatigue strength than most aluminum alloys with particularly good resistance to marine atmosphere and salt water corrosion and hence used extensively in the marine and aircraft components.
Alloy 6011	It is a versatile, heat treatable, highly formable and weldable aluminum alloy with high strength and excellent corrosion resistance.
Alloy 6061	This is the most versatile of the heat treatable aluminum alloys and offers a wide range of mechanical properties and corrosion resistance. Can be easily fabricated and welded and used for steps, walkways, cover plates, platforms trench covers, etc.
Alloy 6063	Architectural grade aluminum with relatively high tensile properties, excellent finish characteristics, high corrosion resistance, good weldability, machinability and best suited for anodizing. Used for trim, moldings, aluminum furniture and parts that need to have a decorative anodized finish.
Alloy 7075	It is one of the highest strength aluminum alloys and its strength-to-weight ratio is excellent. Ideally suited for highly stressed aircraft, transport industry parts and structures where strength is critical. Can be welded, highly polished and used in the plastics mold industry for large parts.

COMMON WORKSHOP METALS

Steels

Hot Rolled Steel: Hot rolling is a mill process which involves rolling steel at a temperature of about 950 degrees C which is above the steel's recrystallization temperature, after which it can be easily shaped and formed. When the steel cools off it will shrink slightly thus giving less control on the size and shape of the finished product when compared to Cold Rolled Steel. Hot Rolled Steel is typically cheaper than Cold Rolled Steels i.e. angles, flats, bars, etc.
Uses: Mainly used in the construction, fabrication and welding trades.

Cold Rolled Steel is essentially Hot Rolled Steel that has been further processed in a cold reduction mill by cold rolling/finishing/drawing/turning/grinding and polishing. This process produces steel with a wider range of surface finishes, higher yield and tensile strengths, closer dimensional tolerances and gets rid of surface imperfections and produces concentricity and straightness, as compared to Hot Rolled Steel. When referring to bar products, the term cold finishing/drawing is used.
Uses: Engineering and Toolroom applications.

Steels are classified into 4 basic groups based on their chemical compositions:

- Carbon Steel
- Alloy Steel
- Stainless Steel
- Tool Steel

Carbon Steel

Carbon Steel is a steel alloy where carbon is the main alloying element. It contains up to 2.0% C by weight, along with a variety of other alloying elements, such as manganese or silicon, depending on the desired attributes and physical properties.

Carbon Steel is grouped into 3 categories:

- Low Carbon Steel (Mild Steel) 0.04% to 0.30% C
- Medium Carbon Steel 0.31% to 0.60% C
- High Carbon Steel (Tool steel) 0.61% to 1.50% C

The higher the carbon content, the harder it is to machine, cut, bend or weld.

Designation	Description and Uses
A36	Commonly used mild steel which has excellent welding properties and is suitable for punching, tapping, drilling and machining purposes
1008	A low-carbon steel containing approximately 0.40% Mn that has excellent weldability, formability and is primarily used for extruded cold headed, cold upset and cold pressed parts and forms.
1010	Soft ductile, easily welded grade and case hardenable. Mainly used for general fabrication work.
1018	General purpose commonly available free machining grade carbon steel. It can be easily formed, machined, welded and fabricated.
1020	Cold finished mild steel for machined parts or shafting, can be case hardened and welds readily.
1026	It is mainly used to manufacture square and round tube by cold rolling flat sections into tubular shapes and then welding it. The tubing is easy to machine and fabricate and it can be bent or drawn, flattened, flared, drilled, punched or welded. Used for structural supports, frames, trailer beds and components, etc.
1030	Medium tensile steel for shafts and medium stressed parts, can be welded if precautions are taken.
1045	A medium carbon steel, suitable for medium stressed parts in machinery and industrial tools. Can be flame or induction hardened.

WORKSHOP METALS

Continued ...

COMMON WORKSHOP METALS

Designation	Description and Uses
1141	It is a medium-carbon steel with higher mechanical properties than other medium carbon steels, as well as free machining properties. It has a fine-grain structure and is used extensively in automatic machines for producing pins, shafts, bolts, studs, etc.
12L14	The addition of lead gives increased machinability. For use in automatic and semi-automatic machines for repetition parts.

Alloy Steel

Alloy steel is steel that has had small amounts of one or more alloying elements (other than carbon) such as manganese, silicon, nickel, titanium, copper, chromium and aluminum added. This produces specific properties that are not found in regular carbon steel. Alloy steels are workhorses of industry because of their economical cost, wide availability, ease of processing, and good mechanical properties. Alloy steels are generally more responsive to heat and mechanical treatments than carbon steels.

Designation	Description and Uses
4130	A chromium molybdenum alloy steel, widely used in the aircraft and fabrication industry because of its ease of weldability, ease of fabrication and mild hardenability. Used for aircraft parts, roll cages for racing cars and other similar applications and available in round bar, hollow tube and plate sections.
4140	Suitable for applications requiring high strength and good toughness, particularly in medium to large cross-sections, medium to high stressed parts and can be nitrided or flame/induction hardened.
4340	The addition of molybdenum for applications requiring high strength and toughness prevents temper brittleness. Used in general engineering, automotive and aircraft applications. It can be flame/induction hardened or nitrided.
6150	A high carbon, chromium-vanadium alloy steel with high strength and hardness. When heat treated, the material has excellent toughness, shock and abrasion resistance and used for hand tools, machinery parts, etc.
8620	A nickel-chromium-molybdenum alloy steel which when hardened by carburization results in a tough core and a hard wear-resistant surface. Used for gears, crankshafts and other similar machine parts.

Stainless Steel

Stainless steel is a steel alloy with excellent corrosion resistance as compared to carbon/alloy steel. Common alloying ingredients include chromium usually between 10 - 20% plus nickel, or molybdenum.

Designation	Description and Uses
303	Free cutting steel for components subjected to corrosion in mechanical and plant engineering, mass produced items such as screws, bolts and nuts.
304	General purpose austenitic stainless steel. Frequently used where welding without subsequent heat treatment is involved.
304L	A commonly used low carbon stainless steel alloy with 18% Cr and 8% Ni. It is easily weldable and is extensively used for drawing and stamping applications.
316	Marine grade stainless steel offering good corrosion resistance and used in food processing, chemical, fishing and hospital equipment.
321	This is a titanium stabilized, corrosion and oxidation resistant stainless steel with very good weldability and used for parts fabrication.

WORKSHOP METALS

Continued ...

COMMON WORKSHOP METALS

Designation	Description and Uses
347	A niobium / tantalum stabilized, corrosion and oxidation resistant stainless steel. Easily weldable and used in the chemical, aircraft and automotive industry.
410	A general purpose stainless steel alloy with mild corrosion resistance. This alloy can be heat treated if required to create an extremely hard surface.
416	This alloy has excellent hardenability and good wear resistant properties but is moderately corrosion resistant to mild acids and alkalies. Used for cutlery, scissors, dental and surgical instruments.
420	It is a high carbon stainless steel alloy with a minimum chromium content of 12% and offers good ductility in its annealed state. Excellent corrosion resistance properties when the metal is polished, surface ground or hardened.
440	A general purpose stainless steel alloy with mild corrosion resistance. This alloy can be heat treated to create an extremely hard surface.

Tool Steel

Tool steel is a term used for a variety of high-hardness, tough, impact and abrasion resistant steels. Used for punching, blanking and mold making dies, bending and impact applications, cutting tools and knives, etc. Tool steels are almost always used to form/mold other metal or plastic components.

Designation	Description and Uses
A Type	A very versatile, all-purpose tool steel that is characterized by a low distortion factor during heat treatment because of its increased chromium content. This tool steel has good machinability and a balance of wear resistance and toughness.
D Type	It is a high carbon/high chromium tool steel with excellent abrasion resistance and has the minimum distortion in heat treatment as compared to other tool steels. Uses include forging, casting and drawing dies.
H Type	This low carbon tool steel with moderately high additional alloys is used to cut materials at high temperatures with added strength and hardness for prolonged exposure to elevated temperatures.
O Type	A general purpose oil hardening tool steel with good abrasion resistance and toughness for a wide range of tool and die applications.
P Type	Used mainly for manufacturing plastic molding dies.
S Type	Medium carbon shock-resisting tool steel with moderate hardness and good impact toughness in both cold and hot work applications.
W Type	It is a cold work high carbon tool steel which is easily machinable, has a high surface case hardness whilst retaining a soft, ductile inner core.

Copper

100 series coppers are considered pure coppers, having between 96% and 99.9%. These grades possess high electrical conductivity and can contain additional alloying elements such as beryllium, cadmium, chromium, cobalt, nickel, or iron which help to improve the strength without affecting electrical conductivity. Copper has a unique property in as much as it softens when it is heated and quenched unlike steel which develops the opposite characteristic and hardens.

Designation	Description and Uses
102	Malleable and easily weldable high purity, high conductivity copper used commonly in the electrical industry.

WORKSHOP METALS

Continued ...

COMMON WORKSHOP METALS

Designation	Description and Uses
C110	It is the most commonly available copper product, electrolytically refined, contains 99.90% Cu and offers excellent electrical and thermal conductivity.
122	This is a high conductivity copper used commonly for oil burners, heating systems and liquid transfer applications.
145	High conductivity copper, which can be easily machined and drilled. It can also be hot forged and has similar properties to alloy 110.
172	Highest strength copper alloy used extensiely for aircraft bearings, drilling equipments, springs, terminals and is non sparking.
182	A chromium copper alloy which has excellent strength, high conductivity and is used for welding tips and holders, electrodes, etc.
1751	This alloy has the highest electrical and thermal conductivity of all copper alloys and is used in the electrical and heat transfer applications.
C14700	A sulphur bearing copper for use in automatic and semi-automatic machines for repetition parts.

Brass

It is an alloy of copper and zinc and is one of the most used materials in the world. It has low friction properties, good strength and electrical conductivity with excellent decorative properties because of its similar-to-gold shine when polished. Extensively used for hardware items like knobs, hinges, wood screws, locks, electrical connectors, etc.

Designation	Description and Uses
230	Commonly known as ***Red Brass***, this material is reddish in color and contains approximately 85% Cu and 15% Zn with a tensile strength of approximately 70,000 psi and a hardness of around 77HRB and normally available in pipe form.
260	Commonly known as ***Cartridge Brass*** because it is generally used for shell casings. Contains approximately 70% Cu and 30% Zn, tensile strength of approximately 61,000 psi and hardness around 77HRB. Not very machinable, but easily worked, normally available in foil, plate, sheet and tube.
330	Contains approximately 66% Cu and 33% Zn with a slight inclusion of lead, tensile strength of approximately 65,000 psi and a hardness of around 70HRB. Good workability and machinability because of a slight lead content and mainly available in tube form.
360	Commonly known as ***Free Machining Brass***, it contains approximately 66% Cu and 33% Zn with a slight inclusion of lead. It has a tensile strength of approximately 65,000 psi and hardness of around 70HRB. The presence of lead in the alloy makes it is highly machinable. Normally available in most solid shapes.
385	Contains approximately 58% Cu and 38% Zn and 4% Pb with a tensile strength of approximately 65,000 psi and hardness of around 80HRB. Used in mass production of brass components in high speed automatic lathes where machining output and long tool life is required and where no further cold forming operations after machining are required like screw thread components, nuts and bolts, etc.
464	Commonly known as ***Naval Brass / Naval Bronze*** and contains approximately 59% Cu and 39% Zn with a slight inclusion of iron and tin and has a tensile strength of approximately 75,000 psi and a hardness of around 82HRB. The tin is added to help deal with corrosion especially in seawater. Normally available in plate, round and sheet form.

COMMON WORKSHOP METALS

Bronze

It is an alloy primarily consisting of copper and tin often with the addition of Al, Mg, Ni or Zn and sometimes metalloids such as arsenic, phosphorus or silicon. These additions produce a range of alloys that may be harder than copper alone and have other useful properties, such as stiffness, ductility or machinability.

Designation	Description and Uses
510	Commonly known as ***Phosphor Bronze*** and used extensively for springs, hardware and thrust bearings because of its capability to be tempered and thus resist fatigue.
954	Commonly known as ***Aluminum Bronze*** and contains approximately 83% Cu, 11% Al, 4% Fe, 1.5% Ni with other slight additives. It is very hard 170 - 180 HRB, abrasive resistant and has excellent strength at elevated temperatures with reasonable machining properties and used for high strength bearings.
LG 2	Commonly known as ***Leaded Gunmetal*** and contains approximately 85% Cu, 5% lead, 5% Sn and 5% Zn with other slight additives. It has excellent machining properties with medium strength, good pressure toughness with reasonable corrosion resistance to sea water and brine. Used for pump, valve components, etc.
PB 1	Commonly known as ***Phosphor Bronze*** and contains approximately 88% Cu, 11% Sn and 1% P with other slight additives. Has good machining properties, high strength, and good corrosion resistance to sea water and brine. Used for heavy load gears, worm wheels and bearings using good lubrication and proper alignment.

Case Hardening Steel

EN-36A	Case hardening chromium-nickel steel for applications requiring excellent wear, high toughness and core strength. Suitable for gear wheels, crankshafts, etc.

High Speed Steel

M2 M35 M42	Tungsten molybdenum high speed steel with excellent toughness and cutting properties. HSS has a wide variety of uses including drills, taps, broaches, milling tools, cold work tools, etc.

Key Steel

EN8 1018	It is manufactured to close tolerances and used extensively in the manufacture of keys for shafts.

Plastic Mold Steel

M200	Suitable for plastic molds, die-casting industries and components for general mechanical engineering. Can be case and flame hardened, gas nitrided and chromium plated.

Silver Steel

AISI 01 DS1407	Accurately ground round bar that is used for a variety of applications such as shafts, pins, etc. Oil/water hardenable.

Cast Iron

35B	Pearlitic grain structure with excellent wear resistance and good machinability. Used for hydraulic and pneumatic piston, end caps, etc.
ASTM A48	This specification covers gray iron castings intended for general engineering use dependent on requirements such as vibration damping, rate of thermal expansion, lubrication retention and machinability. The tensile strength increases as the Class number rises. ASTM A48 Class 20, 25, 30 & 35 are the most commonly available.

COMMON WORKSHOP MATERIAL COMPARISON TABLE

ISO	Country						
	USA	Germany		U.K.	France	Japan	Sweden
	Standard						
	AISI/SAE	Mat. No.	DIN	BS	AFNOR	JIS	SS
P	**PLAIN – UNALLOYED STEELS**						
	1006	1.0201	St 36		Fd 5		1160
	1010	1.1121	Ck 10 (C10E)	045 M 10	XC 10	S 10 C	1265
	1015	1.0401	Ck 15 (C15)	080 M 15	CC 12	S 15 CK	1370
	1020	1.0402	C22	050820	CC 20		1450
	1022	1.1133	GS-20Mn 5	120 M 19	20 M 5	SMnC 420	1410
	1025	1.1158	Ck 25 (C25E)	050 A 20	XC 25	S 25 C	1450
	1035	1.1183	C 35	060 A 35	CC35	S35C	1550
	1039	1.1157	40Mn4	150M36	35M5		
	1040	1.0726	35S20	212M36	35MF4		1140
	1045	1.0503	C45	080M46	CC45		1650
	1050	1.1213	Cf53	060A52	XC48TS	S50C	1674
	1055	1.0535	C55	070M55			1655
	1060	1.0601	C60	080A62	CC55		
	1095	1.1274	Ck101	060A96	XC100		1870
	1115	1.0038	Ck16	030A04			1325
	1140	1.0726	35S20	212M36	35MF4		1957
	1213	1.0715	11 SMn 30	230M07	S250	SUM22	1912
	12L13	1.0718	11 SMnPb 30		S250Pb	SUM22L	1914
	12L14	1.0737	11 SMnPb 37		S300Pb		1926
	1330	1.1170	28Mn 6	150M28	20M5	SCMn1	
	1335	1.1167	36Mn 5		40M5	Smn438(H)	2120
	A204 GrA	1.5415	15Mo 3 / 16Mo3	1501-240	15B3		2912
Steel	A570.36	1.0038	RSt.37-2	4360 40 C	E24 2 Ne	STKM 12A	1311
	A572-60	1.8900	S380N	4360 55 E	NFA35-501		2142
	A573-81	1.0116	St.37-3	4360 40 B	E24-U		1312
	W1	1.1545	C 105 W1	BW 1A	Y105	SK 3	1880
	W210	1.1545	C 105 W1	BW 2	Y120	SUP4	2900
	LOW ALLOY STEELS						
	3135	1.5710	36NiCr6	640A35	35NC6	SNC236	
	3415	1.5732	14NiCr10		14NC11	SNC415	
	4130	1.7218	25CrMo4	1717CDS	25CD4	SCM420	2225
	4137	1.7220	34CrMo4	708A37	35CD4		2234
	4140	1.7225	42CrMo4	708M40	42CD4	SCM440	2244
	4340	1.6582	35CrNiMo6	817M40	35NCD6		2541
	5015	1.7015	15Cr3	523M15	12C3	SCr415	
	5115	1.7131	16MnCr5	527M20	16MC5		2511
	5120	1.0841	St52-3	150M19	20MC5		2172
	5132	1.7033	34Cr4	530A32	32C4	SCr430	
	5140	1.7035	41Cr4	530A40	42C4	SCr440	2245
	5155	1.7176	55Cr3	527A60	55C3	SUP9	
	6150	1.8159	50CrV4	735A50	50CV4	SUP10	2230
	8620	1.6523	21NiCrMo2-2	805M20	20NCD2	SNCM220	2506
	9255	1.0904	55Si7	250A53	55S7		2085
	9840	1.6511	36CrNiMo4	816M40	40NCD3		

WORKSHOP METALS

COMMON WORKSHOP MATERIAL COMPARISON TABLE

ISO	Country						
	USA	Germany		U.K.	France	Japan	Sweden
	Standard						
	AISI/SAE	Mat. No.	DIN	BS	AFNOR	JIS	SS
P Steel	52100	1.3505	100Cr6	534A99	100C6	SUJ2	2258
	A182	1.7380	10CrMo9-10	1501-622	12CD9 10		2218
	A204	1.5415	15Mo3 (16Mo3)	1501-240	15D3		2912
	A573-81	1.0144	St44-2	4360-43C	E28-3	SMH400A	1412
	L1	1.7039	41CrS4	524A14			2092
	L3	1.2067	102Cr6	BL3	Y100C6		
	L6	1.2713	55NiCrMoV6		55NCDV7	SKT4	2550
	O2	1.2842	90MnCrV8	BO2	90MV8		
	S1	1.2542	45WCrV7	BS1		S1	2710
	HIGH ALLOY STEELS (TOOL & DIE STEELS)						
	440C	1.4125	X105CrMo17		Z100CD17	SUS440C	
	446	1.4749	X18CrNi28			SUH446	
	A2	1.2363	X100CrMoV51	BA2	Z100CDV5	SKD12	2260
	A353	1.5662	X8Ni9	1501.509			
	D2	1.2379	X155CrMoV121	BD2	Z160CDV12		2310
	D3	1.2080	X210Cr12	BD3	Z200C12	SKD1	
	D3	1.3343	S 6-5-2	4959BA2	Z40CSD10	SUH3	2715
	D4 (D6)	1.2436	X210CrW12	BD6	Z200CD12	SKD2	2312
	H11	1.2343	X38CrMoV51	BH11	Z38CDV5		
	H12	1.2606	X37CrMoW51	BH12	Z35CWDV5		
	H13	1.2344	X40CrMoV51	BH13	Z40CDV5	SKD61	2242
	H21	1.2581	X30WCrV93	BH21	Z30WCV9	SKD5	
	HW3	1.4718	X45GrSi93	401S45	Z45CS9	SUH1	
	M2	1.3343	S 6-5-2	BM2	Z85WDCV	SKH9	2722
	M35	1.3243	S 6-5-2-5	BM35	6-5-2-5	SKH55	2723
	M42	1.3249	S2-9-2-8	BM34			
	P3	1.2080	X210Cr12	BD3	Z200C12	SKD1	2710
	T1	1.3355	3 18 01	Bt1	Z08WCV	SKH2	
M Stainless Steel	**STAINLESS STEELS - AUSTENITIC**						
	301	1.4310	X8CrNiS18-9	301S21	Z12CN17.07	SUS301	2331
	303	1.4305	X12CrNiS18 8	303S21	Z10CNF18.09	SUS303	2346
	304	1.4350	X5CrNi18 9	304S31	Z6CN18.09	SUS304	2332
	304L	1.4306	X2CrNi19-11	304S12	Z2CrNi18.10	SCS19	2352
	304LN	1.4311	X2CrNiN18-10	304S62	Z2CN18.10	SUS304LN	2371
	308	1.4406	X10CrNi18.08	301S21	Z1NCDU25.20	SCS17	2370
	309	1.4828	X15CrNiSi20-12	309S24	Z15CNS20.12	SUH309	
	310S	1.4845	X12CrNi25-21	310S24	Z12CN25.20	SUH310	2361
	316	1.4401	X5CrNiMo17-12-2	316S16	Z6CND17.11	SUS316	2347
	316L	1.4435	X2CrNiMo18-14-3	316S13	Z2CND17.12	SCS16	2353
	316LN	1.4429	X2CrNiMoN18 13		Z2CND17.13	SUS316LN	2375
	316Ti	1.4571	X10CrNiMoTi1810	320S17	Z6NDT17.12		2350
	317L	1.4438	X2CrNiMo18-15-4	317S12	Z2CND19.15	SUS317L	2367
	318	1.4583	X10CrNiMoNb1812		Z6CNDNb1713B		
	321	1.4541	X6CrNiTi18-10	321S12	Z6CNT18.10	SU321	2337
	347	1.4550	X6CrNiNb18-10	347S17	Z6CNNb18.10	SU347	2338

COMMON WORKSHOP MATERIAL COMPARISON TABLE

ISO	Country						
	USA	Germany		U.K.	France	Japan	Sweden
	Standard						
	AISI/SAE	Mat. No.	DIN	BS	AFNOR	JIS	SS
M Stainless Steel	STAINLESS STEELS - FERRITIC / MARTENSITIC						
	403	1.4000	X6Cr13	403S17	Z6C13	SUS403	2301
	405	1.4724	X10CrAlSi13	403S17	Z10C13	SUS405	
	410	1.4006	X12Cr13	410S21	Z10C14	SUS410	2302
	416	1.4005	X12CrS13	416S21	Z11CF13	SUS416	2380
	430	1.4016	X6Cr17	430S15	Z8C17	SUS430	2320
	430F	1.4104	X14CrMoS17		Z10CF17	SUS430F	2383
	431	1.4057	X17CrNi16-2	431S39	Z15CNi6.02	SUS431	2321
	434	1.4113	X6CrMo17-1	434S17	Z8CD17.01	SUS434	2325
	446	1.4762	X10CrAlSi25		Z10CAS24	SUH446	2322
	CA6-NM	1.4313	X3CrNiMo13-4	425C11	Z4CND13.4M	SCS5	2385
	EV8	1.4871	X53CrMnNiN21.9	349S54	Z52CMN21.09	SUH35/36	
	HNV6	1.4747	X80CrNiSi20	443S65	Z80CSN20.02	SUH4	
K Cast Iron	GREY CAST IRON						
	No. 20B	0.6010	GG10		Ft10D	FC100	0110
	No. 25B	0.6015	GG15	Gd. 150	Ft15D	FC150	0115
	No. 30B	0.6020	GG20	Gd. 220	Ft20D	FC200	0120
	No. 35B	0.6025	GG25	Gd. 260	Ft25D	FC250	0125
	No. 45B	0.6030	GG30	Gd. 300	Ft30D	FC300	0130
	No. 50B	0.6035	GG35	Gd. 350	Ft35D	FC350	0135
	No. 55B	0.6040	GG40	Gd. 400	Ft40D	FC400	0140
	A436 Type 2	0.6660	GGL-NiCr202	L-NiCuCr-202	L-NC-202		0523
	MALLEABLE CAST IRON						
	40010	0.8145	GTS-45	P440/7	MN450	FCMW370	0852
	50005	0.8155	GTS-55	P510/4	MP50-5	FCMP490	0854
	A220-70003	0.8615	GTS-65-02	P570/3	MN650-3	FCMP590	0856
	A220-80002	0.8170	GTS-70-02	P690/2	MN700-2	FCMP690	0862
	NODULAR CAST IRON						
	60-40-18	0.7040	GGG 40	SNG420/12	FCS400-12	FCD400	0717.02
	80-55-06	0.7050	GGG 50	SNG500/7	FGS500-7	FCD500	0727.02
	A43D2	0.7660	GGG-NiCr202	Gd. S6	S-NC202		0776
	100.70.03	0.7070	GGG 70	SNG700/2	FGS700-2	FCD700	0737.01
N Non-Ferrous	ALUMINUM ALLOYS						
	1000	3.0255	Al99-5	L31/34/36	A590 50C		
	7050	3.4345	AlZnMgCu0.5	L86	AZ4GU/9051		
	A360.2	3.2383	G-AlSi10Mg(Cu)	LM9			4253
	A413.0	3.2582	GD-AlSi12				4247
	A413.1	3.2583	G-AlSi12(Cu)	LM20			4260
	A413.2	3.2581	G-AlSi12	LM6			4261
	AZ 81	3.5812	G-MgAl8Zn1	MAG1	G-A9		
	AZ 91	3.5912	G-MgAl9Zn1	MAG7	G-A9Z1		

COMMON WORKSHOP MATERIAL COMPARISON TABLE

ISO	USA	Germany		U.K.	France	Sweden
	AISI/SAE	Mat. No.	DIN	BS	AFNOR	SS
N	EZ33	3.5103	MgSE3Zn2Zr1	MAG6	G-TR3Z2	
	QE22	3.5106	G-MgAg3SE2Zr1	MAG12	G-AG22.5	
	COPPER ALLOYS					
Non-Ferrous	C 18200	2.1293	CuCrZr	CC 102	U-Cr 0.8 Zr	
	C 23000	2.0240	CuZn15	CZ 102	CuZn 15	
	C 26000	2.0265	CuZn30	CZ 106	CuZn 30	
	C 27200	2.0321	CuZn37	CZ 108	CuZn 37	
	C 63000	2.0966	CuAl10Ni5Fe4	CA 104	U Λ 10 N	
	C 81500	2.1292	G-CuCrF 35	CC1-FF		
	C 83600	2.1096	G-CuSn5ZnPb	LG 2	U-E 5 Pb 5 Z 5	
	C 86200	2.0596	G-CuZn34Al2	HTB 1	U-Z 36 N 3	
	C 86500	2.0592	G-CuZn35Al1	HTB 1	U-Z- 36 N 3	
	C 90700	2.1050	G-CuSn10	CT 1		
	C 90800	2.1052	G-CuSn12	Pb 2	U-E 12 P	
	C 93200	2.1090	G-CuSn7ZnPb		U-E 7 Z 5 Pb 4	
	C 93700	2.1176	G-CuPb10Sn	LB 2	U-E 10 Pb 10	
	C 93800	2.1182	G-CuPb15Sn	LB 1	U-Pb 15 E 8	
	C 94100	2.1188	G-CuPb20Sn	LB 5	U-Pb 20	
S	**SUPER ALLOYS (Fe / Ni / Co / Ti - based)**					
Heat-Resistant Super Alloys	Hastelloy B	2.4685	G-NiMo28			
	Hastelloy C	2.4810	G-NiMo30			
	Hastelloy C4	2.4610	NiMo16Cr16Ti			
	Inconel 625	2.4856	NiCr22Mo9Nb	NA 21	NC 22 Fe DNb	
	Inconel 690	2.4642	NiCr29Fe		NC 30 Fe	
	Inconel 718	2.4668	NiCr19FeNbMo		NC 19 Fe Nb	
	Inconel X-750	2.4669	NiCr15Fe7TiAl		NC 15 TNb A	
	Inconel 751	2.4694	NiCr16Fe7TiAl			
	Incoloy 825	2.4858	NiCr21Mo	NA 16	NC 21 Fe DU	
	Monel 400	2.4360	NiCu30Fe	NA 13	NU 30	
	Monel K-500	2.4375	NiCu30Al	NA 18	NU 30 AT	
	N 08020	1.4563	X1NiCrMoCuN31274		Z1NCDU31.27	
	N 08800	1.4558	X2NiCrAlTi3220	NA 15		
	N08031	1.4562	X1NiCrMoCu32287			
	N08330	1.4864	X12NiCrSi36 16	NA 17	Z12NCS35.16	
	Nimonic 75	2.4630	NiCr20Ti	HR 5	NC 20 T	
	R 50250	3.7025	Ti 1	2 TA 1		
	R 52250	3.7225	Ti 1 Pd	TP 1		
	R 54620	3.7145	TiAl6Sn2Zr4Mo2Si			
	R 56400	3.7165	TiAl6V4	TA 10-13	T-A 6 V	
H	**HARDENED METALS (WHITE / HARDENED CAST IRON)**					
Hardened Metals	A532 A25%Cr	0.9650	G-X 260 Cr 27	Grade 3D		0466-00
	A532 A25%Cr	0.9655	G-X 300 CrMo 27 1	Grade 3E		
	Ni-Hard 1	0.9625	G-X 330 NiCr 4 2	Grade 2B		0513-00
	Ni-Hard 2	0.9620	G-X 260 NiCr 4 2	Grade 2A		0512-00

WORKSHOP METALS

COMMON ENGINEERING PLASTICS

The following is a compilation of Common Plastics used in the Engineering Industry for the manufacture/fabrication of components. It is by no means a comprehensive list and as always, it is best to seek the advice of the manufacturer for specific applications.

Thermoplastics

Soften to a flowable state when heated and harden upon cooling. This characteristic allows thermoplastics to be remolded and recycled without affecting the material's physical property.

- Superior Finish
- Highly Recycleable
- Chemical Resistant
- High-impact Resistant
- Eco-friendly Manufacturing
- Remolding / Reshaping Capabilities

Generally more expensive than thermoset plastics.

Thermoset Plastics

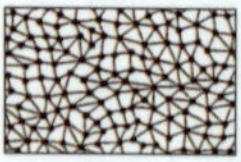

Contains polymers that cross link together during the curing process to form an irreversible chemical bond. Ideal for High-heat applications such as electronics and domestic appliances.

- Cost Effective
- Highly Flexible Design Possible
- Excellent Aesthetic Appearance
- High Levels of Dimensional Stability
- Thick / Thin Wall Design Capabilities
- More Resistant to High Temperatures

Cannot be recycled, remolded or reshaped into original configuration.

Acrylic - Polymethyl Methacrylate PMMA – *Plexiglas® Lucite® Perspex®*

An optically clear transparent plastic having excellent weather resistance, high impact strength, good chemical resistance and is easily thermoformed and fabricated. It is extensively used for instrument covers & enclosures, safety shields, tanks, etc.

Acrylonitrile - Butadiene Styrene ABS – *Magnum® Tecaran® Cycolac®*

A rigid, low-cost thermoplastic which can be easily machined to a fine finish. It can be electroplated and is used in the production of moldings, pipe fittings and products needing long life.

Polyoxybenzylmethyleneglycoanhydride – *Bakelite®*

A hard and stiff plastic and comes with a variety of fillers such as canvas, paper, cloth, etc. Being a good insulator, it is used extensively in the manufacture of base plates, handles and knobs for the electrical industry.

Chlorinated PVC CPVC – *Corzan® Charlotte® Lucalor®*

This is a high heat, corrosion-resistant material with properties similar to PVC and ideally suited for components where high temperatures are necessary. CPVC pipes are preferred over PVC where mechanical strength at higher temperatures is required.

Fiberglass

It has a high strength-to-weight ratio with excellent corrosion resistant and dielectric properties. Used extensively in boat building and tank manufacturing.

High Density Polyethylene HDPE – *Polystone HD 300® Tecafine®*

An economical, high-impact, lightweight thermoplastic with excellent chemical resistance and high tensile strength. It exhibits excellent resilience, stress, crack and abrasion resistance. It is easily machined and requires welded or mechanical assembly.

High Molecular Weight Polyethylene HMWPE – *Polystone HD 500®*

It has high rigidity and creep resistance, ideal to use in the food industry in the manufacture of cutting boards, underlays in food preparation and machine parts.

PLASTICS

COMMON ENGINEERING PLASTICS

Polyacetal (Acetal) POM – *Celcon® Delrin® Ertacetal® Hostaform®*

Possesses high hardness, stiffness, excellent creep resistance & dimensional stability with a low coefficient of friction. Used extensively for the manufacture of precision gears and bearings, electrical insulators, and parts which operate in water between 60° and 90°C.

Polyamide PA – *Nylatron® Nylon® Nylube® Kevlon® Stanyl 46®*

It has high damping properties, good dimensional stability, self-lubricating, high wear resistance with excellent bearing & sliding properties. Used extensively to manufacture gears, bearings, rollers, etc. There are many speciality grades available in this range.

Polycarbonate PC – *Lexan® Makrolon® Hyzod® Merlon® Makrolon®*

Is virtually an unbreakable transparent thermoplastic used extensively in safety and security-related applications. Common uses include safety shields, enclosures, safety spectacles, machine guards and bullet-resistant applications. It is easily glued.

Polyether Sulfone PES – *Victrex® Tecason® Ultrason® Sustason®*

A specially engineered thermoplastic with high thermal, oxidative and hydrolytic stability and good resistance to acids, alkalies, salt solutions, oils and greases.

Polyether-Ether-Ketone PEEK – *Victrex® Ketron® Tecapeek®*

Excellent wear, toughness and impact strength and good dimensional stability. Addition of graphite and carbon fibers results in a bearing grade with low friction and long wear applications. Used in the manufacture of gears, bearings, valves and pumps.

Polyethylene PE – *Polythene® Alathon® Lupolen®*

Possesses excellent abrasion resistance, low coefficient of friction, chemical resistance, impact resistance and damping properties. It is used extensively in the food industry where direct contact with foodstuffs is necessary.

Polyethylene Terephthalate PET – *Tecapet® Ertalyte® Vivak®*

A dimensionally stable thermoplastic used in place of Polyacetal when machining is the main criteria. Commonly used for drink bottle manufacture.

Polylactic Acid (Polylactide) PLA – *Lacton® Ingeo®*

A biodegradable thermoplastic derived from corn starch, tapioca roots, sugarcane, etc. It is one of the materials used for 3D printing with fused deposition techniques.

Polypropylene PP – *Simona® Gehr® Polystone PP® Tecafin®*

An economical, lightweight thermoplastic that offers high corrosion, abrasion and impact resistance. It also exhibits high tensile strength and stiffness at elevated temperatures and is superior to HDPE, easily machined and welded.

Polystyrene PS – *Styrofoam® Thermocol®*

Manufactured in various forms based on application. PS is used to manufacture disposable cutlery, plastic toys, etc. Expanded Polystyrene foam (EPS) is used for packaging to cushion fragile products. Extruded Polystyrene foam (XPS) is used for drinking cups for hot beverages and hot food containers.

Polysulphone PSU – *Ultrason S® Udel® Tecason®*

Possesses excellent radiation stability, good chemical resistance and used wherever higher temperature (in excess of 150°C) is required such as food processing industry, milk machine pumps and medical components subject to sterilisation.

Polytetrafluoroethylene PTFE – *Teflon® Tetron® Fluon® Rulon®*

It has a combination of chemical, electrical, mechanical and thermal properties and the lowest coefficient of friction unmatched in other plastics. It is the most chemically inert of any material. Being non-stick, it is used extensively in the coating of cooking utensils.

COMMON ENGINEERING PLASTICS

Polyurethane PU – *Desmopan® Estane® Avalon® Isothane®*

It has excellent impact and abrasion resistance and resists sunlight and weathering.

Polyvinyl Chloride PVC – *Simona® Gehr® Darvic® Cobex®*

A high corrosion-resistant plastic ideally suited where chemical resistance to acids, alkalies, alcohols and halogens. It has a high strength to weight ratio and is easily fabricated either by welding or gluing. **Unplasticised PVC (uPVC)** is used extensively for pipe & pipe fittings because of its excellent stiffness properties.

Polyvinylidene Fluoride PVDF – *Solf® Kynar® Hylar®*

It is a high performing thermoplastic with high mechanical strength & abrasion resistance. It is highly resistant to most environmental conditions and corrosive materials.

Ultra High Molecular Weight PE UHMWPE – *Simona® Ralloy® Tivar®*

It is high abrasion and chemical resistant when compared to other thermoplastics. It also has exceptional impact resistance even at cryogenic temperatures and self-lubricating.

TYPES OF STRESSES IN ADHESIVE BONDS

Tensile	Shear	Peel	Cleavage
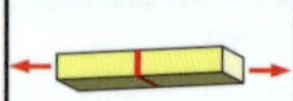	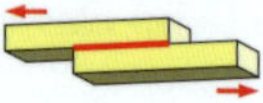	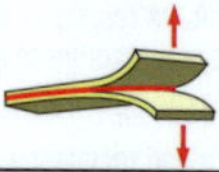	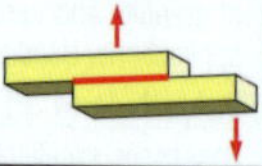

PLASTICS ADHESIVE JOINTS

Poor Joint Design		Preferred Joint Design	
Butt Joint		Butt Joint	
Tee Section Joint		Tee Section Joint	
Basic Lap Joint		Offset Lap Joint	
		Double Lap Joint	
		Inside Taper Lap Joint	
		Outside Taper Lap Joint	
		Scarf Joint	
		Double Scarf Joint	
Offset Lap Joint	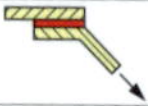	Offset Lap Joint	
Dissimilar Section Joint		Dissimilar Section Joint	
Double Offset Lap Joint	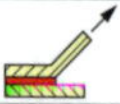	Double Offset Lap Joint	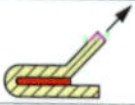
Rod to Sheet Joint		Rod to Sheet Joint	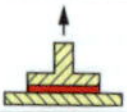

PLASTICS

COMMON ENGINEERING ADHESIVES

The following is a compilation of Common Adhesives used in the Engineering Industry for bonding components. It is by no means a comprehensive list and as always, it is best to seek the advice of the manufacturer for specific applications.

Acrylic / Acrylate: This adhesive is a chemically activated two component structural adhesive and comes in a wide variety of types, e.g., General purpose, High peel and Shock resistant, for use on metal, ceramics and plastic materials such as ABS, Polystyrene and Polycarbonate. Acrylic adhesives are known for their excellent environmental resistance and fast setting time when compared to other resin systems.

Cyanoacrylate *Super Glue®*: An anaerobic compound that comes in a variety of viscosities for gap filling between substrates. May be used with accelerator/primer. Hardening takes place in a few seconds and for this reason is frequently used in the assembly of small parts. It bonds most types of rubber extremely well and has high strength, excellent adhesion to a wide variety of substrates, especially plastics.

Epoxy: Usually come as an Adhesive/Hardener two part compound with a variety of setting times. Epoxy adhesives are widely used for their toughness and resistance to chemical and environmental damage.

Hot Melt: These are solvent-free thermoplastics that melt above 180°F and then rapidly set upon cooling. They can be repeatedly softened by heat which allows parts to be removed, repositioned or replaced during assembly. A hot melt glue gun is used for dispensing the adhesive.

Light Cure: This adhesive cures in seconds on exposure to UV light sources and is specially suitable for bonding glass to glass and glass to metal.

Phenolic, Melamine & Formaldehyde Resins: These are thermosetting adhesives that offer strong bonds and good resistance to high temperatures.

Polysulphides: It is basically a caulking type adhesive that's primarily used for sealing seams. When cured, it is completely waterproof.

Polyurethane: These are mainly available as two part systems. Their flexibility is greater than most other epoxies and exhibit high peel and cohesive strength.

Polyvinyl Acetate (PVA): It is intended primarily for wood and paper bonding and is transparent when dry.

Polyvinyl Chloride (PVC): It is intended primarily for bonding glass, china, porcelain and other ceramic materials. Also used for metal, marble and hard plastics but not recommended for wood.

Pressure Sensitive (Contact Adhesive): These adhesives adhere to most surfaces with very light pressure. They are available in solvent and latex or water-based forms. Both surfaces of the substrate need to be coated with the adhesive and allowed to dry prior to assembly.

Thread Locking & Sealing: Basically an anaerobic sealant intended to produce a waterproof/airtight seal between two mating parts and comes in a variety of strengths ranging from permanent to easily dismantleable.

Silicone: Silicone adhesives have a high degree of flexibility and very high temperature resistance (upto 600°F) when compared to other adhesives but lack the strength of other epoxy or acrylic resins. Room temperature vulcanising (RTV) forms cure through reaction with moisture in the air. It is used extensively in the window glass industry because of its high flexibility, excellent adhesion and high working temperature. This adhesive comes packed mainly in cartridges and is dispensed using a caulking gun.

COMMON ENGINEERING ADHESIVES

Surface Preparation Prior to Bonding

In general, three steps are necessary for cleaning any surface prior to bonding.

1. **De-greasing:**
   Metals - A good quality commercial de-greaser should be used.
   Plastics - A suitable solvent such as Acetone / Methyl Ethyl Ketone is recommended.
2. **Chemical Etching:** Use a suitable acid-based commercial etcher.
   Mechanical Abrading: A Medium grit blasting (sand blasting) using Aluminum Oxide, Silicon Carbide or Garnet grit is recommended.
3. **Final De-greasing:** A good quality commercial de-greaser should be used.

In all cases, parts should be bonded as soon as possible after surface preparation.

Aluminum: De-grease with solvent and dry, clean the surface with suitable Chromic acid solution, rinse thoroughly with clear running water and dry well. Then bond immediately.

Cast Iron: De-grease, Grit blast and final de-grease.

Copper & its Alloys: Immerse in recommended acid, rinse in clear running water, dry and then bond.

Galvanized / Zinc Finished Metals: De-grease, abrade, de-grease, dry & then bond.

Glass: For normal bonding applications, de-greasing alone is sufficient. However, for optimum strength, the section to be bonded should be grit blasted with very fine Silicon Carbide grit and then bonded.

Magnesium & its Alloys: De-grease, immerse in suitable Sodium Hydroxide solution. Rinse in cold water and then immerse in suitable Chromic acid/Chromic Nitrate solution. Rinse again and then immerse again in suitable Hydrofluoric solution. Rinse again. Dry in hot air blast and bond immediately.

Stainless Steel: De-grease, etch, rinse, followed by immersion in suitable Hydrofluoric/ Nitric acid solution. Finally rinse and hot air dry.

Steel: De-grease, medium grit blast, final de-grease and dry.

Teflon: Special etching solutions are recommended prior to bonding.

Phenolic, Polyester, Polyurethane Resins: De-grease, abrade and de-grease again.

Rubber: Surface etching/abrading is recommended followed by thorough rinsing.

Tin: De-grease, abrade and de-grease again.

Common Solvents / Cleaners

Acetone: Used mainly to clean and remove epoxy resins, adhesives, contact cement and ink stains. Dissolves/softens many plastics.

Cyanoacrylate (Super Glue) Adhesive Remover: Used in the removal of unwanted super glue from metal, plastic and rubber surfaces. To disassemble bonded part, soak in solution for extended period of time.

Lacquer Thinner: Used mainly for thinning lacquers (paints) and also used extensively as a general cleaner/de-greaser. It is highly flammable & dissolves/softens many plastics.

Methylene Chloride: When mixed with xylene, makes a strong solvent for cleaning ink, gum, latex, oil and wax. Dissolves/softens some plastics.

Solvent Alcohol (Isopropyl Alcohol): An excellent cleaner for electrical & electronic parts.

Denatured Alcohol (Methylated Spirits): Excellent for cleaning glass and metal.

Turpentine: Mainly used as a thinner/cleaner for oil-based (enamel) paints & varnishes.

Kerosene: Used extensively as a general purpose cleaner/degreaser.

STANDARD GRINDING WHEELS MARKINGS

Standard Wheel Markings: For clarity as well as to avoid mistakes in wheel selection, a standard and universal designation of wheel specification was developed. A typical wheel marking includes the following information:

38A	46	J	8	V	S
Abrasive	Grit	Grade	Structure	Bond Type	Mfr.'s Symbol

Abrasive

Symbol	Characteristics
A	Regular aluminum oxide, hard & tough.
10A, 11A	Premium quality of aluminum oxide.
WA, 38A	White aluminum oxide, hard & brittle
FA, 57A, 89A	Semi-friable, aluminum oxide.
32A, 90A	Strong, sharp mono-crystalline aluminum oxide.
PA, 41A, 88A	Pink aluminum oxide.
RA, 25A, 46A	Ruby aluminum oxide, hard and less brittle than PA.
PSA	Special pink abrasive.
FSA	Special Semi-friable aluminum oxide.
AZ	A modified, high zirconia aluminum oxide.
C, 37C, BC	Black silicon carbide.
GC, 39C	Green silicon carbide.
AC	Blend of A and C

Grit

Coarse			Medium			Fine				Very Fine				Ultra Fine			
10	12	14	30	36	46	70	80	90	100	240	280	320	400	1000	1200	1500	2000
16	20	24	54	60		120	150	180	220	500	600	700	800	2500	3000	4000	6000

Grade: Indicates the relative strength of the bond which holds the abrasive grains in place

Very Soft	Soft	Medium	Hard	Very Hard
D.E.F.G.	H.I.J.K.	L.M.N.O.	P.Q.R.S.	T.U.V.W.X.Y.Z

Structure (Sometimes omitted in wheel markings)

Symbol No.	0	1	2	3	4	5	6	7	8	9	10	11	12	13	14
Abrasive Ratio %	62	60	50	56	54	52	50	48	46	44	42	40	38	36	34

Bond Type

V - Vitrified (Ceramic) : Most common bond used extensively for precision grinding. The porosity & strength of the vitrified bond allows high stock removal & its rigidity helps to maintain consistent precision. Not affected by water, acid, oils or ordinary temperature variations.

B - Resinoid (Organic) : Used for high-speed wheels in fabrication shops & foundries. This bond is excellent for rapid stock removal and heavy work. Resinoid bond is used in most cutting-off and snagging wheels. However, it tends to deteriorate over time and, thus has a limited shelf life.

R - Rubber : Used for centerless grinding feed wheels and for precision ball race grinding wheels. Useful where finish is critical and where burrs and burn need to be held to a minimum.

E - Shellac : Used for roll grinding and other cylindrical grinding operations where fast stock removal and finish are important.

S - Silicate : Used for wheels where heat generated in grinding must be kept to a minimum.

O - Magnesite : Used for wheels designed for spring grinding and thin work pieces.

STANDARD GRINDING WHEEL SHAPES

▼ Primary Grinding Face ▽ Secondary Grinding Face

Shape 1 - Straight

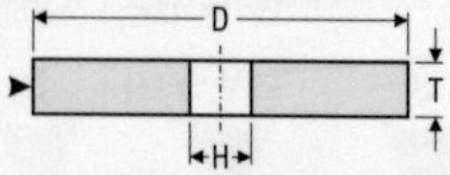

Shape 2 - Cylindrical

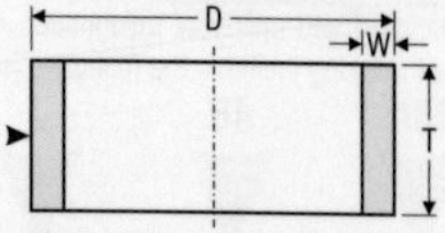

Shape 3 - Tapered One Side

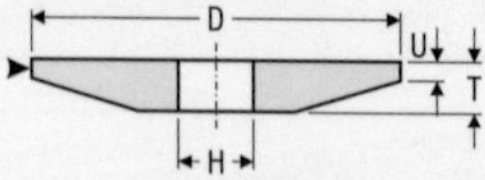

Shape 4 - Tapered Two Sides

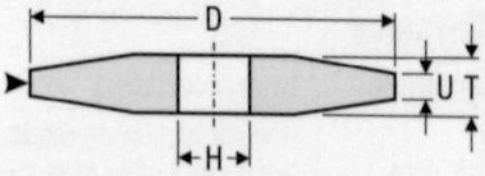

Shape 5 - Recessed One Side

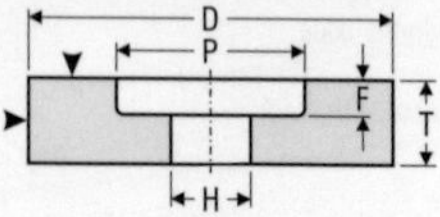

Shape 6 - Straight Cup

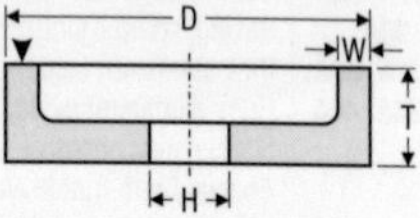

Shape 7 - Recessed Two Sides

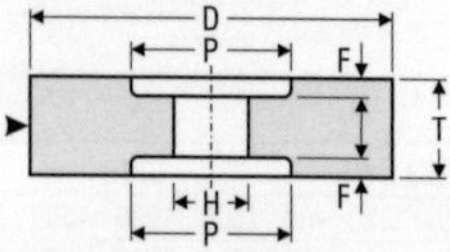

Shape 8 - Relieved Both Sides (Std. Hub)

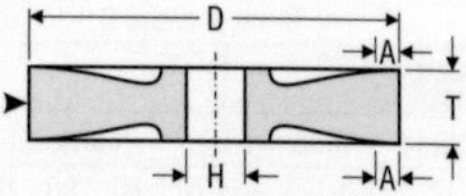

Shape 9 - Double Cup

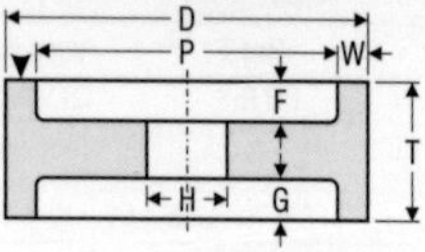

Shape 10 - Raised Dovetail

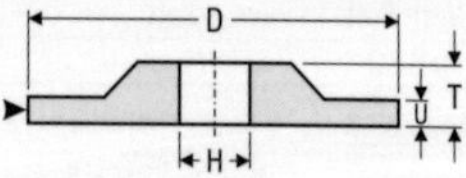

Shape 11 - Flaring Cup

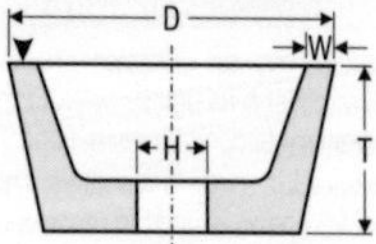

Shape 12 - Dish Type

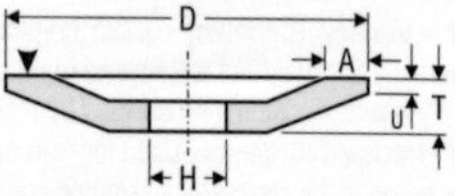

Shape 13 - Saucer

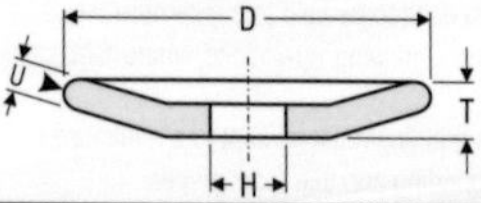

Shape 14 - Depressed Centre (For Valve Seat)

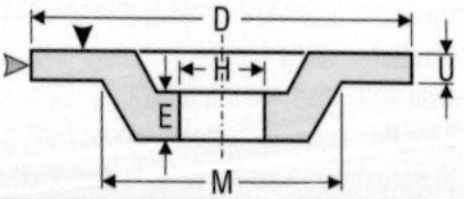

STANDARD GRINDING WHEEL SHAPES

▼ Primary Grinding Face ▽ Secondary Grinding Face

Shape 16 - Cone Curved Side

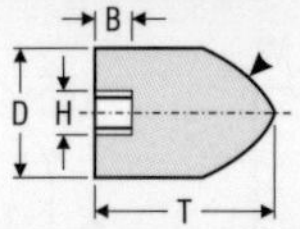

Shape 17 - Cone Square Tip

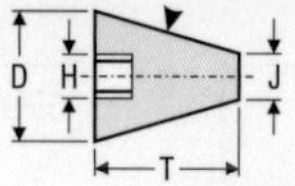

Shape 18 - Plug, Round End

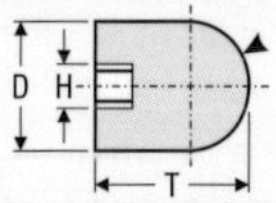

Shape 19 - Plug, Conical End

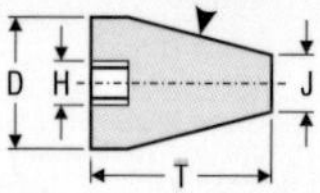

Shape 20 - Relieved One Side

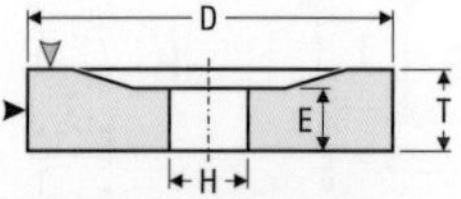

Shape 21 - Relieved Both Sides

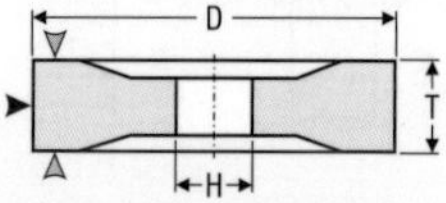

Shape 22 - Relieved One Side Recessed Other Side

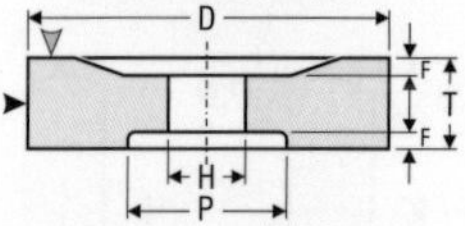

Shape 23 - Relieved & Recessed Same Side

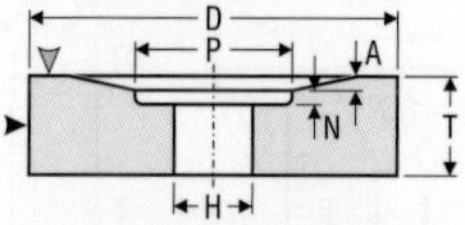

Shape 24 - Relieved One Side Recessed Both Sides

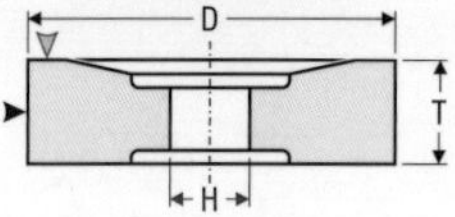

Shape 25 - Relieved Both Sides Recessed One Side

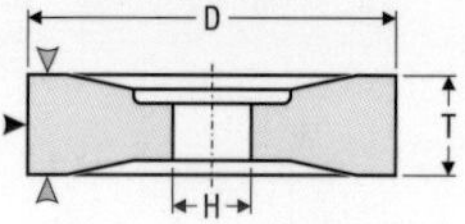

Shape 26 - Relieved Both Sides Recessed Both Sides

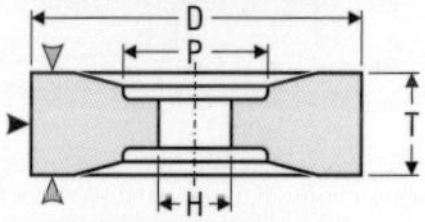

Shape 27 - Depressed Centre (Reinforced)

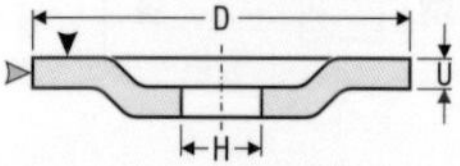

Shape 30 - Cut-Off

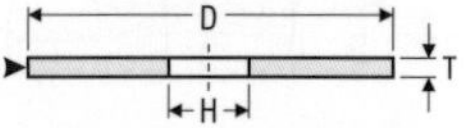

Shape 40 - Mounted

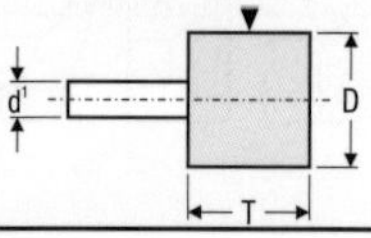

STANDARD GRINDING WHEEL FACES

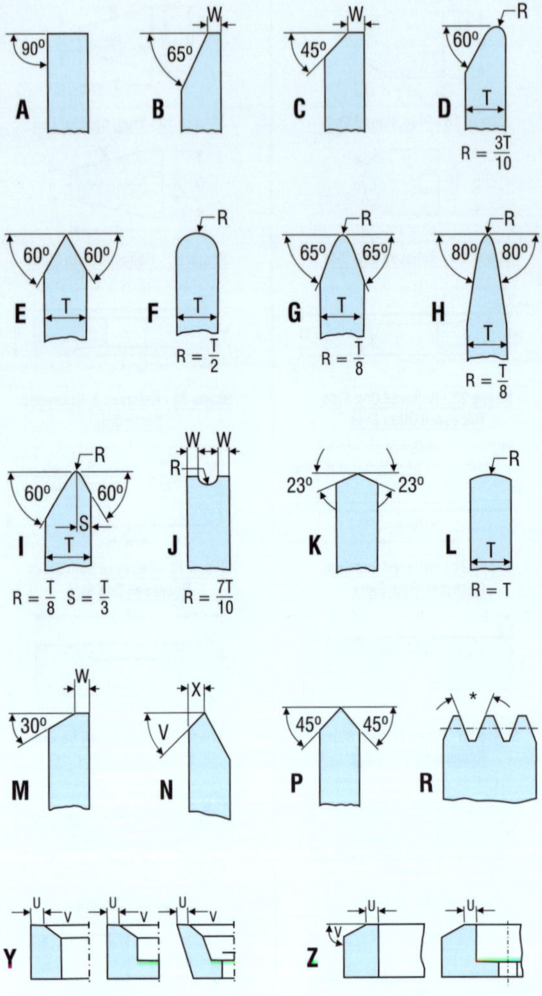

STANDARD MOUNTED WHEEL POINTS SHAPES AND DIMENSIONS

A Shapes - 1/4" Shanks

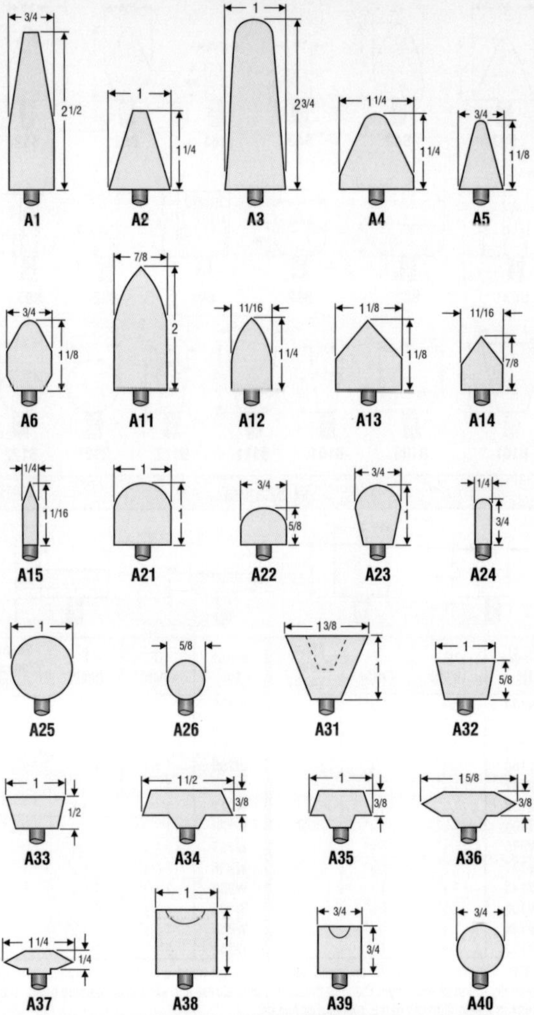

STANDARD MOUNTED WHEEL POINTS SHAPES AND DIMENSIONS

B Shapes - 1/8" Shanks

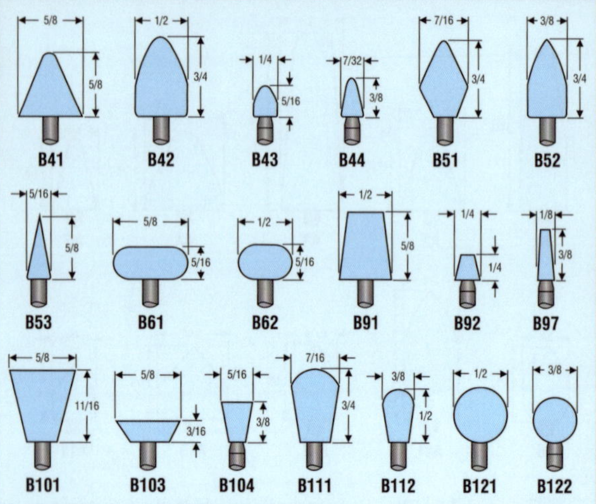

W Shapes

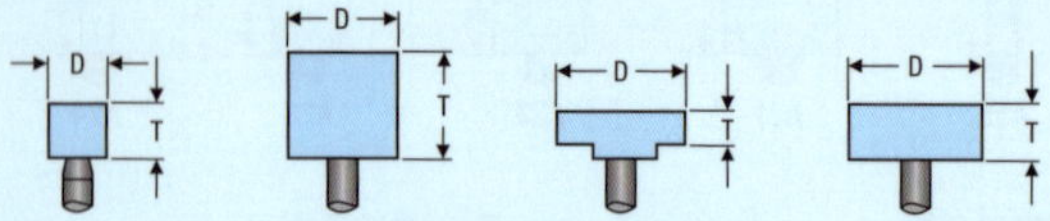

Shape No.	D (inch)	T (inch)	Shank Diameter (inch)	Shape No.	D (inch)	T (inch)	Shank Diameter (inch)
W144	1/8"	1/4"		W195	5/8"	3/4"	
W153	3/16"	3/8"		W196	5/8"	1"	
W154	3/16"	1/2"		W204	3/4"	3/4"	
W160	1/4"	1/4"		W205	3/4"	1"	
W162	1/4"	3/8"		W216	1"	1/4"	
W163	1/4"	1/2"		W217	1"	3/8"	
W164	1/4"	3/4"	1/8"	W220	1"	1	1/4"
W175	3/8"	3/8"		W225	1 1/4"	1/4"	
W177	3/8"	3/4"		W226	1 1/4"	3/8"	
W178	3/8"	1"		W230	1 1/4"	1 1/4"	
W185	1/2"	1/2"		W235	1 1/2"	6	
W186	1/2"	3/4"		W236	1 1/2"	1/2"	
W187	1/2"	1"		W242	2"	1"	

NOTE: Mounted wheels are designed primarily for off-hand grinding applications and are generally made of Aluminum Oxide or Silicon Carbide abrasive in vitrified or resinoid bond. They come in a selection of coarse, medium or fine grit.

TUNGSTEN CARBIDE ROTARY BURRS

Basic Cuts

Standard Cut

General purpose design producing good metal removal and smooth finish. Excellent where the metal is relatively hard and will not clog the flutes.

Uses: Steel Alloys, Stainless Steel, Cast Iron, Copper and Brass.

Double Cut

Produces rapid stock removal in harder materials with minimum tool chatter. This design eliminates loading of the flutes & the production of sharp sliver chips.

Uses: Steel, Cast Iron, Stainless Steel, Titanium, Brass & Copper.

Chip Breaker

Provides the operator with excellent tool control and reduced sliver size chips. Surface finish will be slightly reduced due to the chip breaker pattern.

Uses: Steel, Cast Iron, Stainless Steel, Titanium, Brass and Copper

Diamond Cut

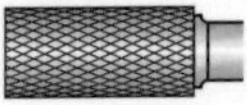

Extremely good tool control producing powder-like chips and increased metal removal. The surface finish, however, will be slightly reduced.

Uses: Heat-treated Steels, Fiberglass, Nickel-Chrome Steel, etc.

Aluminum Cut

Recommended for use on soft materials. The flute design produces rapid metal removal on non-ferrous metals and eliminates clogging of the flutes.

Uses: Aluminum, Plastics, Soft Copper, Rubber & Zinc Die Castings.

Standard Shapes

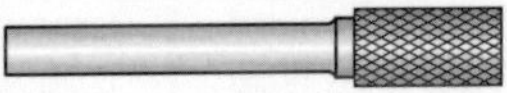

Cylinder Shape - with / without End Cut

Cylinder Shape - Radius End

Tree Shape - Radius End

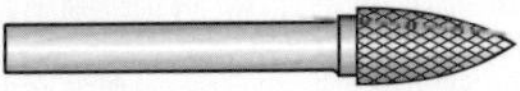

Tree Shape - Pointed End

TUNGSTEN CARBIDE ROTARY BURRS

Oval Shape

Ball Shape

14° Taper Shape - Radius End

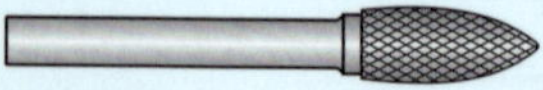

Flame Shape

Cone Shape - ∠14° to 28°

Inverted Cone Shape - ∠14° to 30°

60° Cone Shape

90° Cone Shape

Operating Speed Recommendations (RPM)

Size (Head Diameter) (inch)	(mm)	Standard Cut	Double Cut / Chipbreaker	Diamond Cut	Aluminum Cut
1/16"	1.5	45,000	50,000	60,000	–
1/8"	3	32,000	35,000	42,000	–
3/16"	5	22,000	25,000	30,000	–
1/4"	6	20,000	22,000	25,000	46,000
3/8"	10	16,000	18,000	22,000	31,000
1/2"	12	15,000	16,000	19,000	23,000
5/8"	16	14,000	15,000	18,000	15,000
3/4"	20	13,000	14,000	16,000	12,000
1"	25	10,000	12,000	14,000	8,000

NOTE: The above speeds are recommended for Burrs with Standard Shank Lengths and should be reduced significantly for Burrs with extra long shanks.

The recommended speeds are intended as a General Guide only for Steels, Alloys and Non-ferrous metals.

The Speed for Stainless Steel should be increased by approximately 25%.

Important Note: Because of the high speeds involved, it is most important that approved safety glasses be worn at all times to avoid slivers from entering the eye.

CUTTING TOOL MATERIALS AND APPLICATIONS

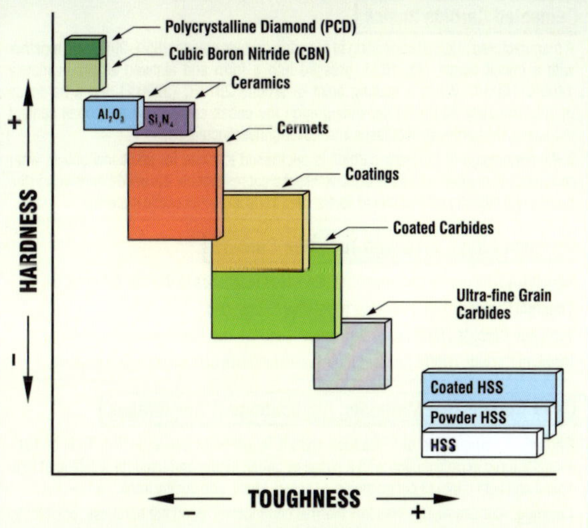

Applications and Productivity Advantages

Material	Principal Characteristics	Typical Applications	Cost/Productivity Advantages
High Speed Steel (HSS)	Tougher Cutting material than carbide but lower wear resistance.	Screw-machine and other low-speed operations, interrupted cuts, low-horsepower machining.	Priced lower than carbide; gives longer tool life, better surface finish.
Uncoated Cemented Carbide	Good combination of abrasive wear resistance and toughness.	Used in moderate to difficult applications related to steel, titanium, cast iron and aluminum.	Higher productivity because of abrasive wear resistance and toughness.
Coated Cemented Carbide	Extremely good combination of wear resistance and toughness	Large variety of grades with hard to tough substrates and used for general machining applications.	Extremely good combination of wear resistance and toughness with long tool life.
Cermet	Chemically stable with a hard and wear resistant substrate.	Finishing operations on irons, steels, stainless steels and aluminum alloys.	Up to 20 times the tool life of conventional Carbide grades.
Ceramics	Wear resistant with good hot hardness	Steel mill-roll resurfacing, finishing operations on cast irons & steels.	Better finishes at higher machining rates.
Cubic Boron Nitride (CBN)	Resists high cutting temperatures at high cutting speeds.	For finished turning of hardened work materials in 45-70 Rockwell C range.	Provides many times the edge life of carbide. Eliminates alternate grinding operations.
Polycrystalline Diamond (PCD)	Extremely wear resistant but sensitie to chipping, dissolves easily in iron.	Roughing to finishing operations on aluminum and other soft or abrasive metallic and non ferrous materials.	Better than 30 times the tool life of carbide, even on high silicon aluminum.

TUNGSTEN CARBIDE, CERAMIC, CERMET, CBN AND PCD CUTTING TOOL CHARACTERISTICS

Cemented Carbide Basics

A manufactured material consisting of tungsten carbide powder (85%-96%) held together with a cobalt binder (4%-15%), pressed into a form and sintered at approximately 1400°C-1600°C. It has a melting point of around 2870°C (5200°F) and a hardness of approximately 93.0 HRa depending upon the cobalt content. As the cobalt content increases, the hardness decreases and the toughness increases.

If the percentage of tungsten carbide is decreased to allow for additional alloys, wear resistance decreases. Because of its excellent wear resistance, it is used extensively in the cutting tool industry for machining steels, cast irons and other exotic materials.

Common Alloys used with Tungsten Carbide

Cobalt (Co): Increases toughness but decreases resistance to thermal deformation.

Titanium Carbide (TiC): Increases resistance to cratering.

Tantalum Carbide (TaC): Increases resistance to thermal deformation and cracking.

Niobium Carbide (NbC): Similar properties to tantalum carbide but less expensive.

Other Cutting Tool Materials, Applications & Advantages

Cermet: A combination of CERamics and METal particles based on TiC, TiCN or TiN - Manufactured in conjunction with a nickel or cobalt binder and used for finishing at high speed and light depth of cut on steels, stainless steels and ductile irons.

Ceramic: Alumina Al_2O_3 - Has low thermal conductivity, good hot hardness, chemically inert with steel and can be used at high speeds and feeds but poor strength and toughness. Alumina ceramics produce the best results when used for high speed finish machining of cast irons.

Ceramic-Composite: Alumina Al_2O_3+TiC, TaC, ZrO_2 - These are designed to overcome problems associated with cracking and fracture weakness and allow a broader range of applications than pure alumina cutting tools and ideal for machining nickel-based alloys.

Silicon Carbide whisker reinforced Alumina: Alumina Al_2O_3+SiC whiskers - It has a fracture toughness and crack resistance similar to tungsten carbide and used to successfully machine nickel-based alloys.

SIALON(Si-Al-O-N) Ceramics: Silicon Nitride based with added aluminum and oxygen - Offers high hardness, and good shock resistance and used for roughing and interrupted cuts on steels and cast iron.

Cubic Boron Nitride (CBN): Polycrystalline cubic boron nitride - Extremely hard and second to diamond and used primarily for hard steels, cast irons and high temperature alloys with hardness greater than 45Rc. CBN tools have excellent abrasion resistance with long and predictable tool life.

Polycrystalline Diamond (PCD): Synthetic diamond - PCD tools offer excellent productivity and tool life and are generally limited to machining nonferrous and nonmetallic abrasive materials such as aluminum-silica alloys, copper, magnesium, silica-filled resins, fiber-reinforced plastics and various non-metals. Not recommended for ferrous metals and nickel alloys.

Single-Crystal Diamond: Natural or synthetic - These are single crystal diamonds with lapped cutting edges and used to produce super fine finishes in the 10RMS range like computer memory chips, optical lenses and other similar materials where no further grinding or polishing is required.

WORKPIECE MATERIAL CLASSIFICATION

Material Groups Classified in accordance with the ISO Standard

In the metal cutting industry there are millions of parts and components that need to be produced from many different materials on a day-to-day basis. The ease of machinability of these components is dependent upon many factors i.e. type of material, subsequent heat treatment and hardness.

The goal of machining operations is to produce more components, faster and cheaper and therefore this requires application of the most aggressive cutting parameters possible.

The term "machinability" refers to how hard or how easy it is to machine a particular material and is classified as ***easy-to-cut*** or ***difficult-to-cut***.

Various factors such as the effects of hardness, adhesiveness, material strength, malleability, thermal conductivity and work hardening properties determine whether these criteria can be achieved.

Machinability Criteria	Easy-to-Cut	Difficult-to-Cut
Tool Life	Long, stable	Short, unstable
Cutting Temperature	Not too high	High
Chips	Easy Controllable	Continuous, vibrations occur
Cutting Resistance	Low	High
Surface Finish	Good	Burrs easily develops
Machining Time	Short	Long

Workpiece materials have been divided into 6 major groups in accordance with the ISO Standard where each group has unique properties regarding machinability.

P	Steel is the most common material in the metal cutting group and ranges from mild steel to high alloy steel and steel castings. The machinability of steel is good and varies according to the carbon content, material hardness, etc.
M	Stainless steels includes ferritic, martensitic and austenitic with a minimum of 12% chromium and includes other alloys such as nickel, molybdenum, etc. The machinability of stainless steel differs depending upon the alloying elements, in general machinability decreases with higher alloy content.
K	Cast iron includes grey, malleable cast iron, nodular cast iron, compact cast iron, austempered cast iron. Because of the inclusion of silicon carbide in most cast irons, the cutting edges of the tool are subject to extreme abrasion.
N	Non ferrous material such as aluminum, copper, brass, etc. These materials are generaly easy to machine with sharp edge tools and long tool life can be expected except for aluminum with silicon content of 13% plus which makes this material very abrasive.
S	Heat resistant super alloys which include high-alloyed iron, nickel, cobalt and titanium based materials. These materials work harden, generate heat and are also sticky which create built-up edges on the tool. Shorter tool life can be expected.
H	Hardened material 45-65 HRc includes chilled cast iron 400-600 HB. These materials generate a lot of heat during the cutting process, are very abrasive and are classified as difficult-to-machine materials. Short tool life is to be expected.

TUNGSTEN CARBIDE GRADE AND APPLICATIONS CHART - ISO TO ANSI

Workpiece Material According to ISO Classification

ISO Class	Workpiece Material	ISO Class	Workpiece Material
P	Steel	N	Aluminum
M	Stainless Steel	S	Heat Resistant Super Alloys
K	Cast Iron	H	Hardened Material

ISO Class	Workpiece Material	ISO Grade	ANSI Grade	Application	Cutting Features
P	Cast Steels Plain Carbon Steel Alloy and Tool Steel Ferritic, Martensitic and Precipitation Hardening Free-Machining Carbon Steel	P10	C7	Light Finishing	↓ Shock Resistance – / – Wear Resistance ↑
		P20	C6	Semi Finishing	
		P30	C6	General Purpose	
		P40	C5	Roughing	
M	Ferritic Stainless Steels Duplex Stainless Steels Stainless Steel Castings Austenitic Stainless Steels Martensitic Stainless Steels Super Austenitic Stainless Steels	M10	C3	Precision Finishing	↓ Shock Resistance – / – Wear Resistance ↑
		M20	C2	Semi Finishing	
		M30	C2	General Purpose	
		M40	C1	Roughing	
K	Cast Iron Gray Cast Iron Nodualr Cast Iron Malleable Cast Iron Compacted Graphite Iron Long Chipping Malleable Iron Short Chipping Malleable Iron	K01	C4	Precision Finishing	↓ Shock Resistance – / – Wear Resistance ↑
		K10	C3	Semi Finishing	
		K20	C2	General Purpose	
		K30	C1	Roughing	
N	Hard Rubber Copper Alloys Magnesium Alloys Aluminum Cast Alloys Aluminum Wrought Alloys Duroplastics, Fiberplastics Carbon & Graphite Composits	N01	C4	Precision Finishing	↓ Shock Resistance – / – Wear Resistance ↑
		N10	C3	Semi Finishing	
		N20	C2	General Purpose	
		N30	C1	Roughing	
S	Heat Resistant Alloys High Temperartue Alloys Titanium & Titanium Alloys Nickel / Cobalt based Alloys Iron based Heat Resistant Alloys Nickel based Heat Resistant Alloys Cobalt based Heat Resistant Alloys	S10	C2	Precision Finishing	↓ Shock Resistance – / – Wear Resistance ↑
		S20	C3	Semi Finishing	
		S30	C6	General Purpose	
		S40	C7	Roughing	
H	Tool Steels Spring Steels Chilled Cast Iron High Speed Steels Ball Bearing Steels Cold and Hot Work Steels Hardened Steels (48-65 HRc)	H01	C4	Precision Finishing	↓ Shock Resistance – / – Wear Resistance ↑
		H10	C3	Semi Finishing	
		H20	C2	General Purpose	
		H30	C1	Roughing	

TUNGSTEN CARBIDE GRADES COMPARISON TABLE

UNCOATED CARBIDE GRADES

ISO Classification		DIJET	ISCAR	KENNAMETAL	KORLOY	MITSUBISHI	SANDVIK	SECO	SUMITOMO	TAEGUTEC	TUNGALOY
P	P10	SRT	IC70	P10	ST10		S1P		ST10P	P10	TX10S
P	P20	SRT DX30	IC70 IC50M	K125M TTM	ST20	UTi20T	SMA		ST20E	P20	TX20 TX25
P	P30	SR30 DX30	IC50M IC54	GK K600 TTR	ST30A	UTi20T	SM30		A30 A30N	P30	TX30 UX30
P	P40	SR30 DX35	IC54	G13			S6		ST40E		TX40
M	M10	UMN		K313	U10		H10A	890	EH510 U10E	M10	TU10
M	M20	DX25 UMS	IC08	K68 KMF K125M TTM	U20	UTi20T	H13A	HX 883	EH520 U2	M20	TU20 UX30
M	M30	DX25 UMS	IC08 IC28	K600 TTR	ST30A	UTi20T	H10F SM30		A30 A30N		UX30
M	M40	UM40	IC128	G13	U40		S6			M40	TU40
K	K01	KG03		K605	H02	UTi05T	H1P		H1 H2		TH03 KS05F
K	K10	KG10 KT9	IC20	K313 K110M THM THM-U	H01	HTi10	H1P H10 HM	890	EH10 EH510	K10	G1F TH10
K	K20	CR1 KG20	IC20 IC10	K715 KMF K600	G10	UTi20T	H13A	890 HX 883	G10E EH20 EH520	K20	G2F KS15F G2 KS20
K	K30	KG30	IC10 IC28	THR	G3	UTi20T		883	G10E		G3
N	N01	KG03		K605			H10 H13A		H1 H2		KS05F
N	N10	KG10 KT9		K313 K110M THM THM-U	H01	HTi10		890 H15	EH10 EH510	K10	TH10 H10T
N	N20	CR1 KG20		K715 KMF K600				HX KX 883 H15 H25	G10E EH20 EH520	K20	KS15F
N	N30	KG30		G13 THR				H25			
S	S01	KG03				RT9005					
S	S10	FZ05 KG10		K10 K313 THM	H01	RT9005 RT9010	H10 H10A H10F H13A	890	EH10 EH510	K10	KS05F TH10
S	S20	FZ15 KG20		K715 KMF		RT9010 TF15		890 883 HX H25	EH20 EH520		KS15F KS20
S	S30	KG30		G13 K600 THR		TF15					

***IMPORTANT NOTE** : The Grade Comparison Tables on pages 169, 170, 171, 172, 173 & 174 are based on published data and intended as a guide only. The data therein should not be assumed to be accurate equivalents. Some grades may be obsolete and are included for reference purposes.

TUNGSTEN CARBIDE GRADES COMPARISON TABLE

COATED CARBIDE GRADES

ISO Classification		INGERSOLL	ISCAR	KENNAMETAL	KYOCERA	SANDVIK	SECO	SUMITOMO	TAEGUTEC	TUNGALOY	WIDIA
P	P01	IN2006	IC428 IC9150 IC8150 IC903	KC9315 KCP05	CA5505 PR730	GC4205	TP1000	AC700G	TT1300 PV3030	T9105	TN7005 TN2505
	P10	IN6520 IN2004 IN2006 IN2540	IC8150 IC9015 IC9150 IC9250 IC507 IC903 IC950	KC9110 KC9315 KCP10 KCP25 KC5010 KC5510 KU10T KC715M	CA5505 CA5515 PR730 PR830 PR915	GC4205 GC4215 GC3115 GC1515 GC1220	TP1000 TP2000 TP2500 T200M CP200	AC700G AC820P AC2000 ACP100 T2000Z ACP200	TT1300 TT7310 TT7400 PV3010 PV3030 TT7080 TT1040	T9105 T9115 AH710	TN7010 TN7015 TN2510 TN2505
	P20	IN6520 IN2004 IN2005 IN2006 IN2015 IN2505	IC520M IC8250 IC9015 IC9025 IC9150 IC308 IC350 IC354 IC507 IC570 IC907 IC908 IC928 IC950	KC9110 KC9215 KC9225 KC9325 KC5020 KC5525 KU25T KC522M	CA5515 CA5525 CA5025 PR630 PR730 PR830 PR915 PR930	GC4215 GC4220 GC2445 GC1220 GC1020 P20	TP2000 TP2500 TP200 TM2000 MP2500 CP250 F25M	AC820P AC2000 ACP200 T2000Z ACP200	TT3500 TT5100 TT7400 KT7300 TT7800 TT7220 TT9020 TT7080 TT9080 TT7070	T9115 T9125 AH120 AH725 AH730 SH730	TN7015 TN7025 TN7525 TN2525 TN6415 TN6435 TN6525
	P30	IN6520 IN6530 IN1030 IN2030 IN2035	IC8350 IC9025 IC9350 IC308 IC350 IC354 IC507 IC570 IC907 IC908 IC1028 IC3028	KC9125 KU30T KC935M KCP30 KC5025 KC5525 KU25T KC725M	CA5525 CA5535 CA5025 PR630 PR660 PR730 PR830 PR930	GC4225 GC4230 GC4235 GC2025 GC2035 GC1020 P20	TP2500 TP3000 TM4000 MP2500 T350M CP500 F30M	AC830P AC3000 AC630M ACP300 T3000Z AC530U ACP200	TT3500 TT5100 TT7400 KT7300 TT9030 TT7030 TT7080 TT9030 TT9080	T9125 T9135 T3130 AH120 AH725 AH740 GH330 AH130 GH130 AH9030 AH3035 AH3135	TN7025 TN7035 TN450 TN7435 TN7535 TN6435 TN6525
	P40	IN6530 IN0545 IN2035 IN2040 IN2540	IC635 IC9350 IC228 IC328 IC528 IC928 IC1008 IC1028 IC3028	KC9140 KC9240 KC9245 KC935M KC735M	CA5535 PR660	GC4235 GC4240 GC3040	TP3000 TP40 TM4000 T350M CP500 F40M T60M	AC830P AC3000 AC630M T3000Z AC530U	TT5100 TT7100 KT7300 TT7800 TT7080 TT8030 TT7070	T9135 AH140	TN7035 TN450 TN7435 TN7535 TN6540
M	M10	IN6520 IN2006 IN2505	IC520M IC9250 IC354 IC507 IC907	KC5010 KC5510 KU10T	CA6015 CA6515 PR730 PR915	GC2015 GC1025	TP2500 TK2000 CP200 CP250	AC610M ACP100 AC510U ACP200	TT5030 PV3010 PV3030 TT9030 TT9080	T9115 AH710	TN7525

*Refer Important Note on page 169

Continued ...

TUNGSTEN CARBIDE GRADES COMPARISON TABLE

COATED CARBIDE GRADES

ISO Classification		INGERSOLL	ISCAR	KENNAMETAL	KYOCERA	SANDVIK	SECO	SUMITOMO	TAEGUTEC	TUNGALOY	WIDIA
M	M20	IN6520	IC4050 IC9025 IC9054 IC9250 IC9350	KC8050 KC9225 KC925M	CA6515 CA6525 CA6015	GC2015 GC2025	TP2500 TP200 TM2000 MP2500 T25M	AC610M ACP100	TT5100	T6120 T9125	TN7025 TN8025 TN7525 TN7535
M	M20	IN1030 IN2005 IN2006 IN2015 IN2030 IN2035	IC250 IC300 IC308 IC354 IC507 IC808 IC830 IC903 IC907 IC928 IC1008 IC1028 IC3028	KC5010 KC5025 KC5510 KC5525 KC715M	PR630 PR730 PR915 PR930	GC1125 GC1220 GC1020 K20	CP250 CP500 F25M	AC520U ACP200	TT5030 PV3030 TT9020 TT9030 TT9080 TT8030	AH630 AH725 AH730 GH330 SH730 GH730	TN6415 TN6435 TN6525
M	M30	IN6520 IN6530	IC635 IC656 IC4050 IC9025 IC9054	KC8050 KC9240 KU30T KC935M	CA6525 CR9025	GC2025 GC2035 GC235 GC2040 GC4240	TP3000 TP40 TM4000 T350M	AC630M AC830P AC3000 ACP300	TT5100 TT7100	T6130 T3130	TN8025 TN450 TN7425 TN7525 TN7535
M	M30	IN1030 IN2030 IN2035 IN2505	IC228 IC250 IC300 IC328 IC908 IC928 IC1008 IC1028 IC3028	KC5025 KC5525 KU25T KC552M	PR630 PR660 PR730	GC1020 GC3040 GC235	CP500 F30M F40M	AC530U ACP300	TT9030 TT9080 TT8030	AH120 AH645 AH725 AH130 GH130	TN6435 TN6525
M	M40	IN6530	IC635 IC656	KC9240 KC9245		GC2035 GC235 GC2040 GC4240	TP40 TM4000	AC630M	TT7100		TN450 TN7435
M	M40	IN1030 IN2015 IN2030 IN2035	IC228 IC250 IC328 IC330	KC725M KC735M	PR660	GC3040 GC235	F40M	AC530U ACP300	TT8010 TT8020 TT8030	AH140 AH4035	TN6540
K	K01		IC428 IC4100 IC5005 IC5010		CA4010	GC3205		AC410K		T5105	TN5505
K	K01		IC810 IC910		PR510	GC4014			PV3030	AH110	TN2505
K	K10	IN6520	IC418 IC428 IC520M IC4100 IC9015 IC9080 IC9150	KC9315	CA4010 CA4115	GC3005 GC3210	TP1000 TK1000 TK2000 MK1500	AC410K ACK200	TT1300 TT7310 TT8115	T5105 T5115 T1115	TN5015 TN2510 TN5415 TN5505 TN5515
K	K10	IN2004 IN2010 IN2505	IC810 IC910 IC950	KC5010 KC5510 KU10T KC510M	PR510 PR905	GC1210 GC1020	CP200	AC510U	PV3010 PC3030 TT6030	GH110 AH110	TN2505 TN6510

*Refer Important Note on page 169

Continued ...

TUNGSTEN CARBIDE GRADES COMPARISON TABLE

COATED CARBIDE GRADES

ISO Classification		INGERSOLL	ISCAR	KENNAMETAL	KYOCERA	SANDVIK	SECO	SUMITOMO	TAEGUTEC	TUNGALOY	WIDIA
K	K20	IN6515 IN6520 IN6530 INDD15	IC418 IC520M IC4050 IC9015	KC9110 KC9315 KC9325 KC915M T250M T200M	CA4115 CA4120	GC3215 GC4220	TP1000 TP2000 TP2500 TK2000 MK1500 T200M	AC700G AC820P ACK200	TT7310 TT8115	T5115 T5125	TN5015 TN5020 TN5415 TN5515 TN5520
		IN1030 IN2005 IN2015 IN2030	IC810 IC830 IC908 IC910 IC928 IC950 IC1008	KC5025 KC5525 KU25T KC520M	PR510 PR905	GC1210 GC1220 GC1020 K20	CP200 CP250	AC510U ACK300	TT6060 TT8020 TT8030	AH120	TN6415 TN6435 TN6520 TN6525
	K30	IN6515 IN6520 IN6530 INDD15	IC520M IC4050	KC8050 KC9125 KC9325 KU30T KC935M	CA4120	GC4220 GC3040	TP2500 TP200 TK2000 MK3000	ACK300		T5125	TN7435
		IN1030 IN2030 IN2035	IC228 IC308 IC950 IC1008	KC5025 KC5525 KU25T		GC1220 GC1020 P20	CP500	ACK300	TT9030 TT6290 TT9030 TT8030	GH130	TN6435 TN6525 TN6540
N	N01							DL1000			TN6501
	N10		IC520	KC5010 KC5410 KC5510 KU10T		GCN20D K20 GC1020	F15M F17M	DL1000	TT6030	DS1100 DS1200	HCK10 TN6403 TN6502
	N20	IN2005	IC808 IC908	KC5025 KC5525 KU25T		GCN20D K20 GC1020	F15M F17M		TT6030 TT8020		
S	S01		IC507 IC907		PR915			AC510U		AH110 AH710 AH905	
	S10	IN2006	IC507 IC903 IC907	KC5010 KC5510 KU10T KC510M	PR915	GC1005 GC1025 GC1105	CP200 CP250	AC510U	TT5030	AH120 AH905	TN6010
	S20	IN2005 IN2006 IN1030 IN2035	IC300 IC808 IC908 IC928 IC3028	KC5025 KC5525 KU25T KC522M	PR915	GC1025 GC1030 GC1125	CP250 CP500 F40M	AC520U	TT5030 TT8020 TT8030	AH730 SH730	TN6025
	S30	IN2005 IN1030 IN2035	IC228 IC328 IC928 IC3028	KC5025 KC5525 KU25T		GC2030	CP500 F40M		TT8020 TT8030		TN6025
H	H01		IC903			GC1010				AH710	TN2505
	H10	IN2006	IC507 IC903 IC907	KC5010 KC5510 KU10T KC635M		GC1020		AC503	TT6030 TT6060 TT1040	AH110 AH120 SH730	TN2505
	H20	IN2005 IN1530	IC808 IC908 IC1008	KC635M		GC1020 GC1025	CP200	AC503	TT6290 TT6060 TT9080	AH120	TN2525
	H30		IC808 IC908 IC1008			GC1025 GC1030 GC1125	F30M		TT6290 TT9030 TT9080		

*Refer Important Note on page 169

TUNGSTEN CARBIDE GRADES COMPARISON TABLE

CERMET GRADES

ISO Classification		KENNAMETAL	KORLOY	KYOCERA	MITSUBISHI	NTK	SANDVIK	SECO	SUMITOMO	TUNGALOY	TAEGUTEC
P	P01	KT315	CC105 CC115 CN1000	TN30 PV720 PV7010	AP25N NX2525	T3N			T110A T1000A T1500Z	GT720 NS710	PV3010 PV3030
P	P10	KT5020 KT125 KT150	CN1500 CN2000 CC125	PV7025 PV60 TN6020	MP3025 UP35N	T15 C30 Q50	CT5005 CT5015 CT525	TP1030 CMP CM	T1500A T1200A T2000Z	GT730 NS520 NS720	CT3000
P	P20	KT1120 KT175	CN20 CN30	TN100M TC60M PV90	VP45N NX99 NX3035	N20 Z15 C50	CT530	TP1020 C15M	T3000Z T130Z	NS730 NS530	CT7000
P	P30				NX4545	Q50 N40			T250A T130A	NS740	
M	M01	KT315	CC105 CC115 CN1000	PV30 TN30 PV7010	AP25N NX2525	T3N			T110A	GT720 NS710	PV3010 PV3030
M	M10	KT5020 KT125 KT150	CN2000 CC125	PV60 TN6020 TN60	MP3025 UP35N	T15 C30 Q50	CT5005 CT5015 CT525	TP1030 CMP CM	T1500A T1200A T2000Z	GT730 NS520 NS720	CT3000
M	M20	KT1120 KT175	CN20 CN30	TN100M TC60M PV90	VP45N NX99 NX3035	N20 Z15 C50	CT530	TP1020 C15M	T3000Z T130Z	NS730 NS530	CT7000
M	M30				NX4545	Q50 N40			T250A T130A	NS740	
K	K01	KT315	CN1000	PV30 PV7005 PV7020	AP25N NX2525	T3N Q15		CM	T110A T1000A T1500Z	NS710 GT720 NS720	PV3030
K	K10	KT125	CN2000	TN60 TN6020		T15 Z15 C7Z	CT5015	C15M	T1200A T2000Z	GT730 NS730 NS530	CT3000

CERAMIC GRADES

Application	Composition	CERAMTEC	KENNAMETAL	KYOCERA	NTK	SANDVIK	SSANG-YONG	SUMITOMO	TAEGUTEC
K	Al_2O_3	SN60 SN80		KA30	HC1 HW2	CC620	SZ200 SZ300		AW120
K	Al_2O_3+TiC	SH2 SH4	KY1615	A65	HC2 HC5 HC6	CC650	ST100 SD200 TC100	NB90S NB90M	AB30
K	SiAlON	SL508 SL606 SL608	KY1310 KY3000		SX9		SN800		AS500
K	Si_3N_4	SL500 SL808	KY1320 KY3500 KYK10	KS6000 KS6050	SX1 SX6 SX8	CC6090 CC6091	SN400 SN500 SN600	SN2000K SN2100K NS260	AS10
K	Si_3N_4+CVD	SL658C SL854C SL858C	KY3400 KYK25	CS7050	SP2 SP9	CC1690		NS260C	SC10
S	Al_2O_3+SiCw		KY4300		WA1 WA5	CC670	SW400 SW800	WX2000	TC430
S	Si_3N_4+TiN		KY2100 KY1540 KYS30	KS6040		CC6060 CC6065	SN800 SN900		AS20
H	Al_2O_3+TiCN	SH2 SH4			HC2 HC5 HC7		ST300 ST500 ST700		AB20
H	Al_2O_3+TiCN		KY4400	A66N PT600M	ZC4 ZC7	CC6050	TC300	NB100C	AB2010

*Refer Important Note on page 169

TUNGSTEN CARBIDE GRADES COMPARISON TABLE

CBN GRADES

ISO Classification		DIJET	ISCAR	KENNAMETAL	KYOCERA	MITSUBISHI	NTK	SANDVIK	SECO	SUMITOMO	TUNGALOY
K	K01	JBN795	IB50 IB85		KBN475	MB710 MB5015	B20 B22 B30	CB7525 CB50 CB7050	CBN050 CBN300P	BNC500 BN7000 BN500	BXM930 BX870 BX910
	K10	JBN330	IB55 IB90	KB1630 KB9610	KBN60M KBN900	MB730 MB4020	B22 B23	CB50 CB7050	CBN200 CBN300 CBN300P CBN400C	BN7000 BN500	BX950
	K20			KB9640	KBN900	MB730 MB 4020	B16 B40		CBN200 CBN300 CBN400C CBN600	BN7000 BNS800	BXC90 BX90S
	K30			KB5630		BC5030 MBS140			CBN500	BNS800	BXC90 BX90S
S	S01					MB730			CBN170	BN7000	BX470 BX480 BX950
H	H01		IB20H IB25H IB50	KB1610	KBN510 KBN05M KBN10M	BC8105 BC8110 MBC010 MB810	B24 B52	CB20	CBN010 CBN100 CBN060K	BNC100 BN1000 BN2000 BNX10	BXM10 BX310
	H10	JBN300 JBN500	IB55 IB25HA	KB1615 KB1620 KB5610 KB9610	KBN525 KBN05M KBN25M	BC8110 MBC020 BC8120 BC8020	B24 B36 B52 B54	CB7015 CB7050 CB50	CBN10 CBN100 CBN150 CBN160C	NC2020 BNC160 BNC200 BN2000	BXM20 BX330
	H20	JBN245		KB1340 KB5625 KB9640	KBN30M KBN35M KBN900	MBC020 BC8120 MB8020 MB8025	B22 B36	CB7025 CB7525	CBN150 CBN160C	BNC2020 BNC200 BNX20	BXC50 BX360
	H30					BC8130 MB835		CB7525		BNC300 BN350 BNX25	BX380

PCD GRADES

ISO Classification		DIJET	ISCAR	KENNAMETAL	KYOCERA	MITSUBISHI	NTK	SANDVIK	SECO	SUMITOMO	TUNGALOY
N	N01	JDA30 JDA735	ID5	PD100 KD1400 KD1405	KPD001	MD205		CD05 CD10	PCD05 PCD10	DA1000 DA150	DX180 DX160
	N10	JDA40 JDA745	ID5	KD100 KD1400 KD1415	KPD010 KPD230 KPD250	MD220		CD10	PCD10 PCD20	DA1000 DA150	DX140
	N20	JDA10 JDA715		KD1425	KPD010 KPD230 KPD250	MD230	PD01		PCD30 PCD30M	DA1000 DA2200	DX110 DX120
	N30	JDA10		KD1400		MD230	PD2		PCD05 PCD30 PCD30M	DA1000 DA2200	DX110 DX120

*Refer Important Note on page 169

TURNING FORMULAS

Cutting Speed: SFM (Surface feet per minute)

$$SFM = \frac{D \times RPM}{3.82}$$

Example: Determine the cutting speed (SFM) required for turning a 4½" diameter part with a spindle speed of 800 RPM.

$$\therefore SFM = \frac{4.5 \times 800}{3.82} = 942.41 \text{ ft/min}$$

Spindle Speed: RPM (Revolutions per minute)

$$RPM = \frac{SFM \times 3.82}{D}$$

Example: Determine the spindle speed (RPM) required for turning a 4½" diameter part with a cutting speed of 400 SFM.

$$\therefore RPM = \frac{400 \times 3.82}{4.5} = 339.56 \text{ rpm}$$

Metal Removal Rate: MRR (in³/min)

$$MRR = SFM \times DOC \times IPR \times 12$$

Example: Determine the metal removal rate (MRR) required for cutting with a depth of cut of 0.020" with a cutting speed of 800 SFM and a feed rate of 0.015 IPR.

$$\therefore MRR = 800 \times 0.020 \times 0.015 \times 12 = 2.88 \text{ in}^3\text{/min}$$

Power Requirement: HP (Horsepower)

$$HP = \frac{SFM \times DOC \times IPR \times CF}{33{,}000}$$

Example: Determine the power requirement (HP) for turning a material with a cutting force of 190,000, a depth of 0.070", a cutting speed of 500 SFM and a feed rate of 0.015 IPR.

$$\therefore HP = \frac{500 \times 0.070 \times 0.015 \times 190{,}000}{33{,}000} = 3.02 \text{ HP}$$

Cutting Time: T (min)

$$T = \frac{L}{IPR \times RPM}$$

Example: Determine the amount of time (T) required for turning to machine an 8" long part with a spindle speed of 500 RPM and a feed rate of 0.012 IPR.

$$\therefore T = \frac{8}{0.012 \times 500} = 1.33 \text{ min}$$

Feed Rate: FR (in/min)

$$FR = Feed \times RPM$$

Example: Determine the feed rate (FR) in inches per minute for turning a shaft at a feed rate of 0.012 inches per revolution at 800 RPM.

$$\therefore FR = 0.012 \times 800 = 9.6 \text{ in/min}$$

Theoretical Surface Finish: TSF (microns)

$$TSF = \frac{IPR^2 \times 1000}{8 \times NR}$$

Example: Determine the theoretical surface finish (TSF) using an insert with a 1/64" nose radius (NR) and a feed rate of 0.012 IPR.

$$\therefore TSF = \frac{(0.012)^2 \times 1000}{8 \times 0.016} = 1.125 \text{ microns}$$

Where:

- *CF* = *Cutting force (lb / in²)*
- *D* = *Diameter of the part (in)*
- *DOC* = *Depth of cut (in)*
- *HP* = *Horsepower*
- *IPR* = *Inches per revolution (in/rev)*
- *L* = *Machined length (in)*
- *MRR* = *Metal removal rate (in/min)*
- *NR* = *Nose radius of insert (in)*
- *RPM* = *Revolutions per minute (rev/min)*
- *SFM* = *Surface feet per minute (ft/min)*
- *T* = *Cutting time (min)*
- *TSF* = *Theoretical surface finish (μ)*
- *FR* = *Feed rate (in/min)*

SURFACE COATINGS FOR CUTTING TOOLS

Physical Vapor Deposition (PVD)

A low-temperature process that produces very tough coatings with high lubricity for use on gummy materials, high temperature alloys, low-carbon steels, and stainless steels at low to medium speeds.

Chemical Vapor Deposition (CVD)

A high-temperature process that produces highly crater resistant coatings with a good balance of toughness and wear resistance for use on a broad range of carbon and stainless steels at medium to high speeds.

Medium Temperature Chemical Vapor Deposition (MTCVD)

A medium-temperature process that produces tougher, slightly less wear-resistant coatings than the traditional CVD process and is recommended for use on gummy materials, ductile irons and high temperature alloys at medium speeds.

Aluminum Chromium Nitride AlCrN)

Mainly used for milling cutters where machining applicaitons involving heavy mechanical and thermal stresses resulting in supreme wear resistance.

Blackening

Blackening is a finishing operation that chemically coats the surface of ferrous metals. Besides enhancing the appearance of the components, the black deposit holds a sealant, usually oil, into the pores of the black oxide coating, where it is held in close contact with the metal and prevents corrosion. There are three types of blackening:

1. **Copper/Selenium Blackening** - This process uses a room temperature acid bath to form a black copper selenium coating in approximately 1-3 minutes.
2. **Caustic Black Oxiding** - A caustic soda bath operating at about 140°C produces a black coating on the surface of the ferrous metal.
3. **Black Dyed Oxylate Coating** - It provides a high quality black corrosion resistant coating on ferrous metals without using heated chemicals. A grey ferrous oxylate coating is applied to the steel and it is then dyed black. This process operates at room temperature.

The performance of any black coating is gauged by its porosity and ability to absorb a subsequent coating of oil which acts as a rust barrier. It is extensively used to coat Tungsten Carbide tool holders, Boring bars and Milling cutters to enhance appearance of the tool.

Blue Finish

Blue Finish is a thin layer of Iron Oxide produced by a carefully controlled steam tempering process. It increases resistance to corrosion and enables the surface of a tool to absorb lubricant in machining operations.

Bronze Finish

The Bronze Finish is a thin oxide layer formed on the tool surface and is applied on High Speed Cobalt Steels.

Chromium Carbide (CrC)

A super hard coating > HRC 80 is applied to the surface using PVD coating process. The hardness of the coating is the same as TiN but with an oxidation resistance that is 20 to 30% higher. Recommended for machining Titanium and Aluminum alloys.

SURFACE COATINGS FOR CUTTING TOOLS

Chromium Nitride (CrN)

A PVD coating and is particularly suitable in non-ferrous applications where titanium based coatings are not successful. It has a surface hardness of approximately 80 Rc and has good corrosion and oxidation resistance. It is mainly used for machining Copper, Die-casting, Brass forgings.

Diamond Coating

An amorphous diamond coating with extremely low coefficient of friction and high hardness. This coating is specially recommended for tapping Aluminum alloys with high Silicon content.

Hard-Chromium Plating (Cr)

The tool is plated with Chromium to improve wear characteristics and reduce sliding resistance and cutting forces. It is especially suitable when tapping Structural grade steels, Copper, Brass, etc.

Nitriding (FeN)

Nitriding is a process which is used to increase the hardness and wear resistance of the surface of a tool. It is particularly suitable for taps that are used on abrasive materials such as castings, bakelite, etc.

Steam Tempering

Steam tempering reduces sliding resistance, thus preventing cold welding, which occurs for example during the machining of Steels that have a low carbon content. Steam tempered tools are mainly used for machining of ferrous materials.

Titanium Aluminum Nitride (TiAlN)

This coating is applied by the PVD process and has a higher hardness than TiN and TiCN. A very thin film of Al_2O_3 (Aluminum Oxide) is formed on the surface of the TiAlN which greatly increases the service life of the tool which can withstand higher temperatures especially when machining Cast iron and tough steels.

Titanium Carbonitride - TiCNite (TiCN)

TiCNite coated tools have a very high surface hardness of approximately 90 Rc. TiCN is harder than other coating materials, hence cutting speeds similar to carbide tools can be achieved with the toughness of high speed-steel. This coating is recommended for use in all steels, cast iron and aluminum.

Titanium Nitride - TiNite (TiN)

A gold colored film of Titanium Nitride with a hardness of approximately 85 Rc is deposited on the tool which extends tool life by reducing friction and enables the tool to be operated at higher speeds and feeds. It is suitable for machining iron based materials, die-casting and plastic mold tooling. Other applications include wear parts, medical and dental instruments, slitting knives, plastic die molds, carbide tools, and high performance forming tools.

Zirconium Nitride - (ZrN)

A ceramic coating applied by the PVD process. Especially recommended for tapping Aluminum and Aluminum alloys.

TUNGSTEN CARBIDE TURNING INSERT IDENTIFICATION (INCH)

TNMG332TM-Grade

T	N	M	G
1	2	3	4

1 Insert Shape

Symbol	Insert Shape
A	Parallelogram 85°
B	Parallelogram 82°
C	Rhombic 80°
D	Rhombic 55°
E	Rhombic 75°
F	Rhombic 50°
H	Hexagon
K	Parallelogram 55°
L	Rectangle
M	Rhombic 86°
O	Octagon
P	Pentagon
R	Round
S	Square
T	Triangle
V	Rhombic 35°
W	Trigon

2 Relief Angle

Symbol	Relief Angle	Symbol	Relief Angle
A	3°	F	25°
B	5°	G	30°
C	7°	N	0°
D	15°	P	11°
E	20°	O	Special

3 Tolerance

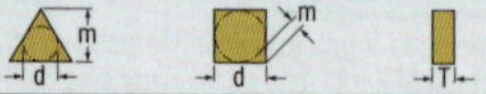

Class	Tol. ± 'm' (inch)	Tol. ± 'T' (inch)	Tol. ± 'd' (inch)
A	0.0002	0.0001	0.0001
C	0.0005	0.0001	0.0001
E	0.001	0.001	0.001
F	0.0002	0.0001	0.0005
G	0.001	0.005	0.001
H	0.0005	0.0001	0.0005
J	0.0002	0.001	0.005
K	0.0005	0.001	0.005
L	0.001	0.001	0.005
M	0.003~0.007	0.005	0.002~0.005
N	0.003~0.007	0.001	0.002~0.005
U	0.005~0.015	0.005	0.003~0.01

4 Fixing / Chip Breaker

Symbol	Feature	Symbol	Feature
A		N	
B		Q	
C		R	
F		T	
G		U	
H		W	
J		X	Special
M			

TUNGSTEN CARBIDE TURNING INSERT IDENTIFICATION (INCH)

3	3	2	TM
5	6	7	8

5 Insert Size (Dia. of insc. circle)

Symbol	Insc. Circle	Fraction
1.2	0.156"	5/32"
1.5	0.187"	3/16"
1.8	0.219"	7/32"
2	0.250"	1/4"
2.5	0.313"	5/16"
3	0.375"	3/8"
4	0.500"	1/2"
5	0.625"	5/8"
6	0.750"	3/4"
8	1.000"	1"

6 Thickness

Symbol	Thickness (T)
1	1/16"
1.2	5/64"
1.5	3/32"
2	1/8"
2.5	5/32"
3	3/16"
3.5	7/32"
4	1/4"
5	5/16"
6	3/8"
8	1/2"

7 Radius

Symbol	Nose Rad.
0	0.004"
0.5	0.008"
1	1/64"
2	1/32"
3	3/64"
4	1/16"
5	5/64"
6	3/32"
7	7/64"
8	1/8"
Round Insert	–

8 Chip Breaker

This Symbol is reserved for Manufacturer's Chip Breaker Design.

ANSI to ISO (Metric) Insert Size Crossover Chart (Common Inserts)

Insert Shape	ANSI	ISO (Metric)
 CNMA 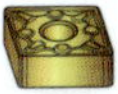 CNMG	432	120408
	433	120412
	643	190612
	644	190616
 SNMA  SNMG	322	090308
	431	120404
	432	120408
	433	120412
	643	190612
	644	190616
	856	250724
	866	250924
 SNUN  SNGN	321	090304
	322	090308
	421	120304
	422	120308
	423	120312
	431	120404
	432	120408
	433	120412
	533	150412
	632	190408
	633	190412
	634	190416
	846	250624
 SPUN  SPMR	321	090304
	322	090308
	421	120304
	422	120308
	423	120312
	532	150408
	533	150412
	632	190408
	633	190412
	634	190416

Insert Shape	ANSI	ISO (Metric)
 TNMA  TNMG	222	110308
	321	160304
	322	160308
	323	160312
	432	220408
	433	220412
	434	220416
	543	270612
	544	270616
	666	330924
 TNUN  TNGN	221	110304
	222	110308
	321	160304
	322	160308
	323	160312
	324	160316
	332	160408
	333	160412
	334	160416
	432	220408
	433	220412
	434	220416
 TPUN  TPMR	221	110304
	222	110308
	321	160304
	322	160308
	323	160312
	432	220408
	433	220412
	434	220416
 WNMA  WNMG	331	060404
	332	060408
	431	080404
	432	080408
	433	080412

CARBIDE INSERT IDENTIFICATION

TUNGSTEN CARBIDE MILLING INSERT IDENTIFICATION (INCH)

SEEN42AFTN-Grade

S	E	E	N
1	2	3	4

1 Insert Shape

Symbol	Insert Shape
A	Parallelogram 85°
B	Parallelogram 82°
C	Rhombic 80°
H	Hexagon
L	Rectangle
M	Rhombic 86°
O	Octagon
P	Pentagon
R	Round
S	Square
T	Triangle
W	Trigon

2 Relief Angle

Symbol	Relief Angle	Symbol	Relief Angle
A	3°	F	25°
B	5°	G	30°
C	7°	N	0°
D	15°	P	11°
E	20°	O	Special

3 Tolerance

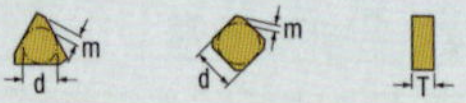

Class	Tol. ± 'm' (inch)	Tol. ± 'T' (inch)	Tol. ± 'd' (inch)
A	0.0002	0.001	0.001
C	0.0005	0.001	0.001
E	0.001	0.001	0.001
G	0.001	0.005	0.001
H	0.0005	0.001	0.0005
J	0.0002	0.001	0.002~0.006
K	0.0005	0.001	0.002~0.006
M	0.003~0.008	0.005	0.002~0.006
U	0.005~0.015	0.001	0.002~0.010

4 Fixing / Chip Breaker

Symbol	Feature	Symbol	Feature
A		Q	
B		R	
F		T	
M		W	
N		X	Special

CARBIDE INSERT IDENTIFICATION

TUNGSTEN CARBIDE MILLING INSERT IDENTIFICATION (INCH)

4	2	A	F	T	N
5	6	7	8	9	10

5 Insert Size (Dia. of insc ⊙)

i.c. i.c.

Symbol	Insc. Circle	Fraction
1.2	0.156"	5/32"
1.5	0.187"	3/16"
1.8	0.219"	7/32"
2	0.250"	1/4"
2.5	0.313"	5/16"
3	0.375"	3/8"
4	0.500"	1/2"
5	0.625"	5/8"
6	0.750"	3/4"
8	1.000"	1"

6 Thickness

T

Symbol	Thickness (T)
1	1/16"
1.2	5/64"
1.5	3/32"
2	1/8"
2.5	5/32"
3	3/16"
3.5	7/32"
4	1/4"
5	5/16"
6	3/8"
8	1/2"

7 Cutting ∠

α

Symbol	Cutting ∠ (α)
A	45°
D	60°
E	75°
F	85°
P	90°
X	65°
Z	Special

8 Relief ∠

β

Symbol	Relief ∠ (β)
A	3°
B	5°
C	7°
D	15°
E	20°
F	25°
G	30°
N	0°
P	11°

9 Edge

Symbol	Figure
E	Hone
F	Sharp
S	Chamfer + Hone
T	Chamfer

10 Direction of Cutting

R	L	N
Right Rotation	Left Rotation	Neutral Right & Left Rotation

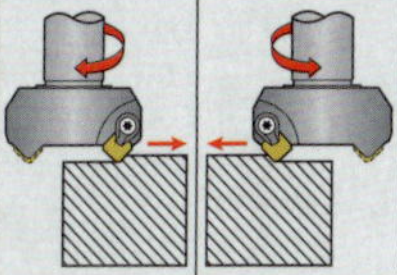

CARBIDE INSERT IDENTIFICATION

INCH & ISO (METRIC) Crossover Chart (Inscribed Circle & Cutting Edge Length)

INCH (Inscribed Circle - inch)								
Inscribed Circle	5/32"	7/32"	1/4"	3/8"	1/2"	5/8"	3/4"	1"
Symbol	5	7	2(8)	3	4	5	6	8
Insert Shape	**ISO (METRIC) (Cutting Edge Length - mm)**							
T	06	09	11	16	22	27	33	44
R S	03	05	06	09	12	15	19	25
D	04	06	07	11	15	19	23	31
C	03	05	06	09	12	16	19	25
W		03	04	06	08			

TUNGSTEN CARBIDE TURNING TOOLHOLDER IDENTIFICATION (INCH)

C	T	G	P	R	16	3	D
1	2	3	4	5	6	7	8

1 Clamping*

Insert Clamping System

- C - Clamp Lock
- D - Clamp+Pin Lock
- M - Clamp+Pin Lock
- P - Pin Lock/Lever Lock
- S - Screw Lock
- W - Pin+Wedge Clamp

External Turning Toolholder

CTGPR163D

3 Approach Angle

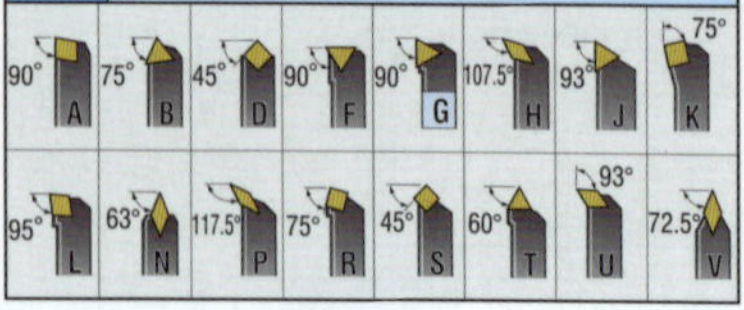

2 Insert Shape

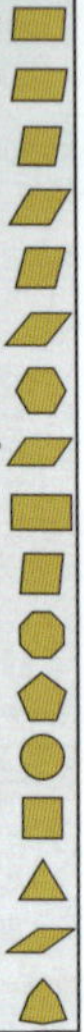

- A - Parallelogram 85°
- B - Parallelogram 82°
- C - Rhombic 80°
- D - Rhombic 55°
- E - Rhombic 75°
- F - Rhombic 50°
- H - Hexagon
- K - Parallelogram 55°
- L - Rectangle
- M - Rhombic 86°
- O - Octagon
- P - Pentagon
- R - Round
- S - Square
- T - Triangle
- V - Rhombic 35°
- W - Trigon

4 Relief ∠

- A - 3°
- B - 5°
- C - 7°
- D - 15°
- E - 20°
- F - 25°
- G - 30°
- N - 0°
- P - 11°
- O - Special

5 Version

R - Right ← L - Left → N - Neutral ↔

6 Holder Size (inches)

Sym.	Width	Height	Sym.	Width	Height
05	5/16	5/16	24	1 1/2	1 1/2
06	3/8	3/8	32	2	2
08	1/2	1/2	64	3/4	1
10	5/8	5/8	66	3/4	1/2
12	3/4	3/4	85	1	1 1/4
16	1	1	86	1	1 1/2
20	1 1/4	1 1/4	91	1 1/4	1 1/2

7 Insert Size (Dia. of insc ⊙)

Symbol	Insc. Circle	Fraction
1.2	0.156"	5/32"
1.5	0.187"	3/16"
1.8	0.219"	7/32"
2	0.250"	1/4"
2.5	0.313"	5/16"
3	0.375"	3/8"
4	0.500"	1/2"
5	0.625"	5/8"
6	0.750"	3/4"
8	1.000"	1"

8 Tool Length (inches)

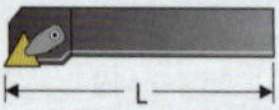

A - 4"	E - 7"
B - 4 1/2"	F - 8"
C - 5"	G - 10"
D - 6"	H - 12"

*Also refer pages 195-199

TUNGSTEN CARBIDE BORING BAR IDENTIFICATION (INCH)

S	24	T	C	T	F	P	R	3	□
1	2	3	4	5	6	7	8	9	10

Boring Bar

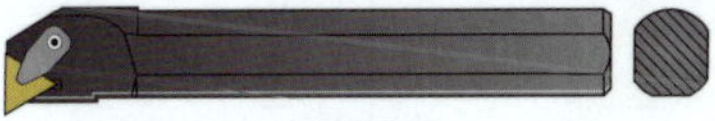

S24TCTFPR3

1 Type

Boring Bar Shank

- A - Steel with Coolant Hole
- C - Carbide Shank
- E - Coolant through Carbide Shank
- [S] - Steel Shank

2 Bar Diameter

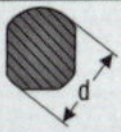

Bar Diameter 'd' (Inches)

Symbol	'd'	Symbol	'd'
03	3/16	16	1
04	1/4	20	1 1/4
05	5/16	24	1 1/2
06	3/8	28	1 3/4
08	1/2	32	2
10	5/8	36	2 1/4
12	3/4	40	2 1/2

3 Tool Length (inch)

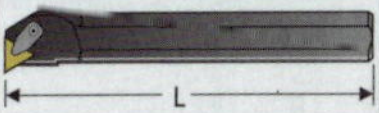

J - 4 1/2"	N - 6 1/2"	S - 10"
K - 5"	P - 6 3/4"	[T] - 12"
L - 5 1/2"	Q - 7"	U - 14"
M - 6	R - 8"	V - 16"

4 Clamping*

Insert Clamping System

- [C] - Clamp Lock
- D - Clamp+Pin Lock
- M - Clamp+Pin Lock
- P - Pin Lock/Lever Lock
- S - Screw Lock

5 Insert Shape

- C - Rhombic 80°
- D - Rhombic 55°
- E - Rhombic 75°
- F - Rhombic 50°
- K - Parallelogram 55°
- R - Round
- S - Square
- [T] - Triangle
- V - Rhombic 35°
- W - Trigon

6 Approach Angle

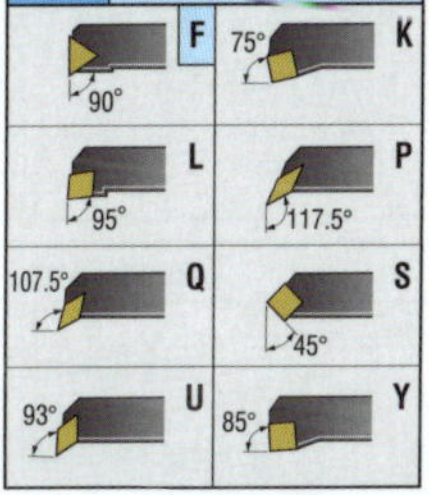

7 Relief ∠

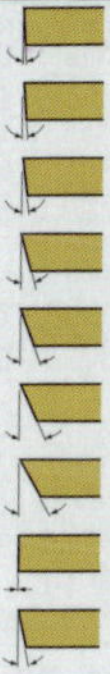

- A - 3°
- B - 5°
- C - 7°
- D - 15°
- E - 20°
- F - 25°
- G - 30°
- N - 0°
- [P] - 11°
- O- Special

8 Version

[R]- Right Hand | L - Left Hand

9 Insert Size (Diameter of inscribed circle)

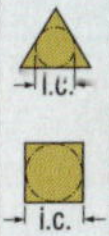

Symbol	Insc. Cir.	Fract.	Symbol	Insc. Cir.	Fract.
1.2	0.156"	5/32"	3	0.375"	3/8"
1.5	0.187"	3/16"	4	0.500"	1/2"
1.8	0.219"	7/32"	5	0.625"	5/8"
2	0.250"	1/4"	6	0.750"	3/4"
2.5	0.313"	5/16"	8	1.000"	1"

10 Manufacturer's Designation

CARBIDE INSERT IDENTIFICATION

*Also refer pages 200-202

TUNGSTEN CARBIDE THREADING INSERT IDENTIFICATION (INCH)

16ERB13UN3MP20

16	E	R	B	13	UN	3M	P20
1	2	3	4	5	6	7	8

1 Edge Length

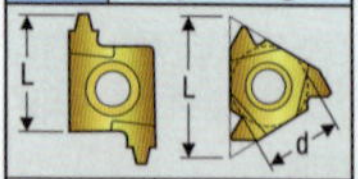

L (mm)	d (inch) - ic
06	5/32"
08	3/16"
11	1/4"
16	3/8"
22	1/2"
27	5/8"

2 Application

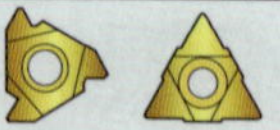

Standard U-Type

Symbol	Application
E	External
I	Internal
U*	U-Type

* UE - External
UI - Internal
UEI - External & Internal

3 Insert Version

External Threading

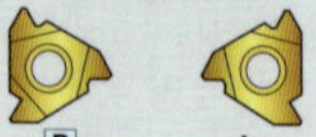

R L

Internal Threading

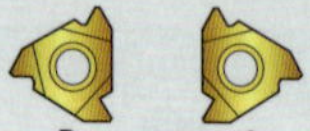

R L

4 Type

M=with Chip Breaker, B=Peripherally Ground & Chip Breaker, □=Regular Type

5 Thread Pitch / tpi

Full Profile (tpi)				Full Profile (mm)				Partial Profile	
48	40	32	24	0.5	0.7	0.75	0.8	A = 0.5 - 1.5 mm	48 - 16 tpi
20	18	16	14	1.0	1.25	1.5	1.75	AG = 0.5 - 3.0 mm	48 - 8 tpi
13	12	11	10	2.0	2.5	3.0	4.0	G = 1.75 - 3.0 mm	18 - 8 tpi
9	8	7	6	4.5	5.0	5.5	6.0	N = 3.5 - 5.0 mm	7 - 5 tpi
5	4.5	4	3	8.0	10.0	12.0	14.0	Q = 5.5 - 6.0 mm	4.5 - 4 tpi
2.5	2							U = 5.5 - 9.0 mm	4.5 - 2.75 tpi

6 Thread Form

Symbol Profile	Symbol Profile	Symbol Profile
60 = V profile 60°	BUT2.5 = Buttress Fig.2.5	RND = Round DIN 405
55 = V profile 55°	BSPT = Whitworth Taper	STACME = Stub Acme
ABUT = American Buttress	EL = Extreme Line Casing	TR = Trapezoidal DIN 103
ACME = Am.Acme-G	H90 = Hughes	UN = American Unified
API = American Pet. Inst.	ISO = ISO, Metric	UNJ = UN-Controlled Root
API RD = API Rd. Casing	MJ = ISO 5855	VAM = Vam Vallourec
BUT = API Buttress	NPT = National Pipe Taper	W = Whitworth (BSW)

7 No. of Teeth / Chip Breaker

Optional: To Manufacturer's Discretion.

8 Carbide Grade

Optional: To Manufacturer's Discretion.

Full Profile

- Workpiece may be slightly oversize as insert 'Tops' the thread.
- No subsequent thread deburring is necessary.

Partial Profile

- Workpiece needs to be accurately pre-machined.
- One insert covers a wide range of pitches.

Single Tooth Insert

Multi Tooth Insert

Refer pages 26-30 for Basic Thread Profile details

TUNGSTEN CARBIDE EXT. THREADING TOOLHOLDER AND INT. THREADING BAR IDENTIFICATION (INCH)

C	E	R	1000	R	16	B
1	2	3	4	5	6	7

External Threading Toolholder

CER1000R16B

C	I	R	0750	P	16	B
1	2	3	4	5	6	7

Internal Threading Bar

CIR0750P16B

1 Clamping*

Insert Clamping System

C - Clamp Lock
S - Screw Lock

NOTE: Some manufacturers use " **D** " for Clamp Lock

2 Application

External / Internal

E - External Threading
I - Internal Threading

3 Version

R	L
Right Hand	Left Hand

4 Shank Size

External Toolholder

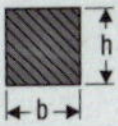

Shank - h x b
0500 = 1/2" x 1/2"
0750 = 3/4" x 3/4"
1000 = 1" x 1"
1250 = 1 1/4" x 1 1/4"
1500 = 1 1/2" x 1 1/2"

Internal Toolholder

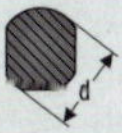

Shank - Diameter d
0500 = 1/2"
0750 = 3/4"
1000 = 1"
1250 = 1 1/4"
1500 = 1 1/2"

5 Tool Length

Symbol Length (inch)

D – 2 1/2"
F – 3 1/4"
H – 4"
K – 5"
L – 5 1/2"
M – 6"
P – 7"
R – 8"
S – 10"
T – 12"
U – 14"
V – 16"

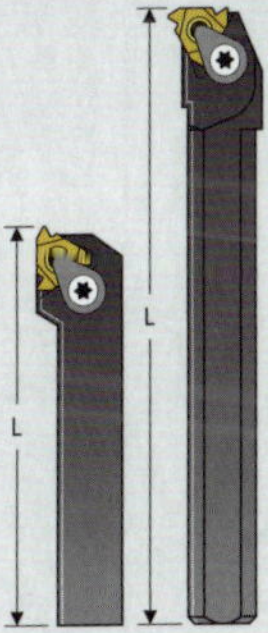

6 Insert Size

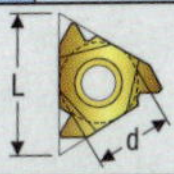

L (mm)	d (inch) - ic
06	5/32"
08	3/16"
11	1/4"
16	3/8"
22	1/2"
27	5/8"

7 Type

B = With Coolant Hole
C = Carbide Shank
V = Vertical (on edge)
O = Offset Style
D = Drop Head
G = Gang Tool
U = U Type

*Also refer pages 195-202

TUNGSTEN CARBIDE TURNING INSERT IDENTIFICATION (METRIC)

TNMG1604082G-Grade

T	N	M	G
1	2	3	4

1 Insert Shape

Symbol	Insert Shape
A	Parallelogram 85°
B	Parallelogram 82°
C	Rhombic 80°
D	Rhombic 55°
E	Rhombic 75°
F	Rhombic 50°
H	Hexagon
K	Parallelogram 55°
L	Rectangle
M	Rhombic 86°
O	Octagon
P	Pentagon
R	Round
S	Square
T	Triangle
V	Rhombic 35°
W	Trigon

2 Relief Angle

Symbol	Relief Angle	Symbol	Relief Angle
A	3°	F	25°
B	5°	G	30°
C	7°	N	0°
D	15°	P	11°
E	20°	O	Special

3 Tolerance

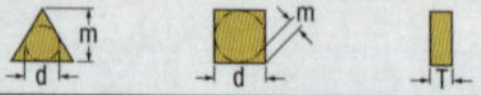

Class	Tol. ± 'm' (mm)	Tol. ± 'T' (mm)	Tol. ± 'd' (mm)
A	0.005	0.025	0.025
C	0.013	0.025	0.025
E	0.025	0.025	0.025
F	0.005	0.025	0.013
G	0.025	0.130	0.025
H	0.013	0.025	0.013
J	0.050	0.025	0.05~0.13
K	0.013	0.025	0.05~0.13
L	0.025	0.025	0.05~0.13
M	0.08~0.18	0.130	0.05~0.13
N	0.08~0.18	0.025	0.05~0.13
U	0.13~0.38	0.130	0.08~0.25

4 Fixing / Chip Breaker

Symbol	Feature	Symbol	Feature
A		N	
B		Q	
C		R	
F		T	
G		U	
H		W	
J		X	Special
M			

CARBIDE INSERT IDENTIFICATION

TUNGSTEN CARBIDE TURNING INSERT IDENTIFICATION (METRIC)

16	04	08	(F)*	(R)#	2G
5	6	7	8	9	10

5 Cutting Edge Length L (mm)

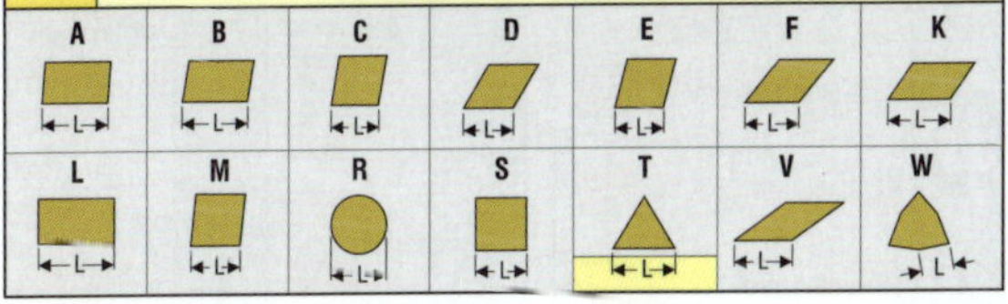

6 Thickness

Symbol	Thickness (S)
01	1.59
T1	1.98
02	2.38
T2	2.78
03	3.18
T3	3.97
04	4.76
05	5.56
06	6.35
07	7.94
09	9.52

7 Radius

Symbol	Nose Rad.
00	Sharp Pt.
01	0.1
02	0.2
04	0.4
08	0.8
12	1.2
16	1.6
20	2.0
24	2.4
28	2.8
32	3.2
00 M0	Round Insert

8 Edge*

Symbol	Figure
E	Hone
F	Sharp
S	Chamfer + Hone
T	Chamfer

* Most manufacturers generally omit this Cutting Edge Symbol.

9 Feed#

Symbol	Figure
R Right	←
L Left	→
N Neutral	↔

Most manufacturers omit this symbol when the insert position is **Neutral**.

10	Chip Breaker	This Symbol is reserved for Mfr.'s Chip Breaker Design.

Recommended Max. Depth of Cut 'd' (mm) for Standard Inserts

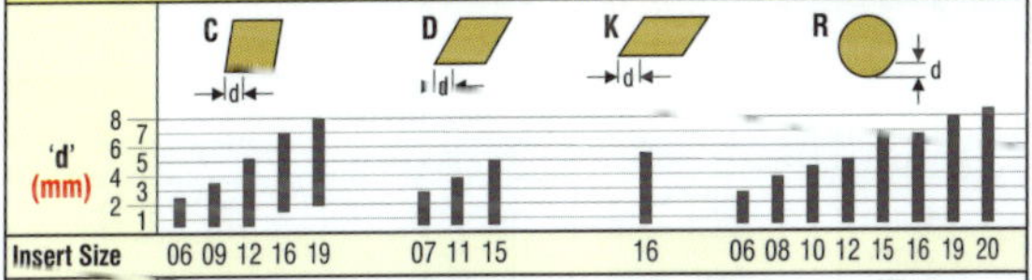

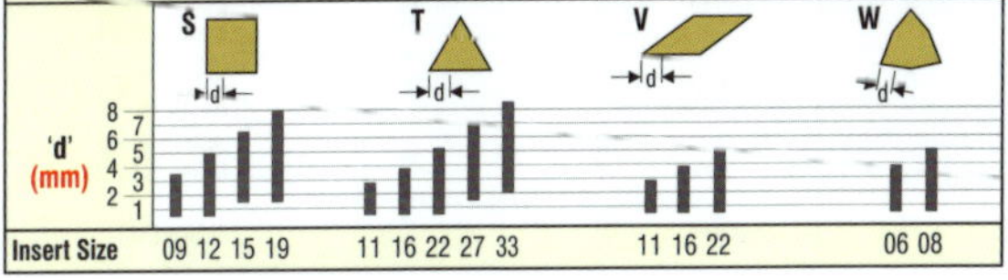

TUNGSTEN CARBIDE MILLING INSERT IDENTIFICATION (METRIC)

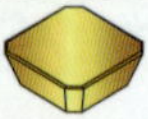

SEKN1203AFTN-Grade

S	E	K	N
1	2	3	4

1 Insert Shape

Symbol	Insert Shape
A	Parallelogram 85°
B	Parallelogram 82°
C	Rhombic 80°
H	Hexagon
L	Rectangle
M	Rhombic 86°
O	Octagon
P	Pentagon
R	Round
S	Square
T	Triangle
W	Trigon

2 Relief Angle

Symbol	Relief Angle	Symbol	Relief Angle
A	3°	F	25°
B	5°	G	30°
C	7°	N	0°
D	15°	P	11°
E	20°	O	Special

3 Tolerance

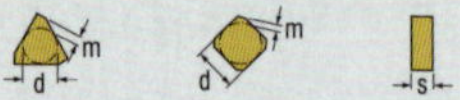

Class	Tol. ± 'm' (mm)	Tol. ± 's' (mm)	Tol. ± 'd' (mm)
A	0.005	0.025	0.025
C	0.013	0.025	0.025
E	0.025	0.025	0.025
G	0.025	0.130	0.025
H	0.013	0.025	0.013
J	0.050	0.025	0.05~0.13
K	0.013	0.025	0.05~0.13
M	0.08~0.18	0.130	0.05~0.13
U	0.13~0.38	0.130	0.08~0.25

4 Fixing / Chip Breaker

Symbol	Feature	Symbol	Feature
A		Q	
B		R	
F		T	
M		W	
N		X	Special

TUNGSTEN CARBIDE MILLING INSERT IDENTIFICATION (METRIC)

12	03	A	F	T	N
5	6	7	8	9	10

5 Cutting Edge Length L (mm)

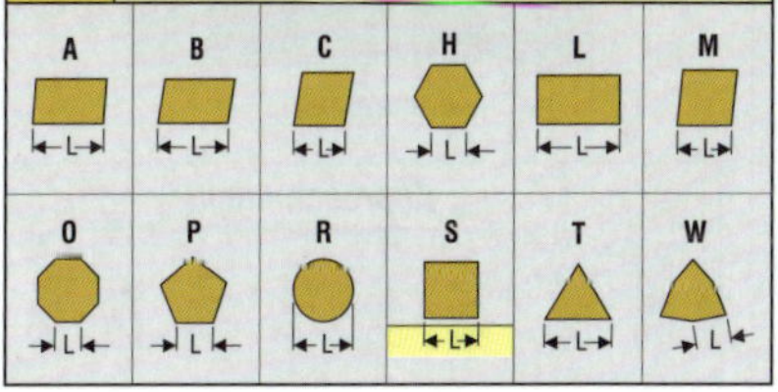

6 Thickness

Symbol	Thickness (S)
02	2.38
03	3.18
T3	3.97
04	4.76
05	5.56
06	6.35

7 Cutting Angle

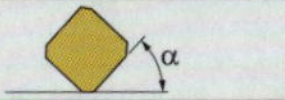

Symbol	Cutting Angle (α)
A	45°
D	60°
E	75°
F	85°
P	90°
X	65°
Z	Special

Metric Series (Nose Radius)

Symbol	Nose Radius
00	Sharp Point
01	0.1
02	0.2
04	0.4
08	0.8
12	1.2
16	1.62
M0	Round Insert

Example:

	1	2	3	4	5	6	7	8	9	10
Std. Series	S	E	K	N	12	03	A	F	T	N
Metric Series	0	E	K	T	06	06	08		S	N

8 Relief ∠

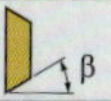

Symbol	Relief ∠ (β)
A	3°
B	5°
C	7°
D	15°
E	20°
F	25°
G	30°
N	0°
P	11°
R	10°
S	14°
T	22°
U	23°
Z	Special

9 Edge

Symbol	Figure
E	 Hone
F	 Sharp
S	 Chamfer + Hone
T	 Chamfer

10 Direction of Cutting

R	L	N
Right Rotation	Left Rotation	Neutral Right & Left Rotation

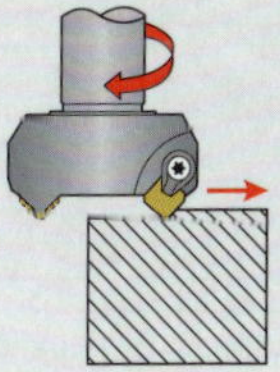

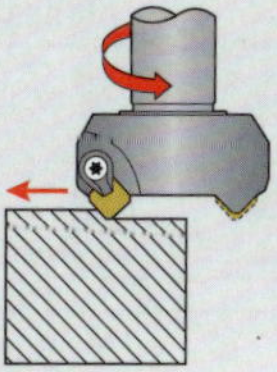

CARBIDE INSERT IDENTIFICATION

TUNGSTEN CARBIDE TURNING TOOLHOLDER IDENTIFICATION (METRIC)

C	T	G	P	R	25	25	M	16	□
1	2	3	4	5	6	7	8	9	10

1 Clamping*

Insert Clamping System

- C - Clamp Lock
- D - Clamp+Pin Lock
- M - Clamp+Pin Lock
- P - Pin Lock/Lever Lock
- S - Screw Lock
- W - Pin+Wedge Clamp

External Turning Toolholder

CTGPR2525M16

2 Insert Shape

- A - Parallelogram 85°
- B - Parallelogram 82°
- C - Rhombic 80°
- D - Rhombic 55°
- E - Rhombic 75°
- F - Rhombic 50°
- H - Hexagon
- K - Parallelogram 55°
- L - Rectangle
- M - Rhombic 86°
- O - Octagon
- P - Pentagon
- R - Round
- S - Square
- T - Triangle
- V - Rhombic 35°
- W - Trigon

3 Approach Angle

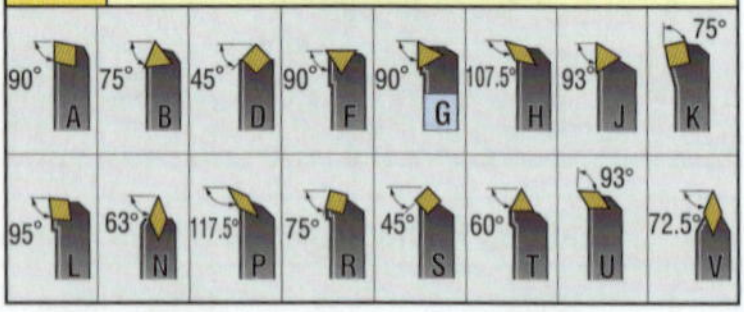

4 Relief ∠

- A - 3°
- B - 5°
- C - 7°
- D - 15°
- E - 20°
- F - 25°
- G - 30°
- N - 0°
- P - 11°
- O - Special

5 Version

R - Right, L - Left, N - Neutral

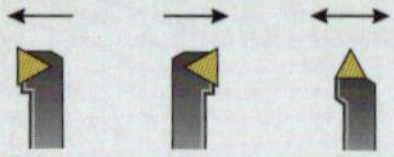

6 Shank

Shank Height – (mm)

7 Shank

Shank Width – (mm)

8 Tool Length (mm)

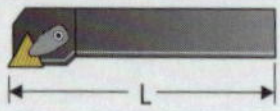

A	32	F	80	P	170
C	50	H	100	R	200
D	60	K	125	S	250
E	70	M	150	T	300

9 Cutting Edge Length L (mm)

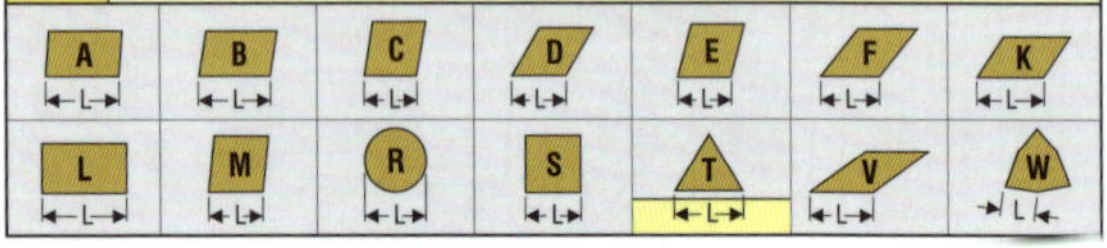

10 Manufacturer's Designation

*Also refer pages 195-199

TUNGSTEN CARBIDE BORING BAR IDENTIFICATION (METRIC)

S	25	T	C	T	F	P	R	16	□
1	2	3	4	5	6	7	8	9	10

1 Type

Boring Bar Shank

- A - Steel with Coolant Hole
- C - Carbide Shank
- E - Coolant through Carbide Shank
- S - Steel Shank

Boring Bar

S25TCTFPR16

2 Shank

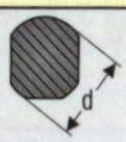

Shank Diameter 'd' (mm)

3 Tool Length (mm)

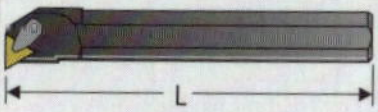

J - 110	N - 160	S - 250
K - 125	P - 170	T - 300
L - 140	Q - 180	U - 350
M - 150	R - 200	V - 400

4 Clamping*

Insert Clamping System

- C - Clamp Lock
- D - Clamp+Pin Lock
- M - Clamp+Pin Lock
- P - Pin Lock/Lever Lock
- S - Screw Lock

5 Insert Shape

- A - Parallelogram 85°
- B - Parallelogram 82°
- C - Rhombic 80°
- D - Rhombic 55°
- E - Rhombic 75°
- F - Rhombic 50°
- H - Hexagon
- K - Parallelogram 55°
- L - Rectangle
- M - Rhombic 86°
- O - Octagon
- P - Pentagon
- R - Round
- S - Square
- T - Triangle
- V - Rhombic 35°
- W - Trigon

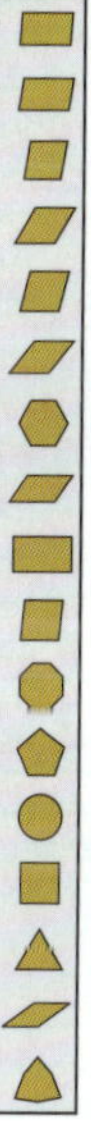

6 Approach Angle

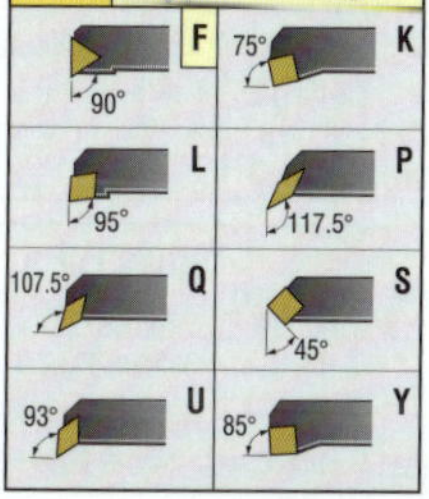

7 Relief ∠

- A - 3°
- B - 5°
- C - 7°
- D - 15°
- E - 20°
- F - 25°
- G - 30°
- N - 0°
- P - 11°
- O - Special

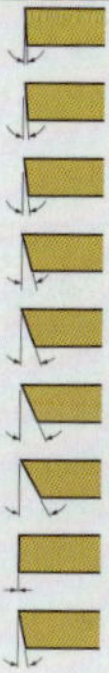

8 Version

R - Right Hand | L - Left Hand

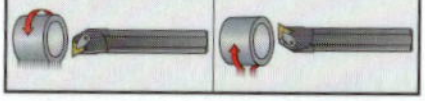

9 Cutting Edge Length L (mm)

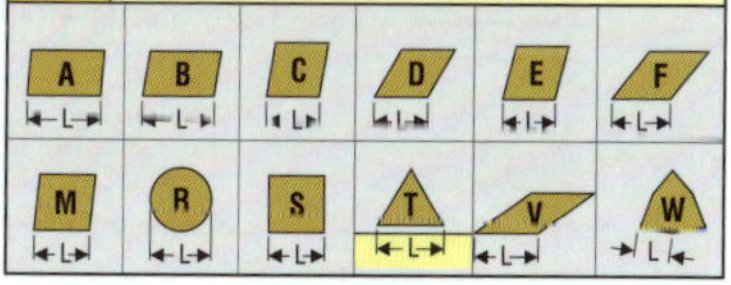

10 Manufacturer's Designation

CARBIDE INSERT IDENTIFICATION

*Also refer pages 200-202

TUNGSTEN CARBIDE THREADING INSERT IDENTIFICATION (METRIC)

16ER1.0ISOA1

16	E	R	1.0	ISO	A1
1	2	3	4	5	6

1 Edge Length

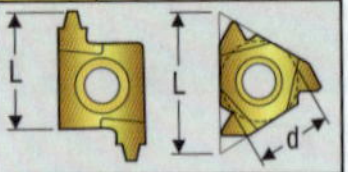

L (mm)	d (mm) - ic
06	3.97
08	4.76
11	6.35
16	9.525
22	12.70
27	15.875

2 Application

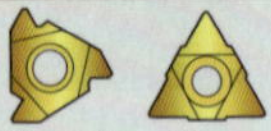

Standard U-Type

Symbol	Application
E	External
N*	Internal
U	U-Type

* Some manufacturers use "I" to denote Internal.

3 Insert Version

External Threading

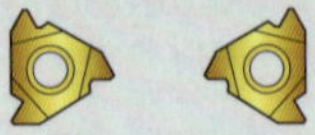

R L

Internal Threading

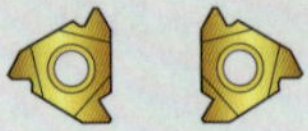

R L

4 Thread Pitch / tpi

Full Profile (mm)				Full Profile (tpi)					Partial Profile
0.5	0.7	0.75	0.8	48	40	32	24	20	A = 0.5 - 1.5 mm 48 - 16 tpi
1.0	1.25	1.5	1.75	18	16	14	13	12	AG = 0.5 - 3.0 mm 48 - 8 tpi
2.0	2.5	3.0	4.0	11	10	9	8	7	G = 1.75 - 3.0 mm 18 - 8 tpi
4.5	5.0	5.5	6.0	6	5	4.5	4	3	N = 3.5 - 5.0 mm 7 - 5 tpi
8.0	10.0	12.0	14.0	2.5	2				K = 5.5 - 8.0 mm 4.5 - 3 tpi

5 Thread Form

Symbol Profile	Symbol Profile	Symbol Profile
60 = V profile 60°	BUT2.5 = Buttress Fig.2.5	RND = Round DIN 405
55 = V profile 55°	BUT2.6 = Buttress Fig.2.6	STACME = Stub Acme
ABUT = American Buttress	BSPT = Whitworth Taper	TR = Trapezoidal DIN 103
ACME = Am.Acme-G	EL = Extreme Line Casing	UN = American Unified
API = American Pet. Inst.	H90 = Hughes	UNJ = UN-Controlled Root
API RD = API Rd. Casing	ISO = ISO, Metric	VAM = Vam Vallourec
BUT = API Buttress	NPT = National Pipe Taper	W = Whitworth (BSW)

6 No. of Teeth / Chip Breaker

Optional: To Manufacturer's Discretion.

Full Profile

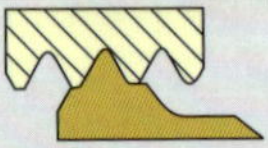

- Workpiece may be slightly oversize as insert 'Tops' the thread.
- No subsequent thread deburring is necessary.

Partial Profile

- Workpiece needs to be accurately pre-machined.
- One insert covers a wide range of pitches.

Single Tooth Insert

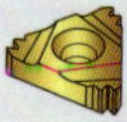

Multi Tooth Insert

Refer pages 26-30 for Basic Thread Profile details

TUNGSTEN CARBIDE EXT. THREADING TOOLHOLDER AND INT. THREADING BAR IDENTIFICATION (METRIC)

C	E	R	2020	K	16	□
1	2	3	4	5	6	7

External Threading Toolholder

CER2020K16

C	I	R	0020	P	16	□
1	2	3	4	5	6	7

Internal Threading Bar

CIR0020P16

1 Clamping*

Insert Clamping System

C - Clamp Lock

S - Screw Lock

2 Application

External / Internal

E - External Threading

I - Internal Threading

3 Version

R Right Hand

L Left Hand

4 Shank Size

External Toolholder

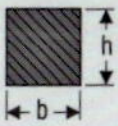

Shank - h x b

Example :
2020 – 20 x 20 mm

Internal Toolholder

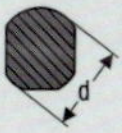

Shank - Diameter d

Example :
0020 = 20 mm dia.

5 Tool Length

Symbol Length (mm)

D – 60
F – 80
H – 100
K – 125
L – 140
M – 150
P – 170
R – 200
S – 250
T – 300
U – 350
V – 400

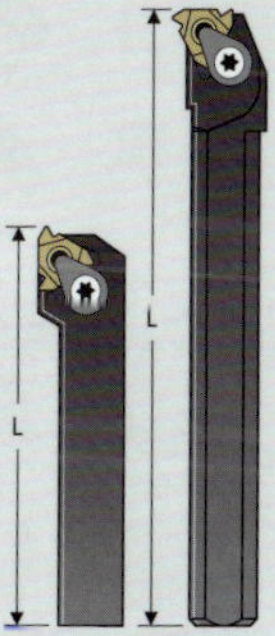

6 Insert Size

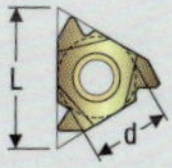

L (mm)	d (mm)-ic
06	3.97
08	4.76
11	6.35
16	9.525
22	12.70
27	15.785

7 No. of Teeth / Chip Breaker

Optional : To Manufacturer's Discretion

*Also refer pages 195-202

QUICK VISUAL IDENTIFICATION OF TUNGSTEN CARBIDE INSERT SHAPES

The following is a list of some of the common basic shapes of Tungsten Carbide Inserts that may be found in most engineering workshops. It is by no means a comprehensive list and is provided as a starting point for Insert identification.

Refer pages 178 to 185 for **INCH** & pages 186 to 193 for **METRIC** for detailed information regarding Insert identification.

NOTE: Chipbreaker shapes and designs vary considerably between manufacturers.

Common – Turning / Milling Inserts

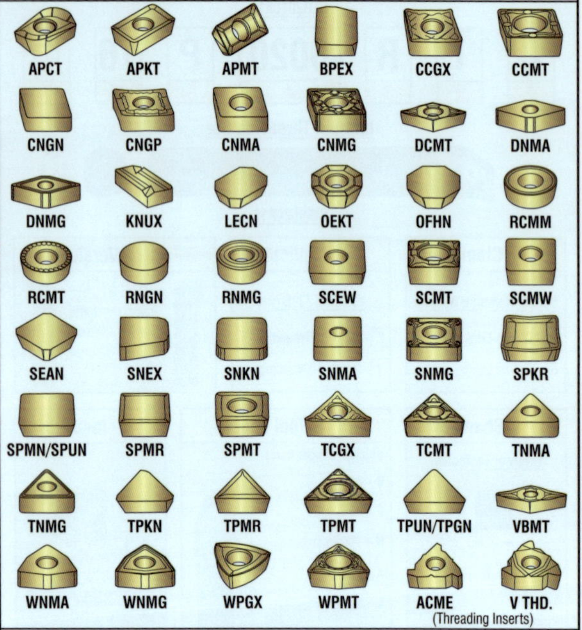

Common – Ceramic / Cermet / PCBN / PCD Inserts

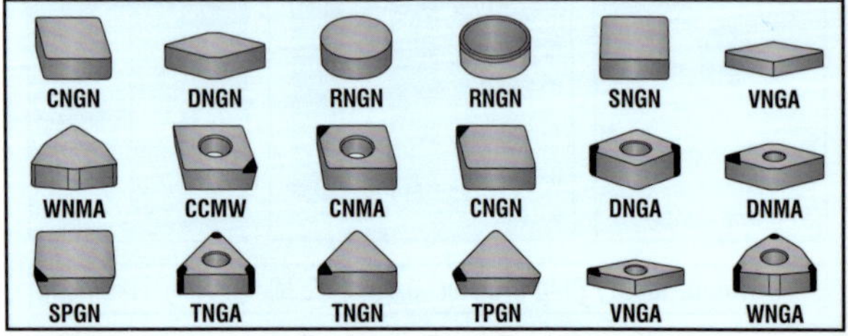

TUNGSTEN CARBIDE TOOLHOLDER SELECTION / APPLICATION

Right Hand Toolholder (R)	Neutral Toolholder (N)	Left Hand Toolholder (L)

Towards Chuck	Direction of Tool Travel (Feed)	Away from Chuck

NOTE: For Toolholder Identification - refer page 182 **(INCH)** & page 190 **(METRIC)**

Insert Clamping Method

C	D	M	P	S	W
				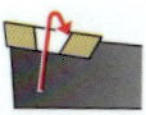	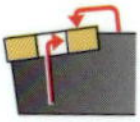
Top Clamping (Inserts w/o Hole) (Clamp Lock)	Top and Hole Clamping (Clamp+Pin Lock)	Top and Hole Clamping (Clamp+Pin Lock)	Hole Clamping (Pin Lock) (Lever Lock)	Screw Clamping (Screw Lock)	Top and Hole Clamping (Pin+ Wedge Clamp)

Toolholders with C - Clamp Lock Assembly

CCBN R/L ____	CCLN R/L ____	CCLP R/L ____	CKJN R/L ____
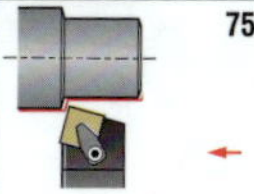 75°	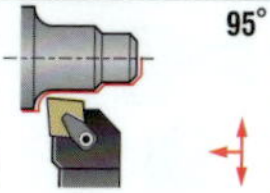 95°	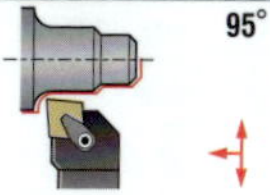 95°	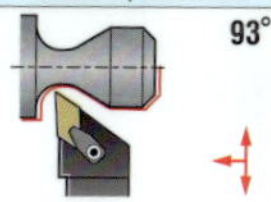 93°
Insert - CNGN, CNMN	Insert - CNGN, CNMN	Insert - CPGN, CPUN	Insert - KNUX
CKNN R/L ____	**CRDCN ____**	**CRDC R/L ____**	**CRDNN ____**
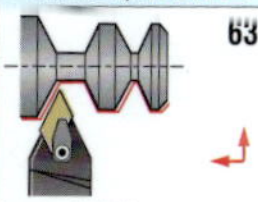 63°	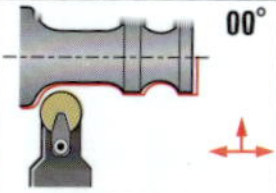 00°	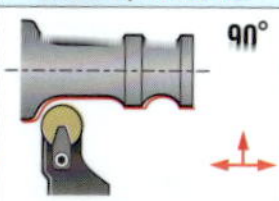 90°	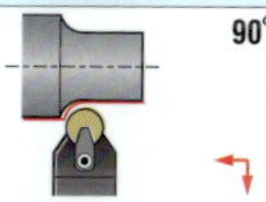 90°
Insert - KNUX	Insert - RCGX, RPGX	Insert - RCGX, RPGX	Insert - RNGN, RNMN
CRGN R/L ____	**CRSN R/L ____**	**CSBP R/L ____**	**CSDPN ____**
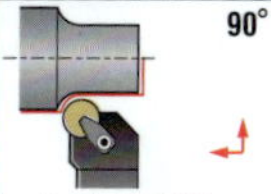 90°	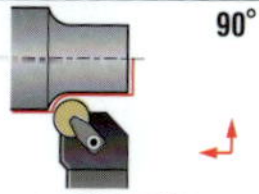 90°	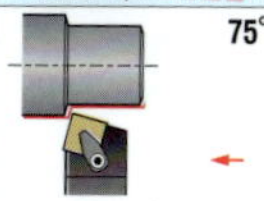 75°	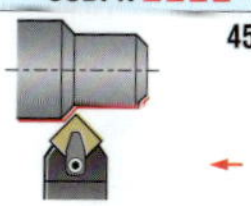 45°
Insert - RNGN, RNMN	Insert - RNGN, RNMN	Insert - SPMR, SPUN	Insert - SPMR, SPUN

Continued ...

TUNGSTEN CARBIDE TOOLHOLDER SELECTION / APPLICATION

Toolholders with C - Clamp Lock Assembly ...continued

CSKP R/L _ _ _ _
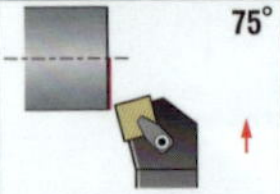

Insert - SPMR, SPUN

CSRP R/L _ _ _ _
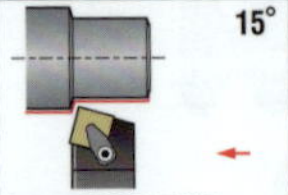

Insert - SPMR, SPUN

CSSP R/L _ _ _ _
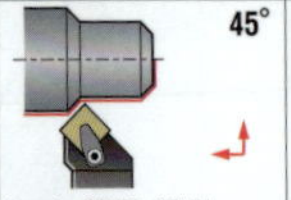

Insert - SPMR, SPUN

CSTP R/L _ _ _ _
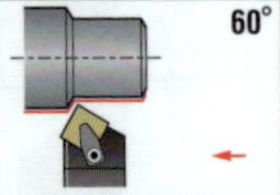

Insert - SPMR, SPUN

CTAP R/L _ _ _ _
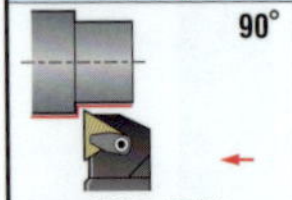

Insert - TPMR, TPUN

CTBP R/L _ _ _ _
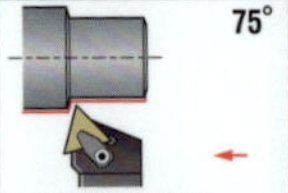

Insert - TPMR, TPUN

CTCON _ _ _ _
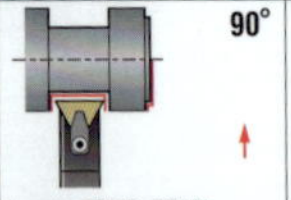

Insert - TPMR, TPUN

CTCP R/L _ _ _ _
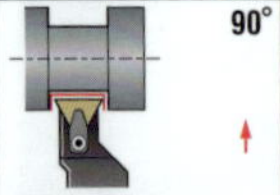

Insert - TPMR, TPUN

CTCPN _ _ _ _
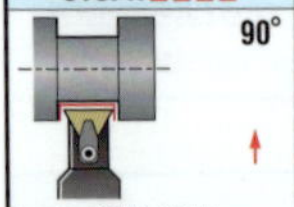

Insert - TPMR, TPUN

CTDP R/L _ _ _ _
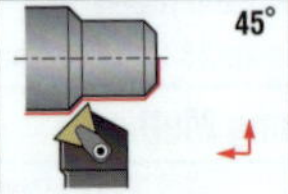

Insert - TPMR, TPUN

CTFP R/L _ _ _ _
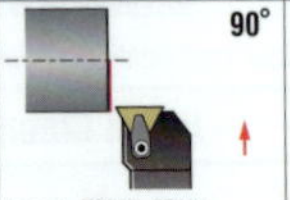

Insert - TPMR, TPUN

CTGP R/L _ _ _ _
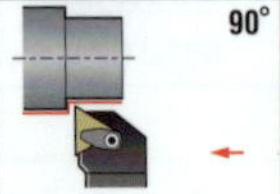

Insert - TPMR, TPUN

Toolholders with D Dual - Top and Hole Clamp

DCBN R/L _ _ _ _
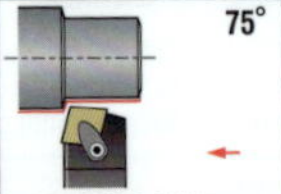

Insert - CNMA, CNMG

DCKN R/L _ _ _ _
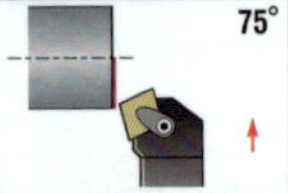

Insert - CNMA, CNMG

DCLN R/L _ _ _ _
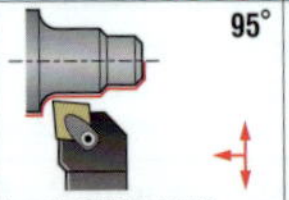

Insert - CNMA, CNMG

DCRN R/L _ _ _ _
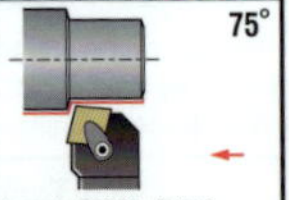

Insert - CNMA, CNMG

DCSN R/L _ _ _ _
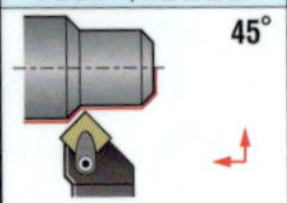

Insert - CNMA, CNMG

DDHN R/L _ _ _ _
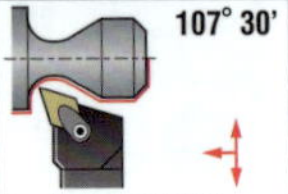

Insert - DNMG, DNMX

DDJN R/L _ _ _ _
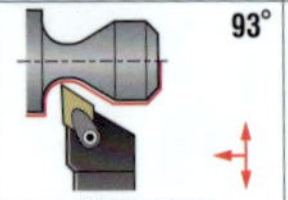

Insert - DNMG, DNMX

DDNNN _ _ _ _
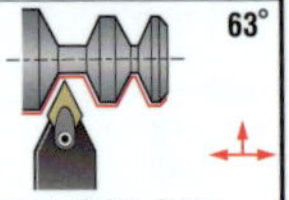

Insert - DNMG, DNMX

DDPNN _ _ _ _
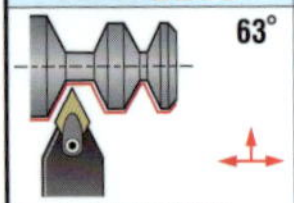

Insert - DNMG, DNMX

DDQN R/L _ _ _ _
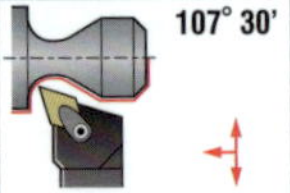

Insert - DNMG, DNMX

DSBN R/L _ _ _ _
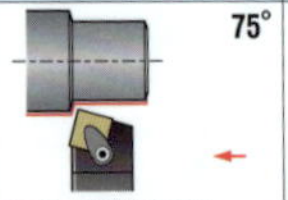

Insert - SNGX, SNMG

DSDNN _ _ _ _
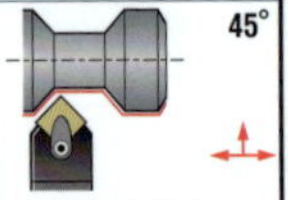

Insert - SNGX, SNMG

DSKN R/L _ _ _ _
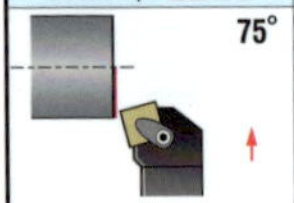

Insert - SNGX, SNMG

DSSN R/L _ _ _ _
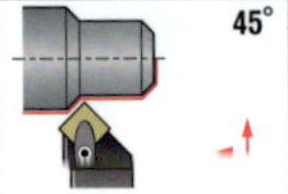

Insert - SNGX, SNMG

DTFN R/L _ _ _ _
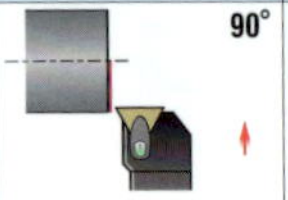

Insert - TNGG, TNMG

DTGN R/L _ _ _ _
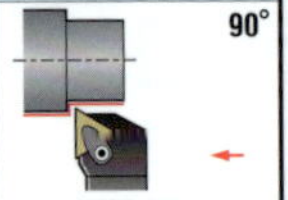

Insert - TNGG, TNMG

Continued ...

TUNGSTEN CARBIDE TOOLHOLDER SELECTION / APPLICATION

Toolholders with D Dual - Top and Hole Clamp ...continued

DVJN R/L ----	**DVQN** R/L ----	**DVVNN** ----	**DWLN** R/L ----
93° 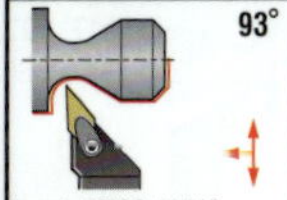 Insert - VNGG, VNMG	117° 30' 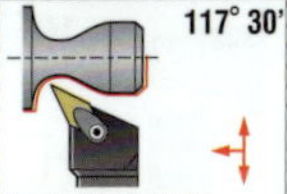 Insert - VNGG, VNMG	72° 30' 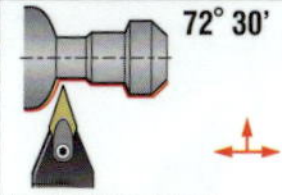 Insert - VNGG, VNMG	95° 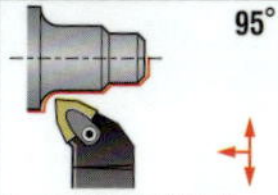 Insert - WNMG, WNMX

Toolholders with M Multiple - Pin and Clamp Lock

MCFN R/L ----	**MCGN** R/L ----	**MCHN** R/L ----	**MCKN** R/L ----
90° 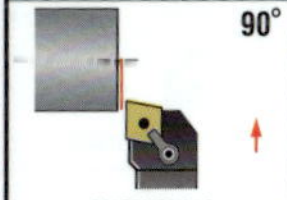 Insert - CNGN,CNMG	90° 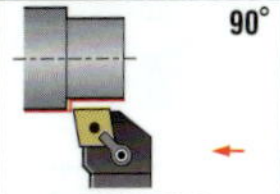 Insert - CNGN,CNMG	50° 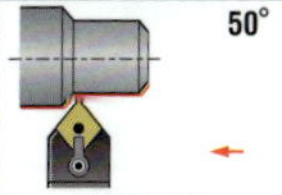 Insert - CNGN,CNMG	15° 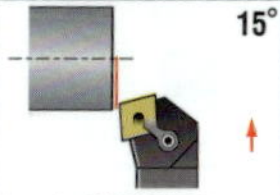 Insert - CNGN,CNMG
MCLN R/L ----	**MCMNN** ----	**MCRN** R/L ----	**MDJN** R/L ----
95° 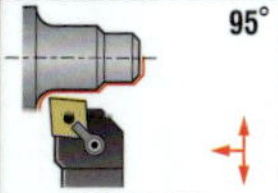 Insert - CNGN,CNMG	50° 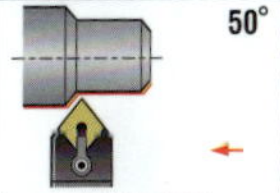 Insert - CNGN,CNMG	75° 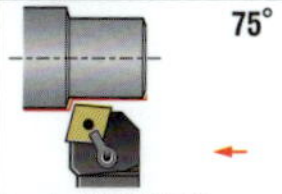 Insert - CNGN,CNMG	93° 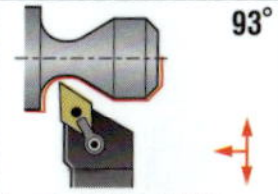 Insert - DNMG, DNMX
MDNNN R/L ----	**MDPNN** R/L ----	**MDQN** R/L ----	**MRGN** R/L ----
62° 30' 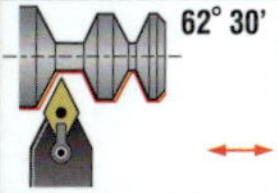 Insert - DNMG, DNMX	117° 30' 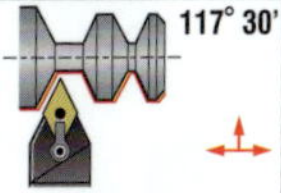 Insert - DNMG, DNMX	107° 30' 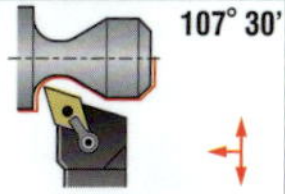 Insert - DNMG, DNMX	90° 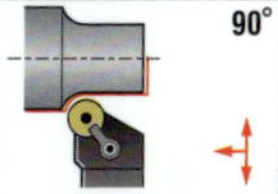 Insert - RNGA, RNMX
MSBN R/L ----	**MSDNN** ----	**MSKN** R/L ----	**MSRN** R/L ----
75° 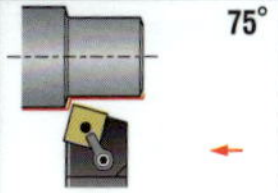 Insert - SNGX, SNMG	45° 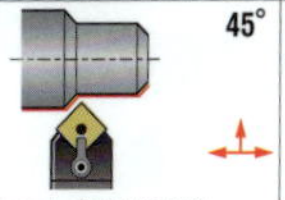 Insert - SNGX, SNMG	75° 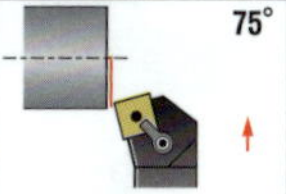 Insert - SNGX, SNMG	75° 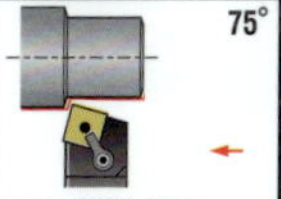 Insert - SNGX, SNMG
MSSN R/L ----	**MTAN** R/L ----	**MTCNN** ----	**MTENN** ----
45° 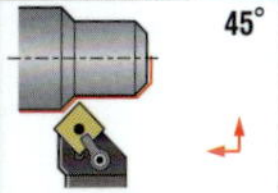 Insert - SNGX, SNMG	90° 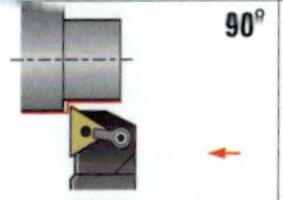 Insert - TNGG, TNMG	90° 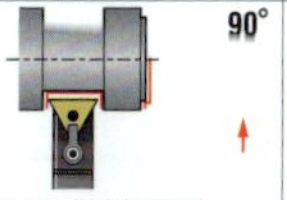 Insert - TNGG, TNMG	60° 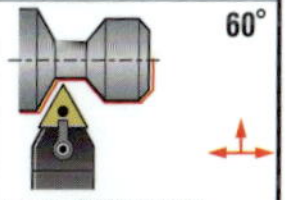 Insert - TNGG, TNMG
MTFN R/L ----	**MTGN** R/L ----	**MTJN** R/L ----	**MTLN** R/L ----
90° 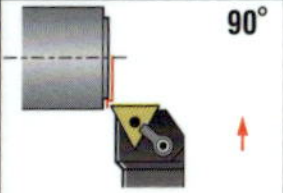 Insert - TNGG, TNMG	90° 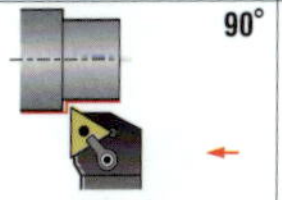 Insert - TNGG, TNMG	93° 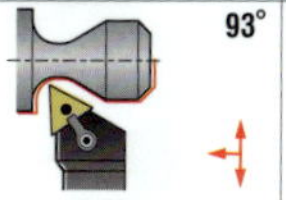 Insert - TNGG, TNMG	95° 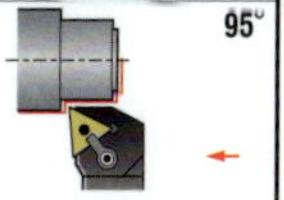 Insert - TNGG, TNMG

TOOL HOLDERS

Continued ...

TUNGSTEN CARBIDE TOOLHOLDER SELECTION / APPLICATION

Toolholders with M Multiple - Pin and Clamp Lock ...continued

MTNNN _ _ _ _ 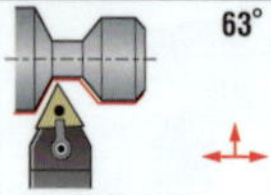 63° Insert - TNGG, TNMG	**MVJN** R/L _ _ _ _ 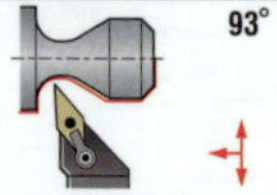 93° Insert - VNGG, VNMG	**MVLN** R/L _ _ _ _ 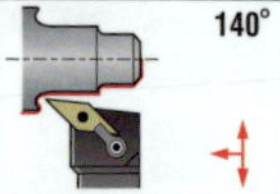 140° Insert - VNGG, VNMG	**MVQN** R/L _ _ _ _ 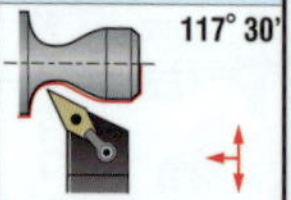 117° 30' Insert - VNGG, VNMG
MVTN R/L _ _ _ _ 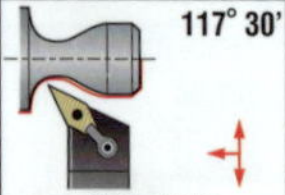 117° 30' Insert - VNGG, VNMG	**MVVNN** _ _ _ _ 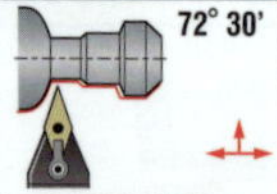 72° 30' Insert - VNGG, VNMG	**MVWN** R/L _ _ _ _ 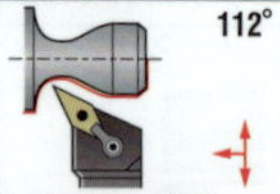 112° Insert - VNGG, VNMG	**MWLN** R/L _ _ _ _ 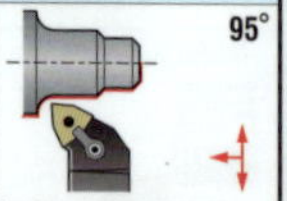 95° Insert - WNMG, WNMX

Toolholders with P - Pin Lock

PCBN R/L _ _ _ _ 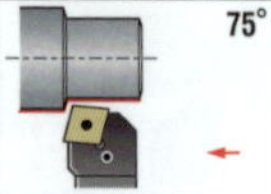 75° Insert - CNGN,CNMG	**PCFN** R/L _ _ _ _ 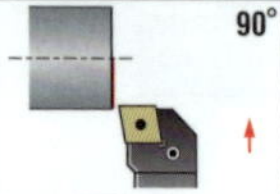 90° Insert - CNGN,CNMG	**PCKN** R/L _ _ _ _ 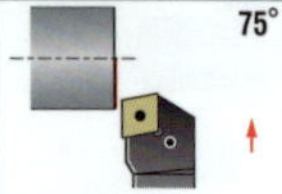 75° Insert - CNGN,CNMG	**PCLN** R/L _ _ _ _ 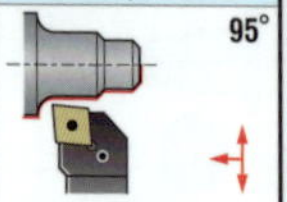 95° Insert - CNGN,CNMG
PCMNN _ _ _ _ 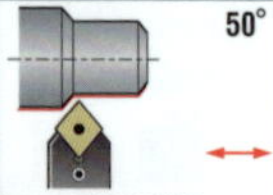 50° Insert - CNGN,CNMG	**PCSN** R/L _ _ _ _ 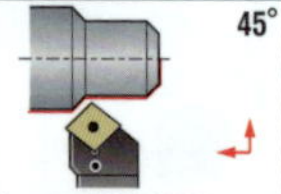 45° Insert - CNGN,CNMG	**PDJN** R/L _ _ _ _ 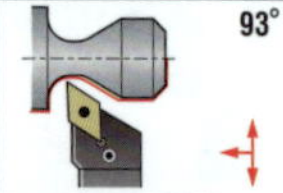 93° Insert - DNMG, DNMX	**PDNN** R/L _ _ _ _ 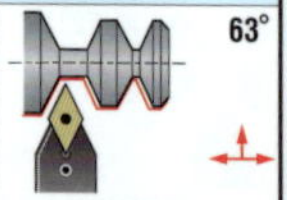 63° Insert - DNMG, DNMX
PRDCN _ _ _ _ 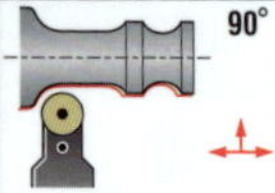 90° Insert - RNGN, RNMN	**PRSC** R/L _ _ _ _ 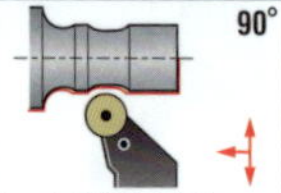 90° Insert - RNGN, RNMN	**PRSN** R/L _ _ _ _ 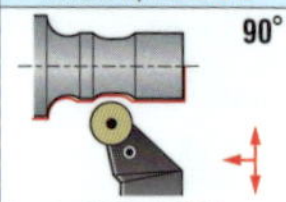 90° Insert - RNGN, RNMN	**PSBN** R/L _ _ _ _ 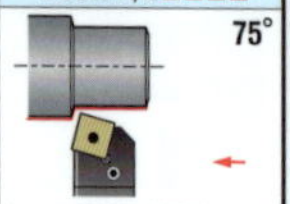 75° Insert - SNGX, SNMG
PSDNN _ _ _ _ 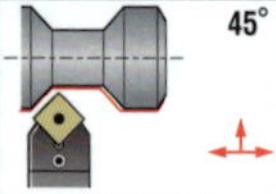 45° Insert - SNGX, SNMG	**PSKN** R/L _ _ _ _ 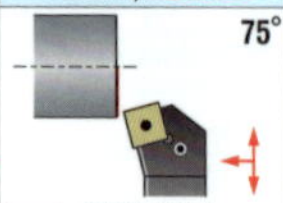 75° Insert - SNGX, SNMG	**PSSN** R/L _ _ _ _ 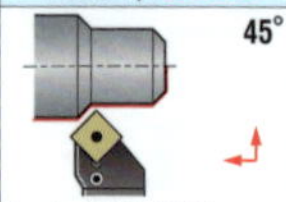 45° Insert - SNGX, SNMG	**PTDN** R/L _ _ _ _ 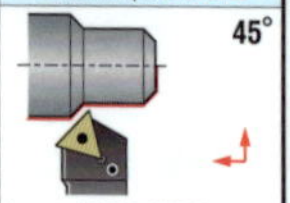 45° Insert - TNGG, TNMG
PTFN R/L _ _ _ _ 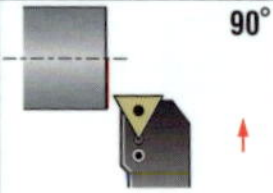 90° Insert - TNGG, TNMG	**PTGN** R/L _ _ _ _ 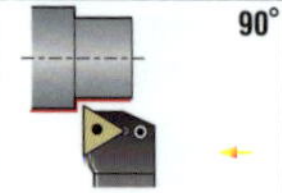 90° Insert - TNGG, TNMG	**PTTN** R/L _ _ _ _ 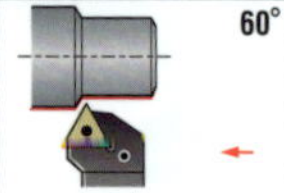 60° Insert - TNGG, TNMG	**PWLN** R/L _ _ _ _ 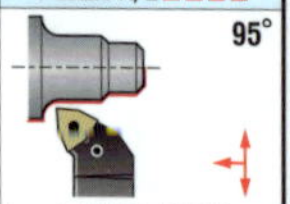 95° Insert - WNMG, WNMX

TOOL HOLDERS

TUNGSTEN CARBIDE TOOLHOLDER SELECTION / APPLICATION

Toolholders with S - Screw Lock

Toolholder	Diagram	Angle	Insert
SCAC R/L ----	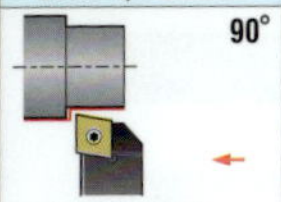	90°	Insert - CCGT, CCMT
SCFC R/L ----	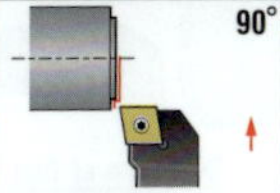	90°	Insert - CCGT, CCMT
SCGC R/L ----	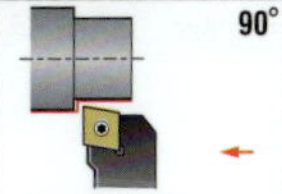	90°	Insert - CCGT, CCMT
SCLC R/L ----	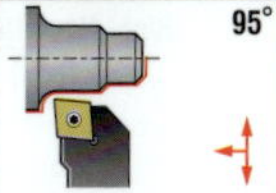	95°	Insert - CCGT, CCMT
SCMCN ----	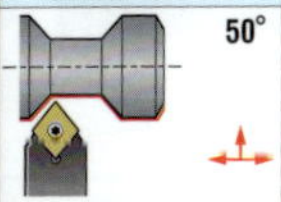	50°	Insert - CCGT, CCMT
SCRC R/L ----	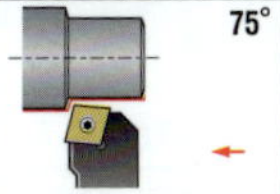	75°	Insert - CCGT, CCMT
SDJC R/L ----	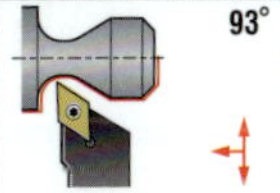	93°	Insert - DCGT, DCMT
SDNCN ----	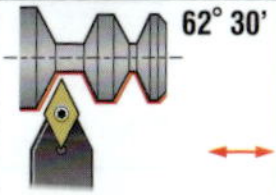	62° 30'	Insert - DCGT, DCMT
SDQC R/L ----	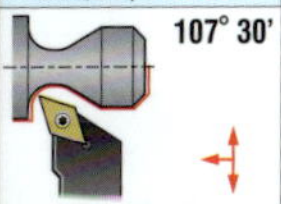	107° 30'	Insert - DCGT, DCMT
SRGC R/L ----	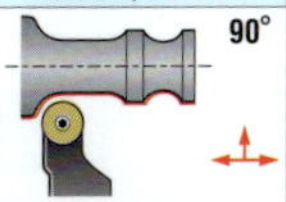	90°	Insert - RCMT, RCMX
SROCN ----	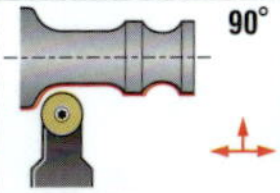	90°	Insert - RCMT, RCMX
SRSC R/L ----	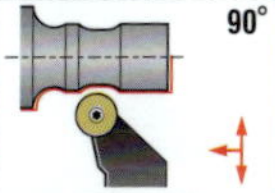	90°	Insert - RCMT, RCMX
SSBC R/L ----	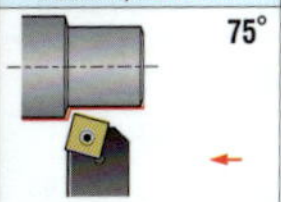	75°	Insert - SCGT, SCMT
SSDCN ----	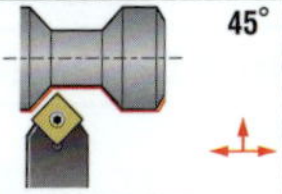	45°	Insert - SCGT, SCMT
STAC R/L ----	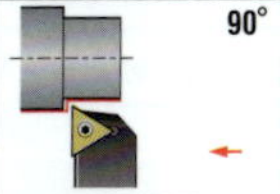	90°	Insert - TCGT, TCMT
STFC R/L ----	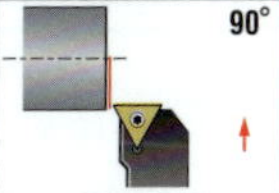	90°	Insert - TCGT, TCMT
STGC R/L ----	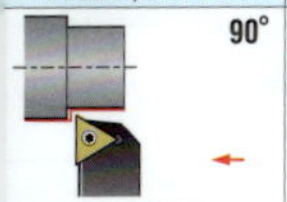	90°	Insert - TCGT, TCMT
STJC R/L ----	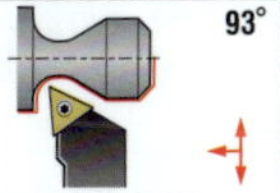	93°	Insert - TCGT, TCMT
SVJB R/L ----	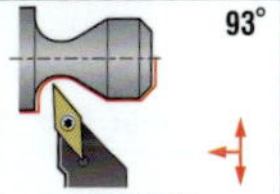	93°	Insert - VBGT, VBMT
SVLB R/L ----	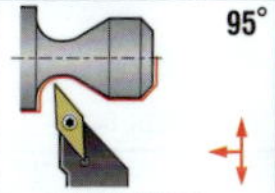	95°	Insert - VBGT, VBMT
SVLC R/L ----	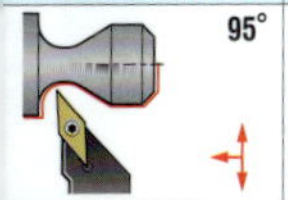	95°	Insert - VBGT, VBMT
SVVB R/L ----	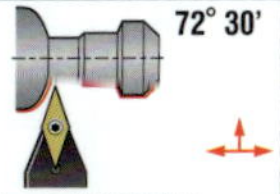	72° 30'	Insert - VBGT, VBMT
SVVC R/L ----	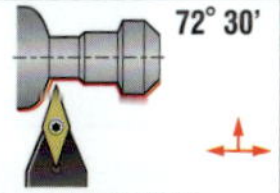	72° 30'	Insert - VBGT, VBMT
SWLC R/L ----	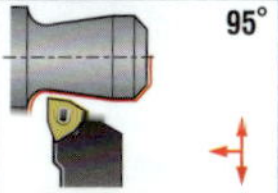	95°	Insert - WBGT, WCGT

Toolholders with W - Wedge Lock

Toolholder	Diagram	Angle	Insert
WTENN	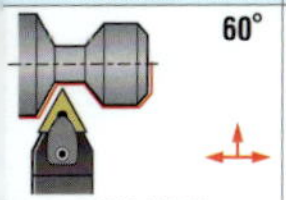	60°	Insert - TNGG, TNMG
WTJN R/L ----	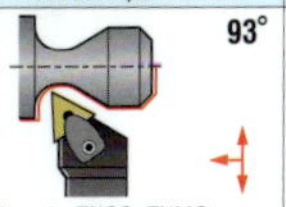	93°	Insert - TNGG, TNMG
WTXN R/L ----	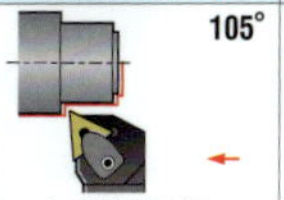	105°	Insert - TNGG, TNMG
WWLN R/L ----	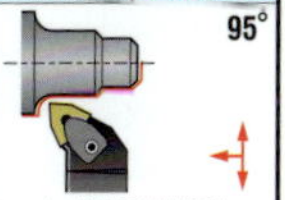	95°	Insert - WNMG, WNMX

TOOL HOLDERS

TUNGSTEN CARBIDE BORING BAR SELECTION / APPLICATION

Boring Bar Feed Direction - Hand of Tool

Right Hand Boring Bar
(R)

Left Hand Boring Bar
(L)

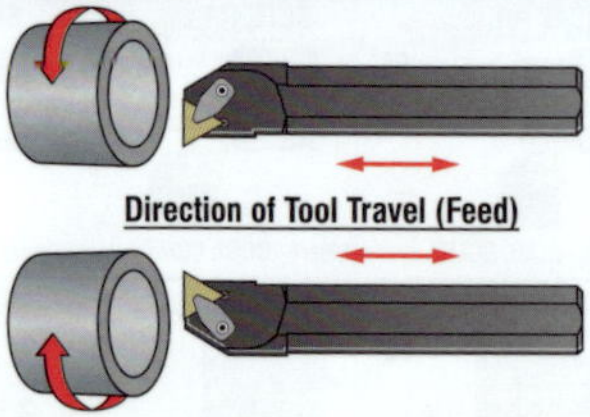

NOTE: For Boring Bar Identification - refer page 183 (INCH) & page 191 (METRIC)

Insert Clamping Method

C

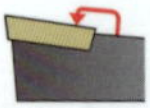

Top Clamping (Inserts w/o Hole) (Clamp Lock)

D

Top and Hole Clamping (Clamp+Pin Lock)

M

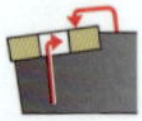

Top and Hole Clamping (Clamp+Pin Lock)

P

Hole Clamping (Pin Lock) (Lever Lock)

S

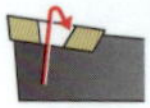

Screw Clamping (Screw Lock)

Boring Bars with C - Clamp Lock Assembly

S 16 T – CCLNR4

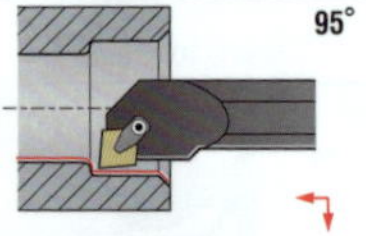

Insert - CNGN, CNMN

_ _ _ – CDUN R/L _ _

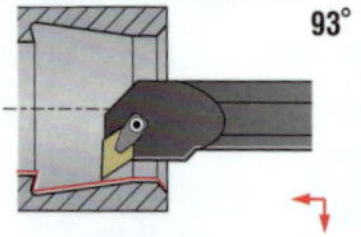

Insert - DNGN, DNGX

_ _ _ – CKUN R/L _ _

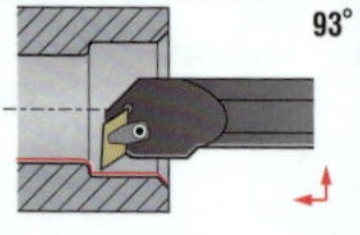

Insert - KNUX

_ _ _ – CSKP R/L _ _

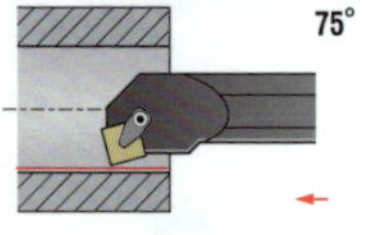

Insert - SPMR, SPUN

_ _ _ – CTFP R/L _ _

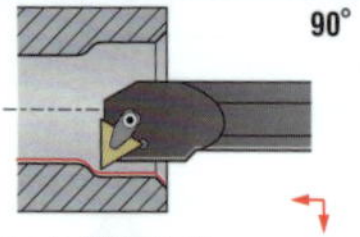

Insert - TPMR, TPUN

_ _ _ – CTUP R/L _ _

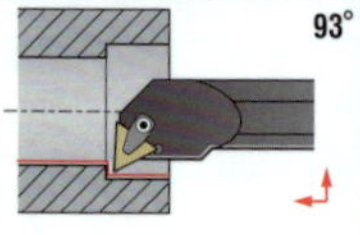

Insert - TPMR, TPUN

Boring Bars with M Multiple - Pin and Clamp Lock

_ _ _ – MCLN R/L _ _

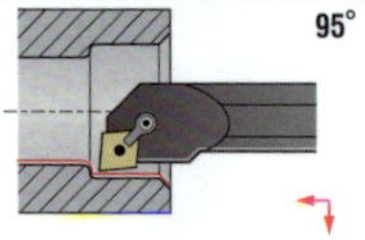

Insert - CNMA, CNMG

_ _ _ – MDQN R/L _ _

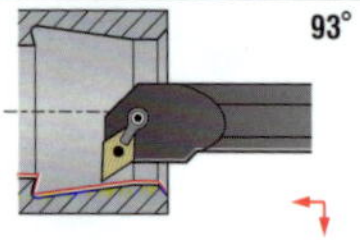

Insert - DNMG, DNMX

_ _ _ – MDUN R/L _ _

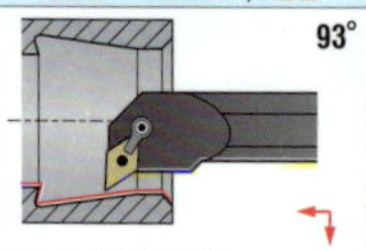

Insert - DNMG, DNMX

BORING BARS

Continued ...

TUNGSTEN CARBIDE BORING BAR SELECTION / APPLICATION

Boring Bars with M Multiple - Pin and Clamp Lock ...continued

--- – MSKN R/L --

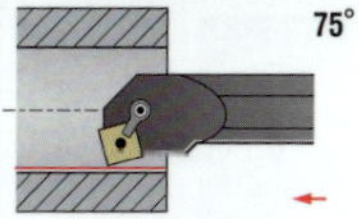

Insert - SNGX, SNMG

--- – MTFN R/L --

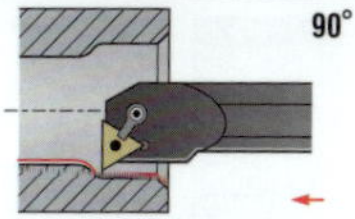

Insert - TNMG, TNMX

--- – MTUN R/L --

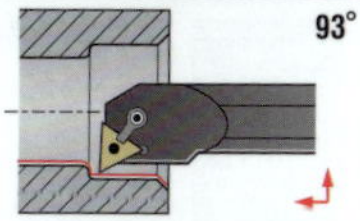

Insert - TNGG, TNMG

--- – MVVN R/L --

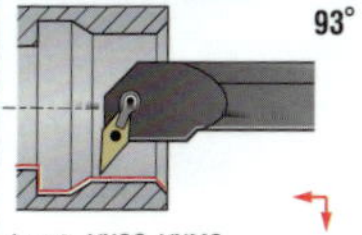

Insert - VNGG, VNMG

--- – MVXN R/L --

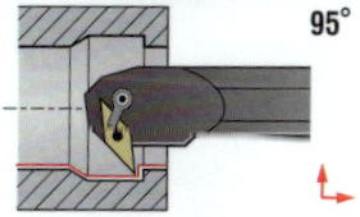

Insert - VNGG, VNMG

--- – MWLN R/L --

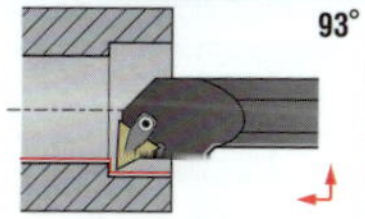

Insert - WNMG, WNMX

Boring Bars with P - Pin Lock

--- – PCKN R/L --

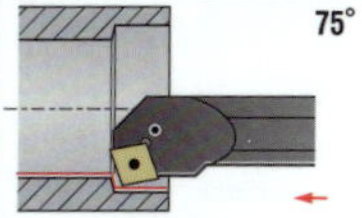

Insert - CNGN, CNMG

--- – PCLN R/L --

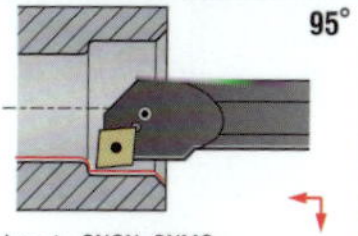

Insert - CNGN, CNMG

--- – PDUN R/L --

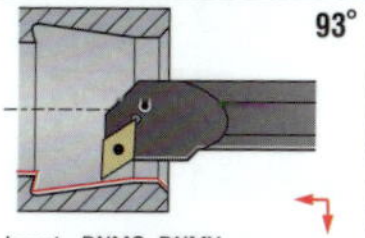

Insert - DNMG, DNMX

--- – PSKN R/L --

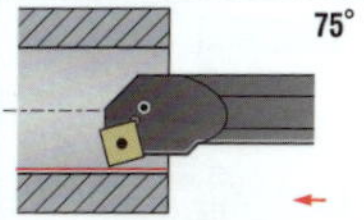

Insert - SNGX, SNMG

--- – PTFN R/L --

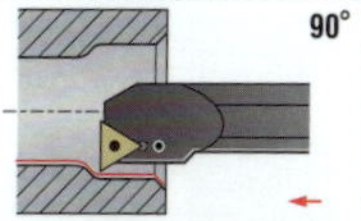

Insert - TNGG, TNMG

--- – PWLN R/L --

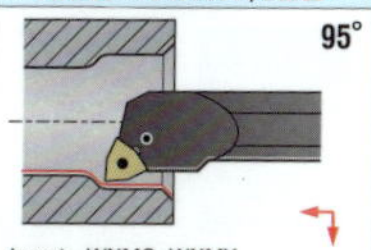

Insert - WNMG, WNMX

Boring Bars with S - Screw Lock

--- – SCFD R/L --

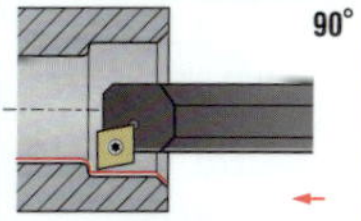

Insert - CCGT, CCMT

--- – SCFP R/L --

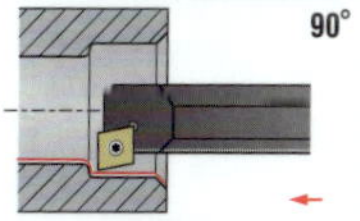

Insert - CCGT, CCMT

--- – SCLC R/L --

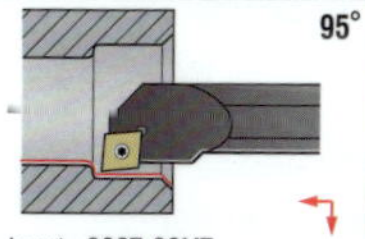

Insert - CCGT, CCMT

--- – SCLD R/L --

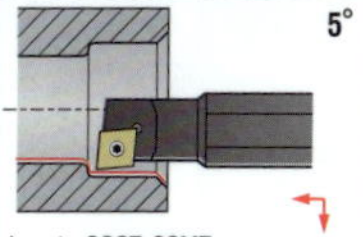

Insert - CCGT, CCMT

--- – SCLP R/L --

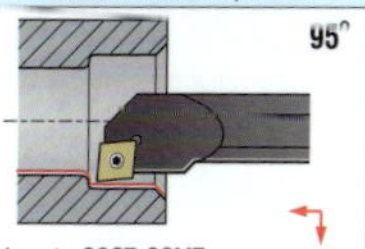

Insert - CCGT, CCMT

--- – SDQC R/L --

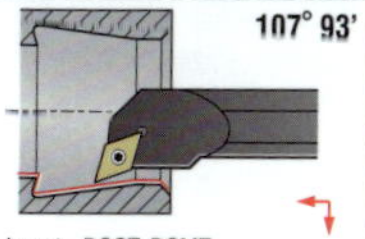

Insert - DCGT, DCMT

Continued ...

TUNGSTEN CARBIDE BORING BAR SELECTION / APPLICATION

Boring Bars with S - Screw Lock ...continued

___ – SDUC R/L __

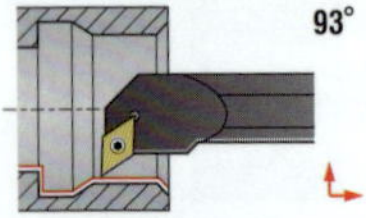

Insert - DCGT, DCMT

___ – SDUP R/L __

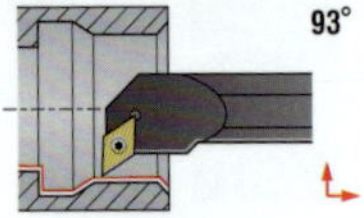

Insert - DCGT, DCMT

___ – SDXP R/L __

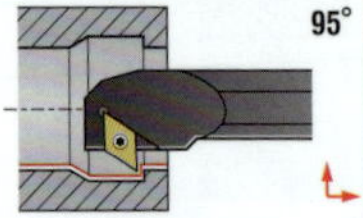

Insert - DCGT, DCMT

___ – SDZC R/L __

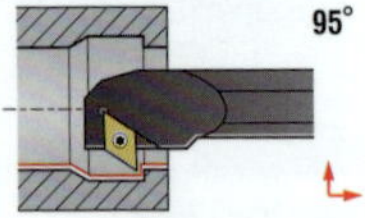

Insert - DCGT, DCMT

___ – SSKC R/L __

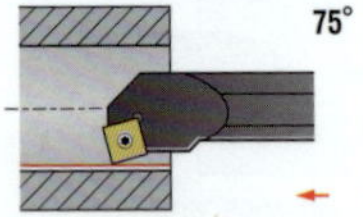

Insert - SCGT, SCMT

___ – SSKP R/L __

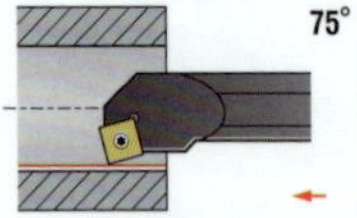

Insert - SPGT, SPMT

___ – STFC R/L __

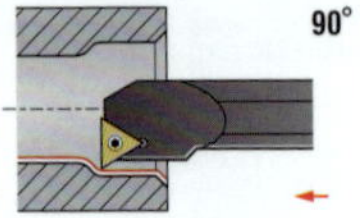

Insert - TCGT, TCMT

___ – STFD R/L __

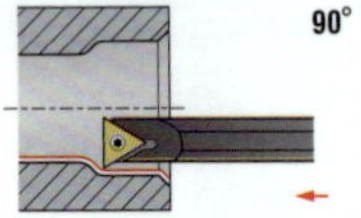

Insert - TCGT, TCMT

___ – STFP R/L __

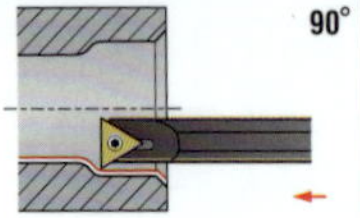

Insert - TCGT, TCMT

___ – STLC R/L __

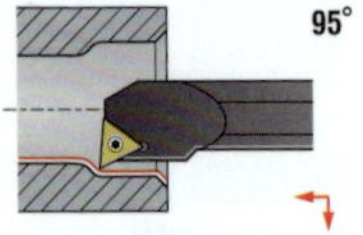

Insert - TCGT, TCMT

___ – STLD R/L __

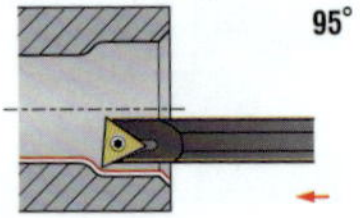

Insert - TCGT, TCMT

___ – STWP R/L __

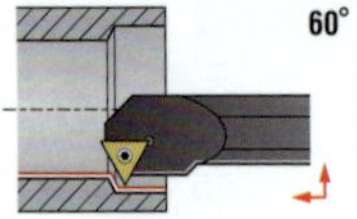

Insert - TCGT, TCMT

___ – SVMB R/L __

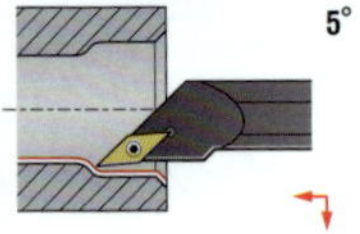

Insert - VBMT, VCGT

___ – SVQB R/L __

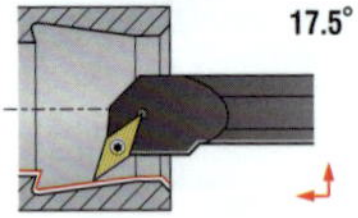

Insert - VBMT, VCGT

___ – SVQC R/L __

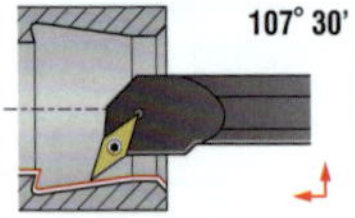

Insert - VBMT, VCGT

___ – SVUC R/L __

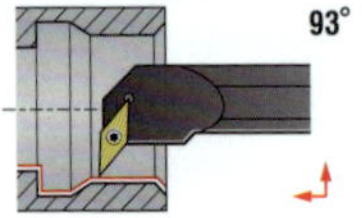

Insert - VBMT, VCGT

___ – SWLC R/L __

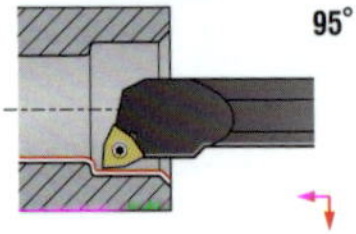

Insert - WBGT, WCGT

___ – SWLP R/L __

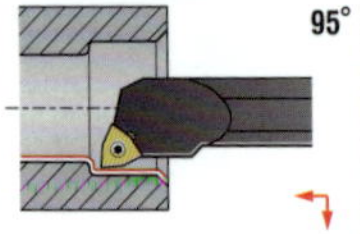

Insert - WBGT, WCGT

BORING BARS

GEOMETRIC DIMENSIONING AND TOLERANCING (GD&T) ACCORDING TO ASME Y14.5-2009

Geometric Dimensioning and Tolerancing **(GD&T)** is a system of symbols, rules and definitions used to define the geometry of mechanical parts.

According to the ASME Y14.5-2009 Standard, the purpose of **GD&T** is to describe the engineering intent of parts and assemblies by using symbols, rules, definitions, tools and examples that convey what degree of accuracy and precision is needed on each controlled feature of the part.

It provides designers the tools to accurately describe the functionality of a part and adds clarity when interpreting engineering drawings, thus avoiding costly errors with fewer outsourcing problems.

GD&T when used correctly is one of the most powerful tools available for the design of parts that can improve quality, reduce cost and shorten delivery time.

Conventional Bilateral Drawing

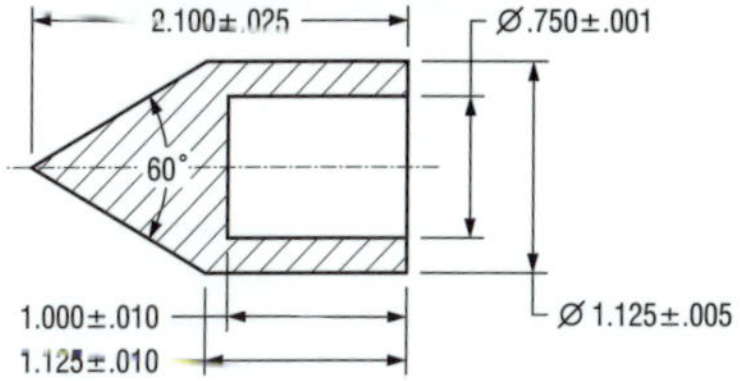

GD&T Drawing

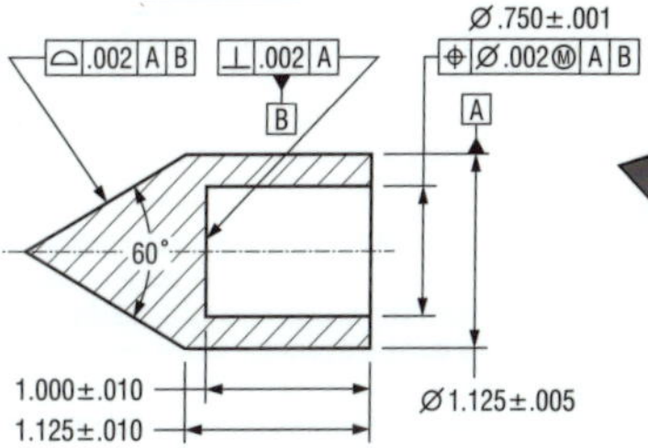

Datum Selection

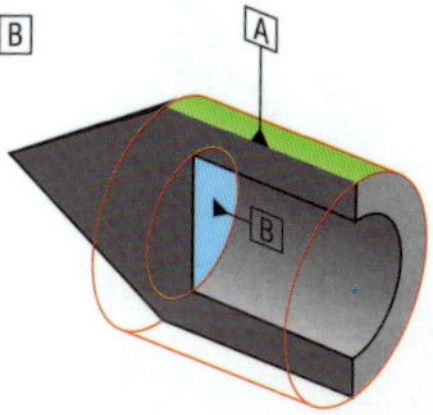

A Datum is theoretically an exact plane, axis or point location where dimensional tolerances are referenced to and can be used for all GD&T symbols except for form tolerances such as Surface Straightness, Flatness, Circularity and Cylindricity.

Feature Control Frame

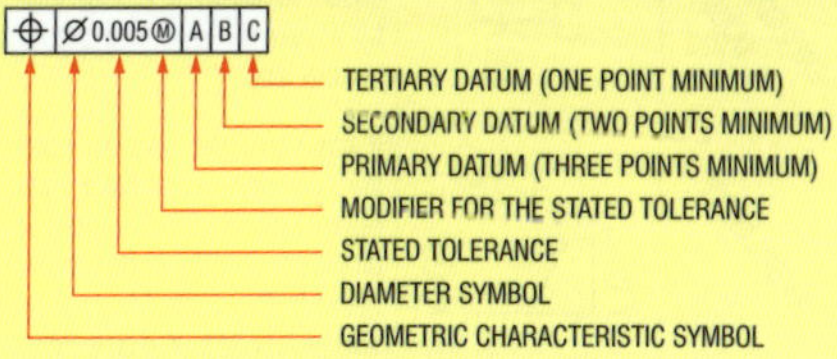

GD&T SYMBOLS AND GUIDELINES

Symbol	Name	On Drawing	Summary Description
FORM			
—	STRAIGHTNESS (SURFACE)	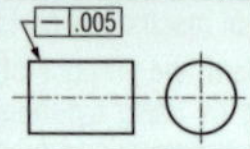	Surface Straightness is a tolerance that controls the form of a line somewhere on the surface or the feature.
— Ⓜ	STRAIGHTNESS (AXIS)	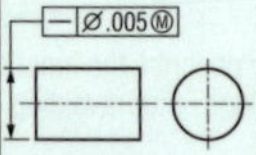	This tolerance controls the central axis of a part and MMC further specifies this by controlling the size of the feature in addition to the allowed "bend" of the axis.
⏥	FLATNESS (SURFACE)	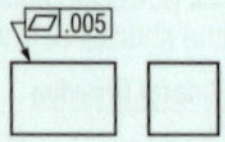	Flatness refers to how flat the surface is regardless of any other datums or features. It refers to two parallel planes that define a zone where the entire reference surface must lie.
○	CIRCULARITY	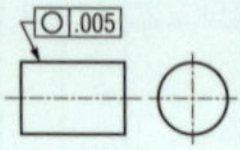	Circularity is used to describe how close an object should be to a true circle. It lies perpendicular to the central axis and bounded by two concentric circles according to the tolerance specified.
⌭	CYLINDRICITY	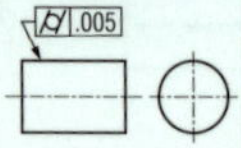	Cylindricity is used to describe how close an object conforms to a true cylinder. It is a tolerance that controls the overall form of a cylindrical feature to ensure that it is round enough and straight enough along its axis.
ORIENTATION			
// Ⓜ	PARALLELISM (SURFACE)	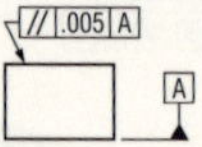	This tolerance controls parallelism between two surfaces or features and is controlled similar to flatness with two parallel planes acting as its tolerance zone.
// Ⓜ	PARALLELISM (AXIS)	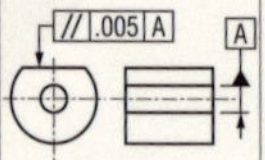	Axis Parallelism is a tolerance that controls how parallel a specific parts central axis needs to be to a datum plane or axis.The axis form is controlled by a cylinder around a theoretical perfectly parallel axis.
⊥ Ⓜ	PERPENDICULARITY (SURFACE)	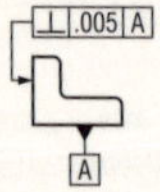	Surface Perpendicularity is a tolerance that controls Perpendicularity between two 90° surfaces and is controlled with two parallel planes acting as its tolerance zone.
⊥ Ⓜ	PERPENDICULARITY (AXIS)	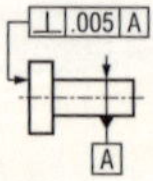	Axis Perpendicularity is a tolerance that controls how perpendicular a specific axis needs to be to a datum and is controlled by a cylinder around a theoretical perfectly parallel axis.
∠ Ⓜ	ANGULARITY	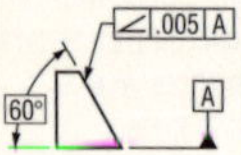	It describes the specific orientation of one surface plane to another datum plane at a referenced angle. The tolerance does not directly control the angle variation but indirectly controls the angle by controlling where the surface can lie based on the datum.

GD&T SYMBOLS AND GUIDELINES

	Symbol	Name	On Drawing	Summary Description
PROFILE	⌒	PROFILE OF A LINE	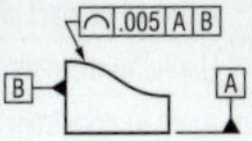	It describes a tolerance zone around any line or feature, usually of a curved shape. Can be applied to any linear tolerance, takes a cross section at any point along the surface and sets a tolerance zone on either side of the profile.
PROFILE	⌓	PROFILE OF A SURFACE	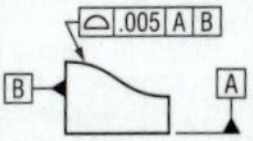	Profile of a surface describes a tolerance zone around a surface, curve or shape. It controls all the points along the surface within a tolerance range that directly matches the profile.
LOCATION	⌖ Ⓢ	(TRUE) POSITION	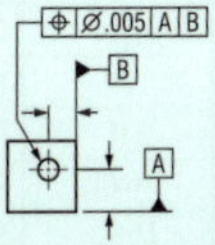	A Position (RFS) tolerance is the total permissible variation in the location of a feature about its exact True Position. When specifying True Position, a datum is usually referenced with x and y coordinates.
LOCATION	⌖ Ⓜ	(TRUE) POSITION	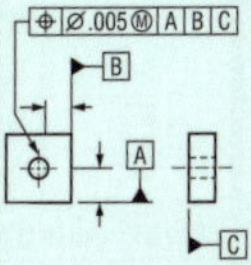	A Position (MMC / LMC) tolerance is the total permissible variation in the location of a feature about its exact True Position. It can be used with (MMC), (LMC), projected tolerances and tangent planes and may apply to everything from points, axes, planes and entire features.
LOCATION	◎ Ⓜ	CONCENTRICITY	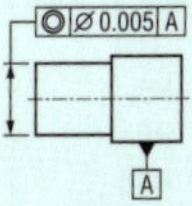	Concentricity is a tolerance that controls the central axis of the referenced feature to a datum feature. It relies on measurements from a derived axis as opposed to a tangible surface or feature.
LOCATION	⌯ Ⓜ	SYMMETRY	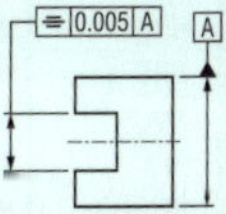	Symmetry is a tolerance that controls how much the median points may deviate from the specified centre plane. For the symmetry to be in tolerance, the median distance between every point on the two surface features needs to fall on that centre plane.
RUNOUT	↗	RUNOUT (RADIAL)	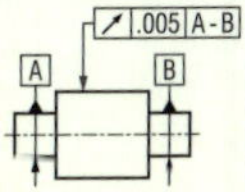	Runout is how much one given reference feature or features vary with respect to another datum when the part is rotated 360° around the datum axis.
RUNOUT	⌰	RUNOUT (AXIAL)	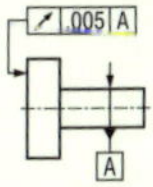	Total Runout tolerance is a combination of how much of variation takes place in the surface of the part when rotated 360° around the datum axis together with the amount of variation in the axial dimension.

GEOMETRIC D&T

GD&T SYMBOLS AND ABBREVIATIONS USED IN ENGINEERING DRAWINGS

Modifier Symbols

Symbol	Modifier	Description
Ⓕ	Free State	Used when part is otherwise restrained
Ⓛ	Least Material Condition (LMC)	Useful to maintain minimum wall thickness
Ⓜ	Maximum Material Condition (MMC)	Used to constrain the tolerance given
Ⓟ	Projected Tolerance Zone	Useful on threaded holes for long studs
Ⓢ	Regardless of Feature Size (RFS)	Default modifier if no other shown
Ⓣ	Tangent Plane	Used where form is not required
Ⓤ	Unequal Bilateral	Refers to unequal profile distribution

Additional GD&T Symbols

Symbol	Term	Symbol	Term
	All round - profile	<u>10</u>	Dimension not to scale
⌒	Arc length		Dimension origin
[35]	Basic dimensions	7X	Number of times/places
⌲	Conical taper	R	Radius
⌴	Counterbore/Spotface	(40)	Reference dimensions
⌵	Countersink	⌳	Slope
↧	Depth/Deep	SØ	Spherical diameter
-[A]-	Datum feature	SR	Spherical radius
Ø6/A1	Datum target	□	Square (shape)
Ø	Diameter	×	Target point

Common Engineering Drawing Abbreviations

Abbreviation	Term	Abbreviation	Term
A/F	Across Flats	GALV	Galvanized
ASSY	Assembly	HB	Hardness Brinell
CL or L	Center line	HRA	Hardness Rockwell A scale
CM	Centimeters	HRB	Hardness Rockwell B scale
CHAM	Chamfered	HRC	Hardness Rockwell C scale
CH HD	Cheese head	HRD	Hardness Rockwell D scale
C'BORE	Counterbore	HRE	Hardness Rockwell E scale
CSK	Countersunk	HV	Hardness Vickers
CSK HD	Countersunk head	HEX	Hexagon
CYL	Cylinder or Cylindrical	HEX HD	Hexagon head
DATUM	Datum system	HYD	Hydraulic
°	Degree (of angle)	IN or "	Inch
Dia or Ø	Diameter	INSUL	Insulated or insulation
DIM	Dimension	INT	Internal
DRG	Drawing	I/D	Internal diameter
EQUI SP	Equally Spaced	KG	Kilogram
EXT	External	LH	Left hand
FIG	Figure	LG	Long
FT or '	Foot	M/C	Machine
GAL	Gallon	MATL	Material

Continued ...

ABBREVIATIONS / MATHEMATICAL SYMBOLS AND INT. ENGINEERING STANDARDS

Common Engineering Drawing Abbreviations ...continued

Abbreviation	Term	Abbreviation	Term
MAX	Maximum	RD HD	Round head
M	Meter	SCR	Screwed
MM	Millimeter	"	Second (of angle)
MIN	Minimum	SK	Sketch
'	Minute (of angle)	SPEC	Specification
NTS	Not to scale	SPH	Spherical
NO.	Number	SQ	Square
O/D	Outside diameter	STD	Standard
PCD	Pitch circle diameter	SWG	Standard (imperial) wire gauge
LB	Pound	THD	Thread
RAD or R	Radius	TPI	Thread per inch
RPM	Revolutions per minute	VOL	Volume
RH	Right hand	WT	Weight

Mathematical Symbols

Symbol	Term	Symbol	Term
$+$	Plus or Positive	$\geq$	Greater Than or Equal To
$-$	Minus or Negative	$\leq$	Less Than or Equal To
$\pm$	Plus/Minus or Pos./Neg.	$\surd$	Square Root Of
X	Multiply By	∞	Infinity
$\div$	Divided By	$\propto$	Proportional To
$=$	Equal To	Σ	Sum Of
$\neq$	Not Equal To	Π	Product Of
$\approx$	Approximately Equal To	Δ	Difference
$\sim$	Of the Order Of / Similar To	$\therefore$	Therefore
$>$	Greater Than	π	Pi
$<$	Less Than	$\parallel$	Parallel To
$\ngtr$	Not Greater Than	$\perp$	Perpendicular To
$\nless$	Not Less Than	$:$	Is To (Ratio)

Common International Engineering Standards

INTERNATIONAL	ISO	International Organisation for Standardization
AUSTRALIA	AS	Australian Standards
FRANCE	AFNOR	Association Française de Normalisation
GERMANY	DIN	Deutsches Institut für Normung
	WR	German Material Number (Werkstoff No.)
ITALY	UNI	Ente Nazionale Italiano di Ulnificazione
JAPAN	JIS	Japanese Industrial Standards
KOREA	KS	Korean Standards
SWEDEN	SS	Swedish Standards
UK	BS	British Standards
USA	AISI	American Iron and Steel Institute
	ANSI	American National Standards Institute
	ASTM	American Society for Testing and Materials
	SAE	Society of Automotive Engineers
	AMS	Aerospace Material Specifications

ADULT PHYSICAL DIMENSIONS

Adult Male Dimensions

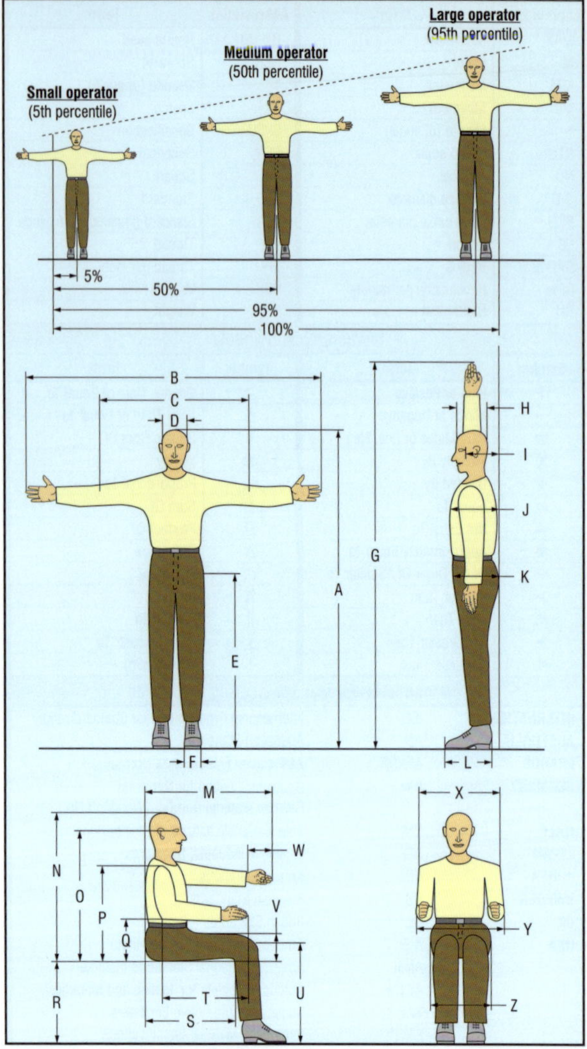

ADULT PHYSICAL DIMENSIONS

(Refer Page 208)

Symbol	Designation	Small Operator (5th Percentile)		Medium Operator (50th Percentile)		Large Operator (95th Percentile)	
		Weight		Weight		Weight	
		121.25 lb	55 kg	168.65 lb	76.5 kg	216.05 lb	98 kg
		(inch)	(mm)	(inch)	(mm)	(inch)	(mm)
A	Height (with shoes)	61.02"	1550	67.52"	1715	74.02"	1880
B	Arm span	62.40"	1585	68.90"	1750	75.59"	1920
C	Arm span, elbows folded	33.46"	850	37.40"	950	41.34"	1050
D	Head width	5.51"	140	5.98"	152	6.50"	165
E	Hip point to floor	32.95"	837	36.50"	927	40.00"	1016
F	Shoe width	3.74"	95	4.13"	105	4.53"	115
G	Overhead reach	74.80"	1900	82.68"	2100	90.55"	2300
H	Head length	6.69"	170	7.40"	188	8.07"	205
I	Eye to back	6.69"	170	7.40"	188	8.07"	205
J	Chest depth	8.27"	210	9.65"	245	11.02"	280
K	Abdominal depth	8.27"	210	10.04"	255	11.81"	300
L	Shoe length	9.84"	250	11.22"	285	12.60"	320
M	Anterior arm reach	29.53"	750	32.48"	825	35.43"	900
N	Chest height	31.50"	800	34.65"	880	37.80"	960
O	Eye height	27.17"	690	30.12"	765	33.07"	840
P	Shoulder height	20.87"	530	23.23"	590	25.59"	650
Q	Elbow height	7.87"	200	9.25"	235	10.63"	270
R	Horizontal sitting height	15.75"	400	17.52"	445	19.29"	490
S	Back to calf	16.54"	420	18.50"	470	20.47"	520
T	Forearm to hand	16.14"	410	18.11"	460	20.08"	510
U	Knee height (with shoes)	19.69"	500	22.05"	560	24.41"	620
V	Thigh thickness	4.72"	120	5.71"	145	6.69"	170
W	Hand length	6.69"	170	7.40"	188	8.07"	205
X	Shoulder width	14.96"	380	17.32"	440	19.69"	500
Y	Elbow to elbow	15.10"	385	17.72"	450	20.28"	515
Z	Hip width	12.60"	320	14.37"	365	16.14"	410

Source : Data compiled from ISO 3411:1999 Human Physical Dimensions of Operators.

NOTE: The above information is intended as a general guide only and varies significantly throughout the world. It is provided to help in the basic design of workstations and equipment.

- The 50th percentile adult male in the United States between 20-65 years old, is 69.1" (1755 mm) tall & weighs 172 lb (78.4 kg). This equates to approx. 2-3% > than the figures in the above table.
- The worldwide average male height is 64.96" (1650 mm) and for women is 60.5" (1537 mm).
- A wheelchair user's eye level is approx. 14.6" (370 mm) below that of the standing person.
- Wheelchair users need a minimum aisle clearance of approx. 31.5" (800 mm) and a turning circle of between 59" (1500 mm) and 67" (1700 mm).

ADULT PHYSICAL DIMENSIONS

Typical Furniture Measurements based on 50th Percentile Male

(Also refer Page 208)

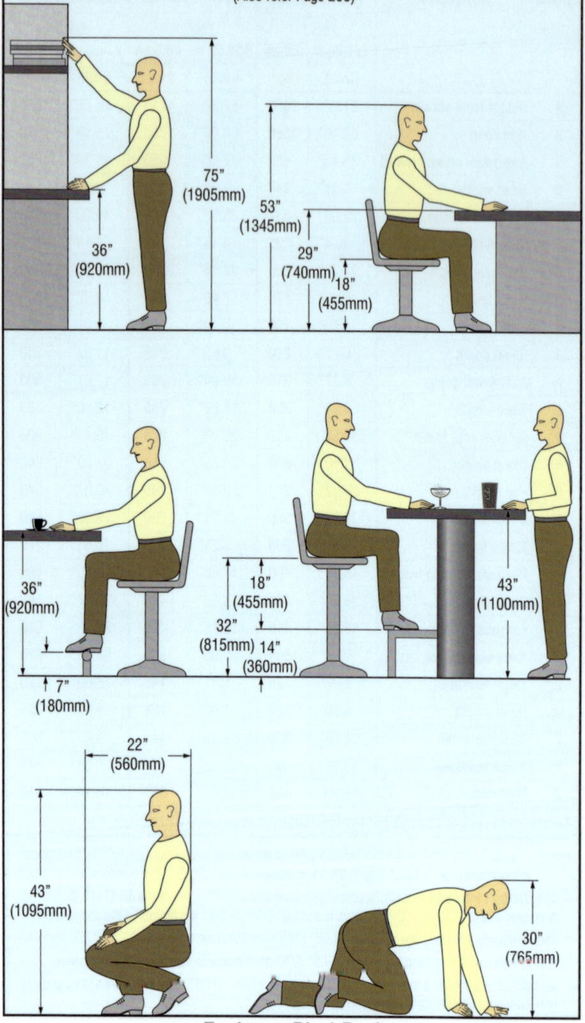

SURFACE ROUGHNESS AND CONVERSION TABLES

Ra - Average roughness of centerline

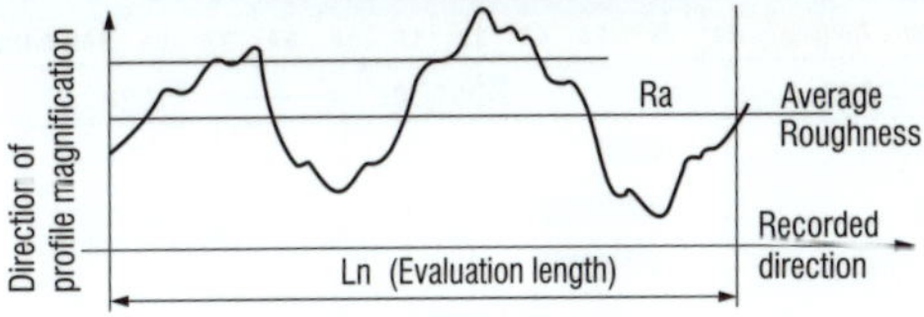

Rmax (Ry) - Maximum height

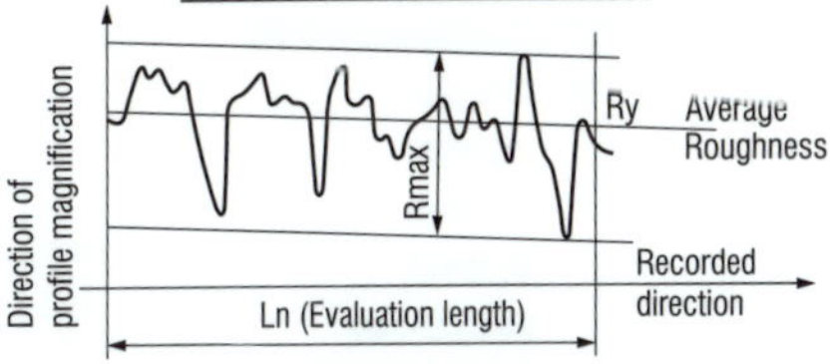

Rz - Average roughness at 10 points

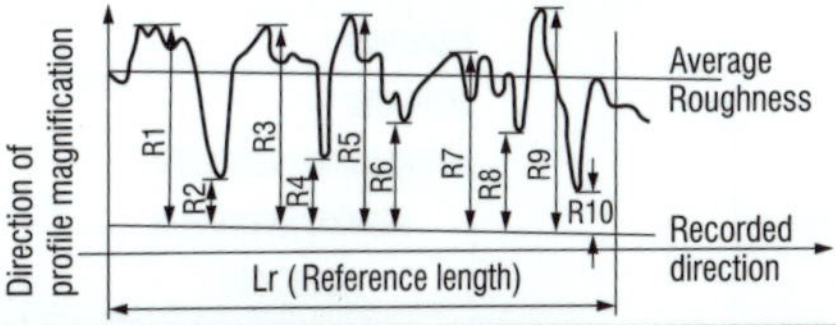

Conversion Tables

Roughness Number	Average Roughness of Centerline		Maximum Height	Average Roughness at 10 Points	Triangle Symbol
N	Ra (μin)	Ra (μm)	Ry (μm)	Rz (μm)	∇ (underlined)
N0	0.50	0.012	0.05	0.063	∇∇∇∇
N1	1	0.025	0.1	0.125	
N2	2	0.05	0.2	0.250	
N3	4	0.10	0.4	0.50	
N4	8	0.20	0.8	1.00	∇∇∇
N5	16	0.40	1.6	2.00	
N6	32	0.80	3.2	4.0	
N7	63	1.6	6.3	7.1	∇∇
N8	125	3.2	12.5	12.5	
N9	250	6.3	25	25	
N10	500	12.5	50	50	∇
N11	1000	25	100	100	
N12	2000	50	200	200	

SURFACE FINISH EXPECTED FROM VARIOUS MANUFACTURING PROCESSES

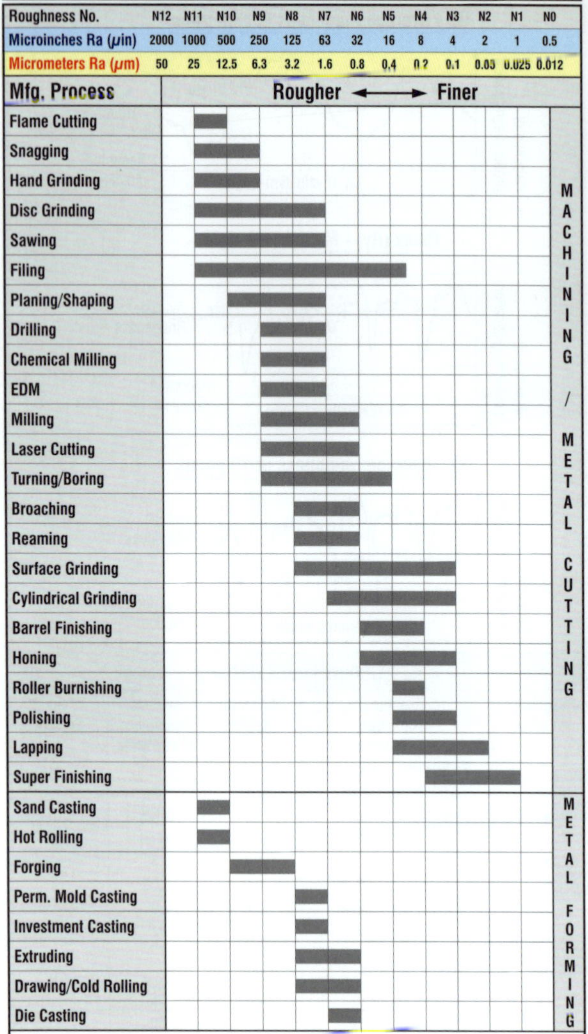

NOTE: The ranges shown above are typical for the processes listed. Higher or lower values may be obtained under special conditions.

COMMON TERMS USED IN THE HEAT TREATMENT OF STEELS

HEAT TREATMENT: A process which involves a combination of heating and cooling of steel in a prescribed manner with respect to time, temperature and rate of heating and cooling so as to induce desired properties in the steel in the solid state.

HARDENING: A process which increases the hardness of steel by changing its chemical and structural makeup.

This usually involves heating the steel to above its critical point and then quenching it in a prescribed medium. Steel with less than 0.3 % carbon cannot be hardened in this manner but its surface can be changed to high carbon steel (by case hardening) which then can be hardened in the normal way.

CASE HARDENING (CARBURIZING): Hardening the surface layer of steel by changing its composition, followed by heat treatment in the normal manner.

The process of case hardening involves heating the low carbon steel in a rich carbon source (powder or gas) at a specific temperature for a set period of time, whereby carbon is absorbed into the surface of the steel to form a high carbon steel casing. The steel is then hardened and tempered in the normal manner.

QUENCHING: An accelerated cooling process by immersion of the heated steel in liquids or gases.

Oils are often preferred to water as a quenching medium as they are less drastic and more uniform in their cooling effect. Quenching in water produces very hard steel whilst quenching in oil produces a slightly softer steel that is less likely to crack or distort.

TEMPERING: Re-heating the quenched steel to a temperature below the transformation range and, if, necessary, holding at that temperature, followed by cooling at a suitable rate to achieve desired mechanical properties.

FLAME HARDENING: Rapid heating of the surface layer of the steel by means of a flame to a temperature above the transformation range, followed by immediate quenching.

INDUCTION HARDENING: A process of heating by electrical induction to the appropriate temperature and quenching in a suitable medium.

NITRIDING: A process of surface hardening by introducing Nitrogen into the surface in a suitable steel by heating and holding it at an appropriate temperature in contact with cracked ammonia or other suitable nitrogenous medium.

This process produces a hard case in the steel without further need for heat treatment.

ANNEALING (NORMALISING): A process of heat treatment for improving mechanical properties brought about by grain refinement and uniformity in structure. The process involves heating and holding for a specified time at a suitable temperature above the transformation range, followed by cooling freely in air.

STRESS RELIEVING (STABILIZING): Heating to and if necessary, holding at a sufficiently high temperature below the transformation range, followed by slow cooling to remove internal stresses only.

STEEL HARDENING / TEMPERING TEMPERATURES JUDGED BY COLOR

Deg. F	Deg. C	Color	
2552	1400	White welding heat	HARDENING
2372	1300	Yellow-White	
2192	1200	Orange-Yellow	
2012	1100	Orange-Red	
1832	1000	Bright Cherry-Red	
1652	900	Cherry-Red	
1472	800	Dull Cherry-Red	
1292	700	Dark Red	
1077	581	Red heat, visible in sunlight	
975	525	Red heat, visible in daylight	
626	330	Greyish Green	TEMPERING
608	320	Grey	
590	310	Light Blue	
570	299	Dark Blue	
560	293	Full Blue	
550	288	Dark Purple	
540	282	Full Purple	
530	277	Light Purple	
520	271	Brown Purple	
510	266	Spotted-Brown	
500	260	Brown Yellow	
490	254	Yellow Brown	
480	249	Dark Yellow	
470	243	Deep Straw Yellow	
460	238	Straw Yellow	
450	232	Pale Straw Yellow	
440	227	Light Yellow	
430	221	Very Pale Yellow	

NOTE: The above colors are intended as a guide only and should not be assumed to be an accurate temperature indicator. Colors vary according to metal composition.

HARDNESS COMPARISON CHART

Brinell Hardness (HB) 3,000 kg	Rockwell Hardness				Vickers Hardness 50 kg	Shore Hardness	Transverse Rupture Strength Tons (short) per inch²
	A Scale 60 kg (2) Brale HRA	B Scale 100 kg (1) 1/16 in. Ball HRB	C Scale 150 kg (2) Brale HRC	D Scale 100 kg (2) Brale HRD			
–	85.6	–	68.0	76.9	940	97	–
–	85.3	–	67.5	76.5	920	96	–
–	85.0	–	67.0	76.1	900	95	–
767	84.7	–	66.4	75.7	880	93	–
757	84.4	–	65.9	75.3	860	92	–
745	84.1	–	65.3	74.8	840	91	–
733	83.8	–	64.7	74.3	820	90	–
722	83.4	–	64.0	73.8	800	88	–
712	–	–	–	–	–	–	–
710	83.0	–	63.3	73.3	780	87	–
698	82.6	–	62.5	72.6	760	86	–
684	82.2	–	61.8	72.1	740	–	–
682	82.2	–	61.7	72.0	737	84	–
670	81.0	–	61.0	71.5	720	83	–
656	81.3	–	60.1	70.8	700	–	–
653	81.2	–	60.0	70.7	697	81	–
647	81.1	–	59.7	70.5	690	–	–
638	80.8	–	59.2	70.1	680	80	–
630	80.6	–	58.8	69.8	670	–	–
627	80.5	–	58.7	69.7	667	79	–
601	79.8	–	57.3	68.7	640	77	–
578	79.1	–	56.0	67.7	615	75	–
555	78.4	–	54.7	66.7	591	73	149.3
534	77.8	–	53.5	65.8	569	71	143.7
514	76.9	–	52.1	64.7	547	70	137.3
495	76.3	–	51.0	63.8	528	68	132.3
477	75.6	–	49.6	62.7	508	66	125.9
461	74.9	–	48.5	61.7	491	65	120.9
444	74.2	–	47.1	60.8	472	63	115.2
429	73.4	–	45.7	59.7	455	61	100.6
415	72.8	–	44.5	58.8	440	59	106.0
401	72.0	–	43.1	57.8	425	58	101.0
388	71.4	–	41.8	56.8	410	56	96.7
375	70.6	–	40.4	55.7	396	54	91.7
363	70.0	–	39.1	54.6	383	52	88.2
352	69.3	(110.0)	37.9	53.8	372	51	85.3
341	68.7	(109.0)	36.6	52.8	360	50	81.8
331	68.1	(108.5)	35.5	51.9	350	48	79.7

HARDNESS COMPARISON CHART

Brinell Hardness (HB) 3,000 kg	Rockwell Hardness				Vickers Hardness 50 kg	Shore Hardness	Transverse Rupture Strength Tons (short) per inch²
	A Scale 60 kg (2) Brale HRA	B Scale 100 kg (1) 1/16 in. Ball HRB	C Scale 150 kg (2) Brale HRC	D Scale 100 kg (2) Brale HRD			
321	67.5	(108.0)	34.3	51.0	339	47	76.8
311	66.9	(107.5)	33.1	50.0	328	46	74.7
302	66.3	(107.0)	32.1	49.3	319	45	73.3
293	65.7	(106.0)	30.9	48.3	309	43	70.4
285	65.3	(105.5)	29.9	47.6	301	–	69.0
277	64.6	(104.5)	28.8	46.7	292	41	66.8
269	64.1	(104.0)	27.6	45.9	284	40	64.7
262	63.6	(103.0)	26.6	45.0	276	39	63.3
255	63.0	(102.0)	25.4	44.2	269	38	61.2
248	62.5	(101.0)	24.2	43.2	261	37	59.7
241	61.8	100.0	22.8	42.0	253	36	58.3
235	61.4	99.0	21.7	41.4	247	35	56.9
229	60.8	98.2	20.5	40.5	241	34	55.5
223	–	97.3	(18.8)	–	234	–	–
217	–	96.4	(17.5)	–	228	33	52.6
212	–	95.5	(16.0)	–	222	–	51.2
207	–	94.6	(15.2)	–	218	32	49.8
201	–	93.8	(13.8)	–	212	31	49.1
197	–	92.8	(12.7)	–	207	30	47.6
192	–	91.9	(11.5)	–	202	29	46.2
187	–	90.7	(10.0)	–	196	–	44.8
183	–	90.0	(9.0)	–	192	28	44.8
179	–	89.0	(8.0)	–	188	27	43.4
174	–	87.8	(6.4)	–	182	–	42.7
170	–	86.8	(5.4)	–	178	26	41.2
167	–	86.0	(4.4)	–	175	–	40.5
163	–	85.0	(3.3)	–	171	25	39.8
156	–	82.9	(0.9)	–	163	–	37.7
149	–	80.8	–	–	156	23	36.3
143	–	78.7	–	–	150	22	35.6
137	–	76.4	–	–	143	21	33.4
131	–	74.0	–	–	137	–	32.7
126	–	72.0	–	–	132	20	31.3
121	–	69.8	–	–	127	19	29.9
116	–	67.6	–	–	122	18	29.2
111	–	65.7	–	–	117	15	27.7

1) In above chart, figures with ( ) are not commonly used.

2) Rockwell scale A, C and D are scaled with a 120° spheroconical diamond brale indentor.

TABLE OF NATURAL SINES

SINE

Deg.	0	10	20	30	40	50	60	
0	0.00000	0.00291	0.00582	0.00873	0.01164	0.01454	0.01745	89
1	0.01745	0.02036	0.02327	0.02618	0.02908	0.03199	0.03490	88
2	0.03490	0.03781	0.04071	0.04362	0.04653	0.04943	0.05234	87
3	0.05234	0.05524	0.05814	0.06105	0.06395	0.06685	0.06976	86
4	0.06976	0.07266	0.07556	0.07846	0.08136	0.08426	0.08716	85
5	0.08716	0.09005	0.09295	0.09585	0.09874	0.10164	0.10453	84
6	0.10453	0.10742	0.11031	0.11320	0.11609	0.11898	0.12187	83
7	0.12187	0.12476	0.12764	0.13053	0.13341	0.13629	0.13917	82
8	0.13917	0.14205	0.14493	0.14781	0.15069	0.15356	0.15643	81
9	0.15643	0.15931	0.16218	0.16505	0.16792	0.17078	0.17365	80
10	0.17365	0.17651	0.17937	0.18224	0.18509	0.18795	0.19081	79
11	0.19081	0.19366	0.19652	0.19937	0.20222	0.20507	0.20791	78
12	0.20791	0.21076	0.21360	0.21644	0.21928	0.22212	0.22495	77
13	0.22495	0.22778	0.23062	0.23345	0.23627	0.23910	0.24192	76
14	0.24192	0.24474	0.24756	0.25038	0.25320	0.25601	0.25882	75
15	0.25882	0.26163	0.26443	0.26724	0.27004	0.27284	0.27564	74
16	0.27564	0.27843	0.28123	0.28402	0.28680	0.28959	0.29237	73
17	0.29237	0.29515	0.29793	0.30071	0.30348	0.30625	0.30902	72
18	0.30902	0.31178	0.31454	0.31730	0.32006	0.32282	0.32557	71
19	0.32557	0.32832	0.33106	0.33381	0.33655	0.33929	0.34202	70
20	0.34202	0.34475	0.34748	0.35021	0.35293	0.35565	0.35837	69
21	0.35837	0.36108	0.36379	0.36650	0.36921	0.37191	0.37461	68
22	0.37461	0.37730	0.37999	0.38268	0.38537	0.38805	0.39073	67
23	0.39073	0.39341	0.39608	0.39875	0.40141	0.40408	0.40674	66
24	0.40674	0.40939	0.41204	0.41469	0.41734	0.41998	0.42262	65
25	0.42262	0.42525	0.42788	0.43051	0.43313	0.43575	0.43837	64
26	0.43837	0.44098	0.44359	0.44620	0.44880	0.45140	0.45399	63
27	0.45399	0.45658	0.45917	0.46175	0.46433	0.46690	0.46947	62
28	0.46947	0.47204	0.47460	0.47716	0.47971	0.48226	0.48481	61
29	0.48481	0.48735	0.48989	0.49242	0.49495	0.49748	0.50000	60
30	0.50000	0.50252	0.50503	0.50754	0.51004	0.51254	0.51504	59
31	0.51504	0.51753	0.52002	0.52250	0.52498	0.52745	0.52992	58
32	0.52992	0.53238	0.53484	0.53730	0.53975	0.54220	0.54464	57
33	0.54464	0.54708	0.54951	0.55194	0.55436	0.55678	0.55919	56
34	0.55919	0.56160	0.56401	0.56641	0.56880	0.57119	0.57358	55
35	0.57358	0.57596	0.57833	0.58070	0.58307	0.58543	0.58779	54
36	0.58779	0.59014	0.59248	0.59482	0.59716	0.59949	0.60182	53
37	0.60182	0.60414	0.60645	0.60876	0.61107	0.61337	0.61566	52
38	0.61566	0.61795	0.62024	0.62251	0.62479	0.62706	0.62932	51
39	0.62932	0.63158	0.63383	0.63608	0.63832	0.64056	0.64279	50
40	0.64279	0.64501	0.64723	0.64945	0.65166	0.65386	0.65606	49
41	0.65606	0.65825	0.66044	0.66262	0.66480	0.66697	0.66913	48
42	0.66913	0.67129	0.67344	0.67559	0.67773	0.67987	0.68200	47
43	0.68200	0.68412	0.68624	0.68835	0.69046	0.69256	0.69466	46
44	0.69466	0.69675	0.69883	0.70091	0.70298	0.70505	0.70711	45
	60	50	40	30	20	10	0	Deg

COSINE

TABLE OF NATURAL SINES

SINE								
Deg	0	10	20	30	40	50	60	
45	0.70711	0.70916	0.71121	0.71325	0.71529	0.71732	0.71934	44
46	0.71934	0.72136	0.72337	0.72537	0.72737	0.72937	0.73135	43
47	0.73135	0.73333	0.73531	0.73728	0.73924	0.74120	0.74314	42
48	0.74314	0.74509	0.74703	0.74896	0.75088	0.75280	0.75471	41
49	0.75471	0.75661	0.75851	0.76041	0.76229	0.76417	0.76604	40
50	0.76604	0.76791	0.76977	0.77162	0.77347	0.77531	0.77715	39
51	0.77715	0.77897	0.78079	0.78261	0.78442	0.78622	0.78801	38
52	0.78801	0.78980	0.79158	0.79335	0.79512	0.79688	0.79864	37
53	0.79864	0.80038	0.80212	0.80386	0.80558	0.80730	0.80902	36
54	0.80902	0.81072	0.81242	0.81412	0.81580	0.81748	0.81915	35
55	0.81915	0.82082	0.82248	0.82413	0.82577	0.82741	0.82904	34
56	0.82904	0.83066	0.83228	0.83389	0.83549	0.83708	0.83867	33
57	0.83867	0.84025	0.84182	0.84339	0.84495	0.84650	0.84805	32
58	0.84805	0.84959	0.85112	0.85264	0.85416	0.85567	0.85717	31
59	0.85717	0.85866	0.86015	0.86163	0.86310	0.86457	0.86603	30
60	0.86603	0.86748	0.86892	0.87036	0.87178	0.87321	0.87462	29
61	0.87462	0.87603	0.87743	0.87882	0.88020	0.88158	0.88295	28
62	0.88295	0.88431	0.88566	0.88701	0.88835	0.88968	0.89101	27
63	0.89101	0.89232	0.89363	0.89493	0.89623	0.89752	0.89879	26
64	0.89879	0.90007	0.90133	0.90259	0.90383	0.90507	0.90631	25
65	0.90631	0.90753	0.90875	0.90996	0.91116	0.91236	0.91355	24
66	0.91355	0.91472	0.91590	0.91706	0.91822	0.91936	0.92050	23
67	0.92050	0.92164	0.92276	0.92388	0.92499	0.92609	0.92718	22
68	0.92718	0.92827	0.92935	0.93042	0.93148	0.93253	0.93358	21
69	0.93358	0.93462	0.93565	0.93667	0.93769	0.93869	0.93969	20
70	0.93969	0.94068	0.94167	0.94264	0.94361	0.94457	0.94552	19
71	0.94552	0.94646	0.94740	0.94832	0.94924	0.95015	0.95106	18
72	0.95106	0.95195	0.95284	0.95372	0.95459	0.95545	0.95630	17
73	0.95630	0.95715	0.95799	0.95882	0.95964	0.96046	0.96126	16
74	0.96126	0.96206	0.96285	0.96363	0.96440	0.96517	0.96593	15
75	0.96593	0.96667	0.96742	0.96815	0.96887	0.96959	0.97030	14
76	0.97030	0.97100	0.97169	0.97237	0.97304	0.97371	0.97437	13
77	0.97437	0.97502	0.97566	0.97630	0.97692	0.97754	0.97815	12
78	0.97815	0.97875	0.97934	0.97992	0.98050	0.98107	0.98163	11
79	0.98163	0.98218	0.98272	0.98325	0.98378	0.98430	0.98481	10
80	0.98481	0.98531	0.98580	0.98629	0.98676	0.98723	0.98769	9
81	0.98769	0.98814	0.98858	0.98902	0.98944	0.98986	0.99027	8
82	0.99027	0.99067	0.99106	0.99144	0.99182	0.99219	0.99255	7
83	0.99255	0.99290	0.99324	0.99357	0.99390	0.99421	0.99452	6
84	0.99452	0.99482	0.99511	0.99540	0.99567	0.99594	0.99619	5
85	0.99619	0.99644	0.99668	0.99692	0.99714	0.99736	0.99756	4
86	0.99756	0.99776	0.99795	0.99813	0.99831	0.99847	0.99863	3
87	0.99863	0.99878	0.99892	0.99905	0.99917	0.99929	0.99939	2
88	0.99939	0.99949	0.99958	0.99966	0.99973	0.99979	0.99985	1
89	0.99985	0.99989	0.99993	0.99996	0.99998	0.99999	1.00000	0
	60	50	40	30	20	10	0	Deg
								COSINE

TABLE OF NATURAL TANGENTS

TAN Deg	0	10	20	30	40	50	60	
0	0.00000	0.00291	0.00582	0.00873	0.01164	0.01455	0.01746	89
1	0.01746	0.02036	0.02328	0.02619	0.02910	0.03201	0.03492	88
2	0.03492	0.03783	0.04075	0.04366	0.04658	0.04949	0.05241	87
3	0.05241	0.05533	0.05824	0.06116	0.06408	0.06700	0.06993	86
4	0.06993	0.07285	0.07578	0.07870	0.08163	0.08456	0.08749	85
5	0.08749	0.09042	0.09335	0.09629	0.09923	0.10216	0.10510	84
6	0.10510	0.10805	0.11099	0.11394	0.11688	0.11983	0.12278	83
7	0.12278	0.12574	0.12869	0.13165	0.13461	0.13758	0.14054	82
8	0.14054	0.14351	0.14648	0.14945	0.15243	0.15540	0.15838	81
9	0.15838	0.16137	0.16435	0.16734	0.17033	0.17333	0.17633	80
10	0.17633	0.17933	0.18233	0.18534	0.18835	0.19136	0.19438	79
11	0.19438	0.19740	0.20042	0.20345	0.20648	0.20952	0.21256	78
12	0.21256	0.21560	0.21864	0.22169	0.22475	0.22781	0.23087	77
13	0.23087	0.23393	0.23700	0.24008	0.24316	0.24624	0.24933	76
14	0.24933	0.25242	0.25552	0.25862	0.26172	0.26483	0.26795	75
15	0.26795	0.27107	0.27419	0.27732	0.28046	0.28360	0.28675	74
16	0.28675	0.28990	0.29305	0.29621	0.29938	0.30255	0.30573	73
17	0.30573	0.30891	0.31210	0.31530	0.31850	0.32171	0.32492	72
18	0.32492	0.32814	0.33136	0.33460	0.33783	0.34108	0.34433	71
19	0.34433	0.34758	0.35085	0.35412	0.35740	0.36068	0.36397	70
20	0.36397	0.36727	0.37057	0.37388	0.37720	0.38053	0.38386	69
21	0.38386	0.38721	0.39055	0.39391	0.39727	0.40065	0.40403	68
22	0.40403	0.40741	0.41081	0.41421	0.41763	0.42105	0.42447	67
23	0.42447	0.42791	0.43136	0.43481	0.43828	0.44175	0.44523	66
24	0.44523	0.44872	0.45222	0.45573	0.45924	0.46277	0.46631	65
25	0.46631	0.46985	0.47341	0.47698	0.48055	0.48414	0.48773	64
26	0.48773	0.49134	0.49495	0.49858	0.50222	0.50587	0.50953	63
27	0.50953	0.51319	0.51688	0.52057	0.52427	0.52798	0.53171	62
28	0.53171	0.53545	0.53920	0.54296	0.54673	0.55051	0.55431	61
29	0.55431	0.55812	0.56194	0.56577	0.56962	0.57348	0.57735	60
30	0.57735	0.58124	0.58513	0.58905	0.59297	0.59691	0.60086	59
31	0.60086	0.60483	0.60881	0.61280	0.61681	0.62083	0.62487	58
32	0.62487	0.62892	0.63299	0.63707	0.64117	0.64528	0.64941	57
33	0.64941	0.65355	0.65771	0.66189	0.66608	0.67028	0.67451	56
34	0.67451	0.67875	0.68301	0.68728	0.69157	0.69588	0.70021	55
35	0.70021	0.70455	0.70891	0.71329	0.71769	0.72211	0.72654	54
36	0.72654	0.73100	0.73547	0.73906	0.74447	0.74900	0.75355	53
37	0.75355	0.75812	0.76272	0.76733	0.77196	0.77661	0.70120	52
38	0.78129	0.78598	0.79070	0.79544	0.80020	0.80498	0.80978	51
39	0.80978	0.81461	0.81946	0.82434	0.82923	0.83415	0.83910	50
40	0.83910	0.84407	0.84906	0.85408	0.85912	0.86419	0.86929	49
41	0.86929	0.87441	0.07966	0.88473	0.88992	0.89515	0.90040	48
42	0.90040	0.90569	0.91099	0.91633	0.92170	0.92709	0.93252	47
43	0.93252	0.93797	0.94345	0.94896	0.95451	0.96008	0.96569	46
44	0.96569	0.97133	0.97700	0.98270	0.98843	0.99420	1.00000	45
	60	50	40	30	20	10	0	Deg

COTANGENT

TABLE OF NATURAL TANGENTS

TAN								
Deg	**0**	**10**	**20**	**30**	**40**	**50**	**60**	
45	1.00000	1.00583	1.01170	1.01761	1.02355	1.02952	1.03553	44
46	1.03553	1.04158	1.04766	1.05378	1.05994	1.06613	1.07237	43
47	1.07237	1.07864	1.08496	1.09131	1.09770	1.10414	1.11061	42
48	1.11061	1.11713	1.12369	1.13029	1.13694	1.14363	1.15037	41
49	1.15037	1.15715	1.16398	1.17085	1.17777	1.18474	1.19175	40
50	1.19175	1.19882	1.20593	1.21310	1.22031	1.22758	1.23490	39
51	1.23490	1.24227	1.24969	1.25717	1.26471	1.27230	1.27994	38
52	1.27994	1.28764	1.29541	1.30323	1.31110	1.31904	1.32704	37
53	1.32704	1.33511	1.34323	1.35142	1.35968	1.36800	1.37638	36
54	1.37638	1.38484	1.39336	1.40195	1.41061	1.41934	1.42815	35
55	1.42815	1.43703	1.44598	1.45501	1.46411	1.47330	1.48256	34
56	1.48256	1.49190	1.50133	1.51084	1.52043	1.53010	1.53986	33
57	1.53986	1.54972	1.55966	1.56969	1.57981	1.59002	1.60033	32
58	1.60033	1.61074	1.62125	1.63185	1.64256	1.65337	1.66428	31
59	1.66428	1.67530	1.68643	1.69766	1.70901	1.72047	1.73205	30
60	1.73205	1.74375	1.75556	1.76749	1.77955	1.79174	1.80405	29
61	1.80405	1.81649	1.82906	1.84177	1.85462	1.86760	1.88073	28
62	1.88073	1.89400	1.90741	1.92098	1.93470	1.94858	1.96261	27
63	1.96261	1.97681	1.99116	2.00569	2.02039	2.03526	2.05030	26
64	2.05030	2.06553	2.08094	2.09654	2.11233	2.12832	2.14451	25
65	2.14451	2.16090	2.17749	2.19430	2.21132	2.22857	2.24604	24
66	2.24604	2.26374	2.28167	2.29984	2.31826	2.33693	2.35585	23
67	2.35585	2.37504	2.39449	2.41421	2.43422	2.45451	2.47509	22
68	2.47509	2.49597	2.51715	2.53865	2.56046	2.58261	2.60509	21
69	2.60509	2.62791	2.65109	2.67462	2.69853	2.72281	2.74748	20
70	2.74748	2.77254	2.79802	2.82391	2.85023	2.87700	2.90421	19
71	2.90421	2.93189	2.96004	2.98868	3.01783	3.04749	3.07768	18
72	3.07768	3.10842	3.13972	3.17159	3.20406	3.23714	3.27085	17
73	3.27085	3.30521	3.34023	3.37594	3.41236	3.44951	3.48741	16
74	3.48741	3.52609	3.56557	3.60588	3.64705	3.68909	3.73205	15
75	3.73205	3.77595	3.82083	3.86671	3.91364	3.96165	4.01078	14
76	4.01078	4.06107	4.11256	4.16530	4.21933	4.27471	4.33148	13
77	4.33148	4.38969	4.44942	4.51071	4.57363	4.63825	4.70463	12
78	4.70463	4.77286	4.84300	4.91516	4.98940	5.06584	5.14455	11
79	5.14455	5.22566	5.30928	5.39552	5.48451	5.57638	5.67128	10
80	5.67128	5.76937	5.87080	5.97576	6.08444	6.19703	6.31375	9
81	6.31375	6.43484	6.56055	6.69116	6.82694	6.96823	7.11537	8
82	7.11537	7.26873	7.42871	7.59575	7.77035	7.95302	8.14435	7
83	8.14435	8.34496	8.55555	8.77689	9.00983	9.25530	9.51436	6
84	9.51436	9.78817	10.07803	10.38540	10.71191	11.05943	11.43005	5
85	11.43005	11.82617	12.25051	12.70620	13.19688	13.72674	14.30067	4
86	14.30067	14.92442	15.60478	16.34986	17.16934	18.07498	19.08114	3
87	19.08114	20.20555	21.47040	22.90377	24.54176	26.43160	28.63625	2
88	28.63625	31.24158	34.36777	38.18846	42.96408	49.10388	57.28996	1
89	57.28996	68.75009	85.93979	114.58865	171.88540	343.77371	Infinity	0
	60	**50**	**40**	**30**	**20**	**10**	**0**	**Deg**
								COTANGENT

INDEX

INDEX

INDEX

INDEX

NOTES

1 •
2 •
3 •
4 •
5 •
6 •
7 •
8 •
9 •
10 •
11 •
12 •
13 •
14 •
15 •
16 •
17 •
18 •
19 •
20 •
21 •

NOTES

1 •
2 •
3 •
4 •
5 •
6 •
7 •
8 •
9 •
10 •
11 •
12 •
13 •
14 •
15 •
16 •
17 •
18 •
19 •
20 •
21 •

NOTES

NOTES

NOTES

NOTES

NOTES

1/16"

NOTES

1/8"

Thank you for purchasing the Engineers Black Book

What makes the Engineers Black Book unique ?

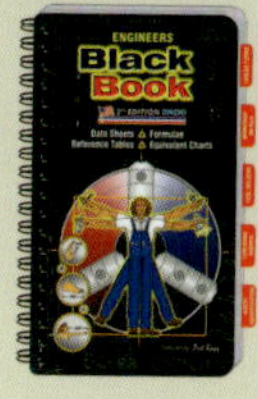

- ✓ Contains very useful engineering information
- ✓ Every page is Grease and Tear Proof
- ✓ It is Wiro Layflat bound so it stays open unassisted
- ✓ Full Color for easy reading
- ✓ It's portable and fits into the pocket - Take it everywhere!
- ✓ It includes a set of Self Adhesive Index Tabs

We would like to hear your feedback !

If you have any suggestions for improvements or concerns about this book we would like to hear from you? Tell us what you think and leave a review by going to: www.EngineersBlackBook.com/review or scan the adjacent QR code with your smartphone. Enter in your details, along with the unique serial number printed on the below hologram label.

In appreciation for doing so, we would like to send you a Drill Size and Decimal Equivalent Chart (aka. E30 Technical Hot Sheet) that includes Tapping Drill Sizes at no charge - anywhere in the world. Please allow 4 weeks for delivery.

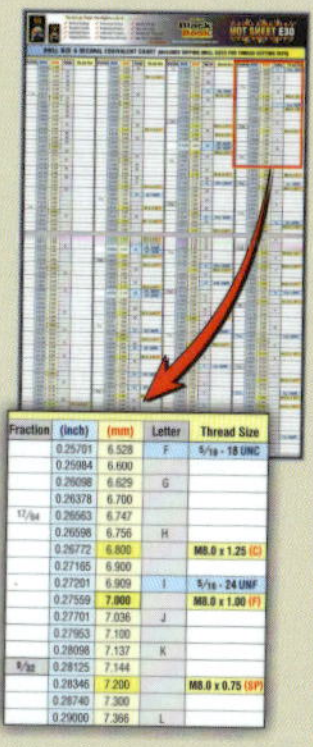

The specifications are as follows.

- ✓ Contains Drill Size & Decimal Equivalents: Up to 3 decimal places in Millimeters and 5 decimal places in Inches.
- ✓ Drill Sizes for Tapping - UNC, UNS, UNF, Metric Coarse, Metric Fine, Metric Special.
- ✓ Exact Chart Size: When folded out 16.54 inches (420.00mm) High x 11.00 inches (279.40mm) Wide.
- ✓ Glare Proof and Full Color.
- ✓ Range 0.00394 inches / 0.1mm - 4.0 inches / 1011.60mm

Thank you once again for giving us your time. Your feedback is important to us and will help us to improve future publications.

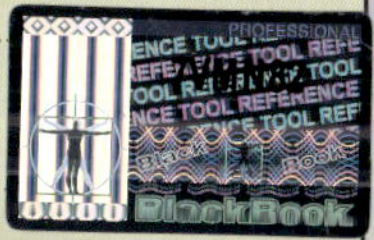

Please visit us online and see our other titles:

THIS BOOK BELONGS TO

....

....

....

....

....

Additional copies of this book may be purchased from your local Industrial Equipment & Engineering Supplies Company. **If you are having difficulty in obtaining additional copies, please e-mail the Publisher/Distributor listed on the front inside cover of this book**.

In view of the fact that this book would be frequently used in a workshop environment, **every page has been Matt Laminated to be Glare-free and Grease-proof**. Stains may be removed using a clean cotton rag lightly dampened with soap to wipe the stain away.